A C S   S Y M P O S I U M   S E R I E S **418**

Marine Toxins

Origin, Structure, and Molecular Pharmacology

Sherwood Hall, EDITOR
U.S. Food and Drug Administration

Gary Strichartz, EDITOR
Brigham and Women's Hospital
Harvard Medical School

Developed from a symposium held
in Woods Hole, Massachusetts, under the auspices
of the International Union of Pure and Applied Chemistry,
and sponsored through an interagency agreement
between the U.S. Army Medical Research
Institute of Infectious Diseases
and the Center for Food Safety and Applied Nutrition,
U.S. Food and Drug Administration

American Chemical Society, Washington, DC 1990

Library of Congress Cataloging-in-Publication Data

Marine Toxins: Origin, Structure, and Molecular Pharmacology
   Sherwood Hall, editor, Gary Strichartz, editor

   p.   cm.—(ACS Symposium Series, 0097–6156; 418).

   "Developed from a symposium held in Woods Hole, Massachusetts, under the auspices of the Commission of Food Chemistry, Applied Chemistry Division, International Union of Pure and Applied Chemistry, and sponsored through an interagency agreement between the U.S. Army Medical Research Institute of Infectious Diseases and the Center for Food Safety and Applied Nutrition, U.S. Food and Drug Administration"

   Symposium held on Aug. 27–30th, 1987.

   Includes bibliographical references.

   ISBN 0–8412–1733–5

   1. Marine pharmacology—Chemistry—Congresses. 2. Marine toxins—Congresses.

   I. Hall, Sherwood, 1944–   . II. Strichartz, Gary. (Gary R.), 1943–
   III. International Union of Pure and Applied Chemistry. Commission on Food Chemistry. IV. U.S. Army Medical Research Institute of Infectious Diseases. V. Center for Food Safety and Applied Nutrition (U.S.) VI. Series

   [DNLM: 1. Marine Toxins—congresses. QW 630 M338 1987]

RS160.7.M37   1990 615.9'5—dc20
DNLM/DLC
for Library of Congress   89–18505
                          CIP

The paper used in this publication meets the minimum requirements of American National Standard for Information Sciences—Permanence of Paper for Printed Library Materials, ANSI Z39.48–1984.

Foreword

The ACS SYMPOSIUM SERIES was founded in 1974 to provide a medium for publishing symposia quickly in book form. The format of the Series parallels that of the continuing ADVANCES IN CHEMISTRY SERIES except that, in order to save time, the papers are not typeset but are reproduced as they are submitted by the authors in camera-ready form. Papers are reviewed under the supervision of the Editors with the assistance of the Series Advisory Board and are selected to maintain the integrity of the symposia; however, verbatim reproductions of previously published papers are not accepted. Both reviews and reports of research are acceptable, because symposia may embrace both types of presentation.

Contents

PALYTOXIN

Preface

O<small>N</small> THURSDAY AFTERNOON, July 30, 1987, people began to fall ill in Champerico, a small town on the Pacific coast of Guatemala. During the next three days, more than 200 people in this town and neighboring areas were affected; 26 died. All those afflicted had consumed meals prepared from fresh clams. Upon investigation, this was found to be an outbreak of paralytic shellfish poisoning, due to marine toxins accumulated by the clams from phytoplankton growing in the waters along the coast.

Why study marine toxins? Episodes like the one above are one reason; they command attention because of the loss of life and health when they occur or because of the costs of preventing and avoiding them. Much research on marine toxins is directed toward moderating their impact by understanding the various syndromes and developing effective strategies to deal with them.

However, there are positive reasons for studying marine toxins and their actions. Marine toxins play a significant role in structuring ecological relationships, both as offensive weapons for predators and as defensive weapons by potential prey. The very high potencies of some marine toxins, many of which seem clearly to have evolved to optimize their performance in these roles, are attained through strong, highly selective interactions, frequently with specific sites on excitable membranes. Investigating the mechanisms of toxin action has revealed a great deal about the physiology of affected tissues and systems. For example, the concept of a membrane channel arose from investigations with tetrodotoxin and saxitoxin, marine toxins that block sodium channels. Marine toxins themselves have proven challenging subjects for chemical study—structure elucidation, characterization, and synthesis.

This field by its nature draws from several distinct traditional disciplines. One major goal of this book is to present work from these diverse disciplines, embracing the major areas of inquiry and the investigational tools with which information is obtained, in a form accessible to the non-specialist as well as the specialist. Although an exhaustive survey was impractical, great care was taken to include representations from the full range of organisms and compounds—from

ix

procaryotes to vertebrates, from small molecules to peptides. The scope of the book goes slightly beyond the title, including blue-green algae that occur in fresh or brackish waters.

Three themes have been emphasized in assembling this book. The first is a description of the sources of natural toxins, with particular emphasis on the metabolic pathways responsible for their synthesis. The second theme concerns the structures of toxins and, in particular, which features determine both the distribution of toxic materials among the organs of intoxicated species and the stability, disposition, and biotransformation of those toxins. The third theme proceeds on a molecular scale to investigate the interactions between toxins and their primary sites of action, not only to understand these interactions better, but to gain a clearer view of the normal function of the target site. The goal central to these themes is the understanding, on a molecular level, of the toxins, their origins, and the structures with which they interact. It is hoped that this volume will aid in the pursuit of that understanding.

As this book goes to press, two events have occurred that bear mention. First, the structure of ciguatoxin has just been published in the *Journal of the American Chemical Society* (1989, *111*, 8929–8931). Despite the determined efforts of many scientists over several years, the structure of this important marine toxin, discussed in Chapters 8, 11, and 13, has resisted elucidation. Its publication marks a notable achievement, attained through the cooperation of investigators with different, complementary skills.

Second, there has been another severe outbreak of paralytic shellfish poisoning along the Pacific coast of Central America, with very high levels of shellfish toxicity detected in Guatemala. There have, however, been no deaths and very few illnesses reported there, due to a timely response based on information gained from detailed investigations of the first outbreak. This time they were ready.

Acknowledgments

This book has been developed from a conference titled "Natural Toxins from Aquatic and Marine Environments," which was in its final planning stages at the time of the Guatemala outbreak. The conference was held August 27–30, 1987, at the Marine Biological Laboratory in Woods Hole, Massachusetts. Funding for the conference and for the production of this volume were provided by the U.S. Army Medical Research Institute of Infectious Diseases. We wish to thank this organization and particularly Col. David L. Bunner, M.D., for their support and encouragement.

Finally, and perhaps most significantly, we wish to thank the people at the ACS Books Department for their cooperation and the excellent quality of their work. Particular thanks are due to Robin Giroux, whose professional skill, enduring optimism, and unshakable determination are largely responsible for the successful publication of this volume.

SHERWOOD HALL
U.S. Food and Drug Administration
Washington, DC 20204

GARY STRICHARTZ
Anesthesia Research Laboratories
Brigham and Women's Hospital
Harvard Medical School
Boston, MA 02115

November 20, 1989

GENERAL   CONSIDERATIONS

Chapter 1

Pharmacology of Marine Toxins

Effects on Membrane Channels

Gary Strichartz and Neil Castle

Anesthesia Research Laboratories, Harvard Medical School, Brigham and Women's Hospital, 75 Francis Street, Boston, MA 02115

Ion channels in plasma membranes are primary targets of marine toxins. These channels are important regulators of a cell's physiology, and many of the pathophysiological effects of toxins arise from actions on ion channels. In this chapter we present the voltage-gated Na^+ channel, as it exists in excitable cells, as an example of a receptor with multiple binding sites for different types of toxins. The toxins are classified according to their physiological effects and described by their chemistry. Occluders, activators and stabilizers are considered as modes for toxins binding to and acting directly on the ion channel. Other, indirect actions of toxins, mediated by cellular metabolism are also discussed to provide a broad overview of the many possible modes for action of marine toxins.

"Channels" is the word used to denote the entities that account for passive ion transport across all membranes (1). Channels operate to produce very small, transient changes in ion concentrations within cells. Ionic balances are important for the regulation of cellular activities through control of intracellular pH, ion concentration, and membrane potential. For example, cytoplasmic Ca^{2+} levels critically determine the activities of a variety of enzymes, many of which are elements in extensive "cascades" that amplify the effects of single "trigger" molecules, such as hormones (2). Internal Ca^{2+} is often elevated by influx of extracellular Ca^{2+} through Ca^{2+}-selective channels (3). The example of several ion channels in mediating some rapid physiological events in neuromuscular systems is diagrammed in Figure 1.

A short primer on electrophysiology will benefit the subsequent description of toxin actions. Membrane potential is the voltage of the inside of the cell (or organelle) relative to the outside and is most often dominated by the K^+ diffusion potential; typical values range from -90 to -40 mV (negative inside) and result from the selective permeability to K^+ of the resting membrane. This permits a few K^+ ions to flow passively out from the cell, leaving the interior with a residual net negative charge. This is an example of net outward ionic current (charge flow) **polarizing** the membrane; conversely, net inward current **depolarizes** the membrane. Thus treatments that reduce outward K^+ current at rest (e.g., K^+ channel blockers) tend to depolarize cells, as do treatments that increase inward Na^+ currents (e.g., Na^+ channel activators). Cells are also depolarized by indirect dissipation of the ion (K^+) gradients through metabolic poisons or by direct inhibition of the active pumps, either the Na^+/K^+ ATPase or other exchange pumps that move one species of ion in trade for another (4). From a biochemical perspective, it is often helpful to view ion channels as the enzymes that catalyze the reactions of passive ion transport. Indeed, these

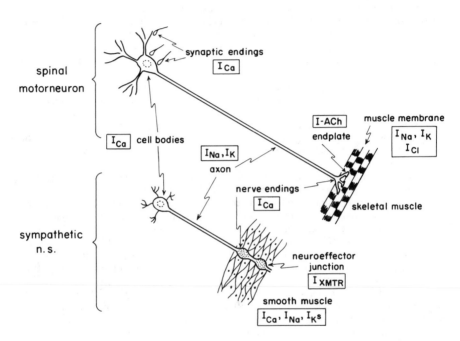

Figure 1. A depiction of the several different ionic currents necessary for the acute function of neuromuscular transmission in the skeletal motor and the efferent autonomic nervous system. The boxed current designations are associated, by the arrows, with those cellular regions where their physiological role is most evident, although these currents often exist in other regions of the cell. I_{XMTR} = neurotransmitter-activated current; I_{ACh} = acetylcholine-activated current; I_X = ion selective current where X = Na^+, Ca^{2+}, K^+, Cl^-, as noted.

enzymes have turned out, when analyzed, to be proteins, and they express the classical enzyme attributes of substrate specificity (ion selectivity), inhibition, and regulation. That many of the toxins to be described here act as inhibitors or regulators of ion channel activities further consolidates this analogy.

Consequences of channel activity often feed back on the channel itself, through reciprocally antagonistic actions that may be biochemical or physiological. A good electrophysiological example is posed by certain neurons that act as "bursting pacemakers", firing impulses in "bursts" for brief periods interrupted by impulse-free intervals (5). The initial burst results from an underlying, low frequency "pacemaker depolarization", due to the membrane potential-dependent opening of Ca^{2+} channels restricted to the soma; the subsequent activation of similarly "voltage-gated" Na^+ channels produces the superimposed spiking burst discharge which propagates along the nerve's axon. The accompanying entry of Ca^{2+} ions of itself activates "calcium-dependent" K^+ channels (6), which catalyze K^+ efflux and thus repolarize the soma membrane, shutting off the Ca^{2+} channels and the pacemaker potential. Restoration of "initial" conditions, by uptake and transport of Ca^{2+} (7), witnesses a reappearance of the pacemaker Ca^{2+} current and re-initiation of the bursting output. Intracellular Ca^{2+} may also trigger the catalyzed phosphorylation of proteins (8), release of internally sequestered Ca^{2+} (9), and itself may be elevated through the activation of other "second messenger" systems (10). The possibilities of reciprocal, complex interactions are numerous, and a single response to ion channel activation is probably a rare physiological event.

Although most examples of ion channel actions are drawn from neurobiology, their importance goes well beyond cells specifically designated as neuronal. Ion channels exist in every animal and plant cell membrane, plasmalemma and organelle included. The voltage-gated variety, previously considered hallmarks of "excitable cells" of nerve, heart, and muscle, are widespread among plasma membranes. Fibroblasts (11), endocrine cells (12-14), and those of the immune system (15) all possess and utilize ion channels in their repertoire of physiological responses. Therefore, systemically distributed toxins may directly affect many types of tissues not merely those traditionally termed "excitable".

Our goal in this chapter is to describe mechanisms through which marine toxins may act on ion channels. Examples often cite the actions of specific toxins, but sometimes a potential target will be noted or the effects of a non-marine toxin described. Since new toxins are reported regularly, a broad basis for the actions of future discoveries seems to us appropriate here.

It is not our intention to compile an exhaustive survey, but rather to categorize and to organize toxins into mechanistically identifiable classes. At first this seems a simple, direct task, but we will quickly demonstrate the complexity of actions. From the present understanding of ion channel function, no single toxin appears to exert only one simple action. One of the reasons for this is that ion channels are complex proteins whose functions involve subtle, concerted conformational changes, and, at present, our understanding of the structural basis for these functions is inadequate for a proper, molecular description of toxin-induced phenomena. Nevertheless, the observed effects are important to describe and understand, for our comprehension of channel mechanics is paralleled by our evolving knowledge of their molecular toxinology.

Classification of Toxin Actions

The actions of toxins may be classified according to the our current perspective of ion channel function. Channels open, or "gate", in response to a range of "stimuli", variables that perturb the population distribution among a set of possible channel conformations.   Recognized perturbations include changes in membrane potential, ions themselves, neurotransmitters binding directly to the channel and a range of "regulatory" substances which modify the response to other perturbations through covalent or non-covalent association with the channel itself (*1*).   Thus, a channel performs the operations of ion **conductance**, when it is "open", of **gating** between open and closed states (of which there are often several), and of **regulating** between "activatable" (gateable = openable) states and other, non-activatable (non-gateable or "inactivated") states.

Ions may carry current through the channel's "pore" (*16*), bind in the channel's pore and occlude it (*17*), bind to or associate with structural features beyond the pore and thereby modify gating (*18*), and alter the actions of other ligands (*19*).   There are many different types of ion channels, exhibiting different ion selectivities (thus, "Na^+" channels, "K^+" channels, etc.), and a variety of gating and regulatory mechanisms. Rather than describe all of them here, we will use as an example the voltage-gated Na^+ channels.   Although certain quantitative physiological and pharmacological properties of Na^+ channels differ among neuronal, skeletal muscular, and myocardial Na^+ channels, we will treat them here as a single entity, with exceptions as noted.   The advantage of using the Na^+ channel as an example rests with its elaborate and well-documented toxinology.   Many animals have evolved toxins as offensive or defensive weapons that are directed, respectively, against their prey's or predator's Na^+ channels, and these have been characterized extensively (*20*).   The disadvantage of exemplifying the Na^+ channel lies in its apparent insensitivity to endogenous regulatory mechanisms, either reversibly binding ligands (hormones, peptides) or covalent modification (e.g., phosphorylation), which are commonly seen with potassium and calcium channels (*see* below).   These activities may or do exist, respectively, on Na^+ channels but their physiological importance remains unknown.

The toxinology of Na^+ channels is best understood in the context of their normal physiology.   Sodium channels exist in several different conformations or states, most of which do not conduct ions across the membrane (i.e., are "closed") and some which do conduct ions (i.e., are "open").   Transitions among these different states are termed "gating" reactions.   A very simple scheme for Na^+ channel gating and its resultant kinetic output is diagrammed in Figure 2.   The channel has two non-conducting states — one existing in resting membranes (*R*, resting) and one in membranes held depolarized (*I*, inactivated) — as well as one conducting state (*O*, open).   Numbers at the arrows denote the relative rate constants for the transitions among different states.   In normal, toxin-free channels responding to membrane depolarization, the $R \rightarrow O$ process, called "activation", is markedly faster than inactivation ($O \rightarrow I$) and is moderately reversible; if inactivation did not occur, 80% of the channels would populate O at equilibrium.   However, inactivation reactions do occur and are almost totally irreversible, 99.9% of the channels populating the I state at the end of long depolarizations. The transient population of O is shown by the heavy dashed curve in Figure 2, which graphs the dynamics of the fraction of Na^+ channels in state O after a depolarizing step in membrane potential.   In contrast to this "macroscopic" population function, the "microscopic" behavior of a single Na^+ channel is depicted by the horizontal traces displayed on the same time scale.   Each of these traces shows the probability of one sodium channel being open.   For the toxin-free channel, the first opening occurs shortly after depolarization; temporary reversal of opening ("deactivation") accounts for many of the early closings, but eventually the virtually irreversible $O \rightarrow I$ reaction traps the channel in the inactivated state.   Inactivation must also be occurring directly from the resting state (i.e., $R \rightarrow I$) in order to assure microscopic reversibility, but normally that process is slow and competes ineffectively with the rapid activation gating.

At a different membrane potential the gating rate constants would differ.

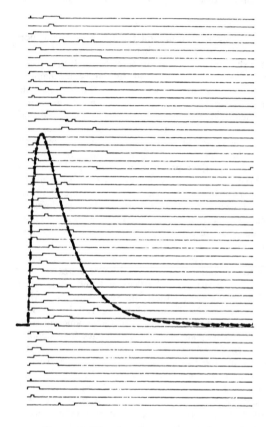

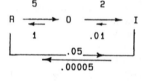

Figure 2. Kinetic schemes for Na^+ channel gating (right) and the graphed time-course for single channels (solid lines, the higher position is "open") and for the population of many channels (broken line, the fraction "open" increases upwardly). Numbers at the arrows of the kinetic scheme are the rate constants, in 10^3 sec^{-1}. The period of simulation is 5 msec. Computerized model courtesy of Dr. Daniel Chernoff.

Activation is slower in less depolarized membranes and inactivation drains the open (and resting) state more effectively. In fact, real Na^+ channels gate by more complex pathways, including several closed states intermediate between R and O, as well as multiple inactivated states. Inactivation from these intermediate states is probably faster than from R, and the entire activation process, in its fully branched entirety, is rich with kinetic possibilities. However, the effects of toxins may be understood in general by the simpler scheme presented in Figure 2.

A real protein underlies these kinetic abstractions. Sodium channels are large (ca. 200 k daltons) glyco-peptides (*21*), heavily acylated with fatty acids (*22*) and carrying a net charge or dipole moment (*23*). Anchored within the plasma membrane by hydrophobic, electrostatic, and covalent bonding, the channel protein's electrically charged regions will respond to changes of the membrane potential by conformational changes to the most stable, energetically accessible disposition. These changes constitute the gating reactions we have simplified for Figure 2. Binding of toxins to the channels modifies these conformational changes, altering the energetics and changing the apparent gating kinetics.

The known toxinology of the Na^+ channel classifies compounds into three categories: **activators, stabilizers**, and **occluders**. Although these terms describe the apparent major effect of the toxin, it should be realized that multiple effects are more often the rule than the exception.

Activators

These toxic agents increase the probability that Na^+ channels will open at membrane potentials where such openings are usually quite rare. As noted above, the normal, voltage-gated process of opening is called "activation", thus the title of this class of drugs. Most of the known activators are not derived from marine organisms, but instead are alkaloids extracted from terrestrial plants or amphibians, e.g., aconitine, veratridine (VTD: veratrum spp.) or batrachotoxin (BTX), respectively (Figure 3; ref *24*). These relatively lipophilic, organic molecules, typically of mass less than 10^3 daltons, produce their full pharmacological effects whether applied intracellularly or from without the cell, and probably act from within the membrane. The effects are inhibited noncompetitively by occluder toxins (e.g., tetrodotoxin, *see* below), and result from binding at a specific site on the channel (*25, 26*).

Activators alter the opening and the closing probability of Na^+ channels, tending to activate channels at voltages near the cell's resting potential and to prevent their normal, complete inactivation. The result is a long-lasting increase in Na^+ permeability and a corresponding depolarization from the steady inward Na^+ current (*27*). But if the membrane potential is held constant by laboratory manipulations (voltage-clamp), then the drug-modified channel's activation and inactivation kinetics may be examined under more controlled conditions (*28–30*).

One example of channel activation by a marine toxin is shown in Figure 4. Here the effects of a brevetoxin (BvTX) activator on Na^+ currents under voltage-clamp are characterized by three aspects, according to the traditional description of such phenomena; the amplitude of currents is reduced, the voltage range for activation is "shifted", and the kinetics of inactivation, shown as the declining phase of the currents, are slowed (*31*). It should be noted that BvTX and a related organic marine toxin, ciguatoxin (*32*), act at one site on the channel which is separate from the other activator (e.g., BTX) binding site (*33–35*), yet the physiological effects are largely indistinguishable (*see* chapter by Baden et al., this volume).

At first these effects of activators may appear complicated, requiring multiple factors to explain the total response, but reference to the kinetic model for channel gating (Figure 2) clarifies the complexity. Activators may produce spontaneous opening

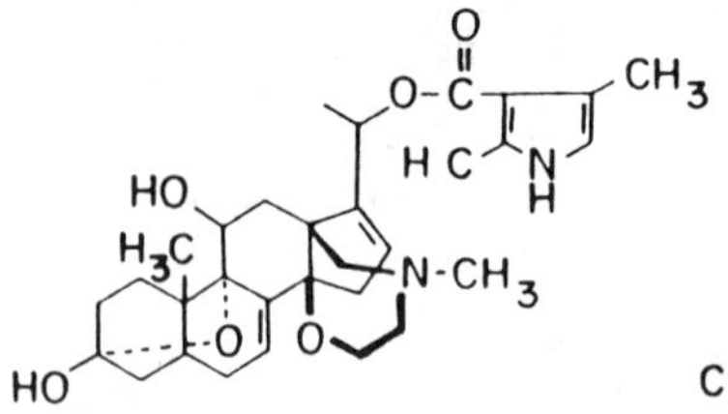

Batrachotoxin

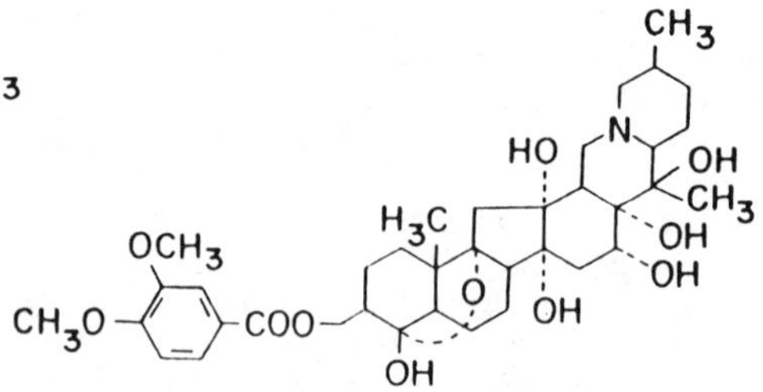

Veratridine

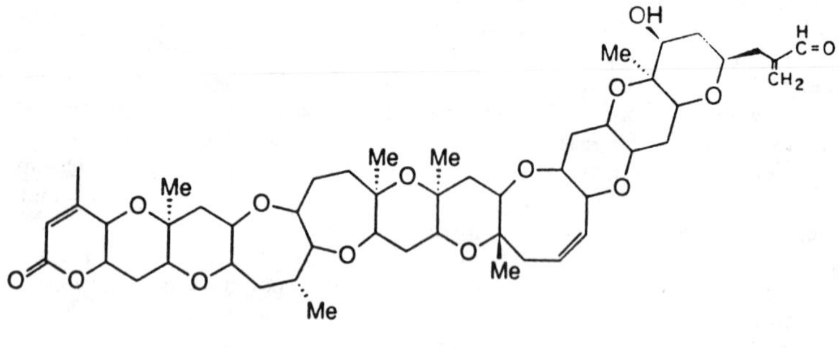

Brevetoxin-B

Figure 3. The chemical structures of two "classical" alkaloid activators (batrachotoxin and veratridine) and of a recently characterized marine toxin [brevetoxin B (BvTX-B)], that acts at a different site on the Na$^+$ channel.

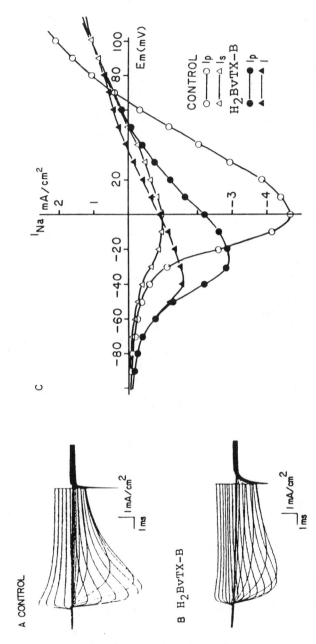

Figure 4. Effects of dihydro-brevetoxin B (H$_2$BvTX-B) on Na$^+$ currents (I$_{Na}$) in crayfish axon under voltage-clamp. (A) A family of Na$^+$ currents in control solution; each trace shows the current kinetics responding to a step depolarization (ranging from -90 to +100 mV in 10 mV increments). Incomplete inactivation at large depolarizations is normal in this preparation. (B) Na$^+$ currents after internal perfusion with H$_2$BvTX-B (1.2 μM). I$_{Na}$ inactivation is slower and less complete than in the control, and the current amplitudes are reduced. (C) A plot of current amplitudes at their peak value (I$_p$; $\circ$, $\bullet$) and at steady-state (I$_s$; $\triangle$, $\blacktriangle$ for long depolarizations) shows that toxin-modified channels (filled symbols) activate at more negative membrane potentials (E$_m$) and correspond to a reduced peak Na$^+$ conductance of the axon (Reproduced with permission from Ref. 31. Copyright 1984 American Society for Pharmacology and Experimental Therapeutics).

by shifting channels at rest from R to intermediate closed (C) and even open (O) states. Activation of activator-modified channels therefore requires smaller membrane depolarization (leading to increased spontaneous openings) and occurs without the kinetic delay seen in unmodified channels (36), consistent with few or no intermediate states to pass through before opening. Activator binding stabilizes the open or pre-open states and thus inhibits their inactivation reaction, slowing the net decline of the Na^+ current.

The shift in activator-bound channels from resting to intermediate closed forms is equivalent to having a higher affinity of these closed forms for the activators. This behavior is characteristic of "modulated receptor" models, described later in this chapter. Channels react faster with most activators if the membrane is repetitively depolarized, again consistent with the drug's binding to activated channels ($28, 29, 37$). When open, the drug-modified channels have a lower conductance and a different ion selectivity compared to unmodified channels, i.e., the detailed energetics of ion transport are altered ($30, 36, 38$). Finally, modified channels may close and re-open many times in a depolarized membrane, in contrast to normal channels which rapidly inactivate (38).

One activator molecule binding to a Na^+ channel influences the total function, including all the gating and the ion permeation. It is pharmacologically noteworthy that these lipophilic ligands, of a modest size compared to the large channel protein, exert such extensive influence. Either the structural features subserving channel functions are localized in a small, central domain of the protein, or concerted, molecule-wide structural changes attending these functions can be affected through one integrative site, or the activator molecule can move in milliseconds among multiple points on the Na^+ channel.

The summed effects of activators are to increase the population of open channels (albeit of reduced Na^+ conductance) over a broad range of membrane potentials. Under normal conditions this results in slow depolarization of the membrane attended by a period of rapid, spontaneous impulse firing. Nerve, muscle, and heart all appear to be sensitive to activators, which probably also penetrate the blood brain barrier, affecting the central nervous system; the pathophysiology of activator intoxication is extensive and interactive. Activator effects are antagonized by local anesthetics and thus, ironically, an antidote might be found in the intentional, transitory administration of systemic local anesthetic in conjunction with artificial respiration, at least until the activators can be cleared from the body. By partially blocking activator-modified Na^+ currents, the often toxic systemic anesthetics can reverse the excited responses of whole organs and prevent the persistent depolarization that may otherwise result in cell death.

Stabilizers

A second category of toxins has an apparently simple action. These drugs prolong and inhibit the inactivation phase of Na^+ currents during depolarization, thereby stabilizing channels in the open state without affecting the activation process (Figure 5). The best examples of such toxins from marine sources are small peptides isolated from nematocysts of a variety of anemones (39) and larger proteins found in venoms of Conidae (40). While some of these toxins at high concentrations do reduce the amplitude of Na^+ currents, the more typical effect is prolongation of the period of high Na^+ conductance. In non-voltage clamped membranes, the effect of such stabilizers is to greatly prolong the action potential, generating a "plateau" depolarization lasting from tenths of seconds to seconds (41). This long depolarization greatly increases the amount of Ca^{2+} that will enter a cell through voltage-gated Ca^{2+} channels, and also serves to generate "local circuit" currents that spread passively to

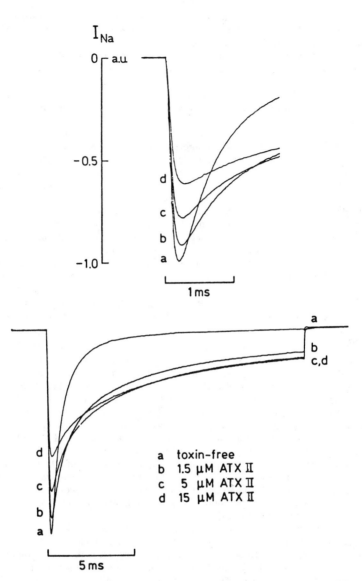

Figure 5. Multiple actions of toxin II from *Anemonia sulcata* (ATX II) on voltage-clamped Na^+ currents (I_{Na}) from amphibian myelinated nerve. This "stabilizer" toxin works in a dose-dependent manner to inhibit channel inactivation (*see* bottom panel) and, as a consequence, delay the time of peak current (*see* top panel). The reduction of peak current amplitude does not result directly from these kinetic alterations and is not observed with all stabilizers (Reproduced with permission from Ref. 39. Copyright 1981 SPPIF).

adjacent, toxin–free regions of a cell (e.g. along an axon) and initiate repeated impulses there. The net result in either case is an increase in depolarization-related activities such as secretion, contraction, etc.

Stabilizers bind at a site separate from those of traditional activators and of ciguatoxin-brevetoxin, but they exert a synergistic action on both types of activators (35, 42). This action potentiates the activators and generally increases their efficacy, yielding larger depolarizations at lower doses (42); it occurs uniquely with the peptide stabilizers and not with ions or oxidants that also slow the inactivation of Na^+ current (37).

The actions of stabilizers present one of the classic voltage-dependent toxin effects. Depolarizations of the membrane sufficient to energetically drive toxin-bound channels to an inactivated conformation also lead to dissociation of the stabilizer (43-45). The energetics of this process are related to its kinetics in an interesting way. For a tightly binding stabilizer toxin, of $K_D \sim 10^{-8}M$, small depolarizations (e.g., to -20 mV) result only in slow dissociations that require minutes to complete, whilst large depolarizations (e.g., to + 50 mV) result in complete dissociation within 10 msec. Stabilizers that bind less strongly, with $K_D \sim 10^{-6}M$, reverse more rapidly at lower depolarizations. Toxin dissociation results not from a direct effect of the membrane potential on the binding reaction per se, but on a voltage-dependent modulation of the toxin binding site from a high to a low affinity form. Apparently the inactivated state has an affinity for stabilizers that is at least an order of magnitude less than that of the resting or open states. Sufficiently positive potentials force the stabilizer-bound channel into an inactivated conformation, leading to rapid relaxation of toxin binding to the new equilibrium corresponding to the low affinity, inactivated conformation.

Occluders

Tetrodotoxin (TTX) and the saxitoxins (STX) are classic examples of marine toxins acting as the third class of agents, occluders. These small (300–450 daltons), organic cations act only from the external surface of Na^+ channels to produce a readily reversible, yet usually high affinity $(K_D \sim 10^{-9}M)$ block of the channel (see Figure 6). This block is antagonized by Na^+, Ca^+, and H^+ (19, 47) and was originally thought to correspond to a "plugging" of the channel's pore (48). The rapid kinetics of activation and inactivation of conducting channels are unaffected by TTX/STX, and the movement of charge associated with activation remains unchanged (23). The overall effect appears equivalent to reducing, reversibly, the number of channels that have patent conductance pathways (49; see Figure 6). But more recent evidence questions this simple plugging model; the structural dependence of toxin activity is more complicated than necessary for simple pore blockade (50), and the physiological behavior extends beyond simple occlusion (47).

In mammalian cardiac Na^+ channels, an unusually low affinity for TTX $(K_D\ 10^{-6}M)$ is accompanied by an apparent "voltage-dependent" action — the toxin block is altered by the membrane potential, probably through some conformational change of the channel (51). Channel gating is also subtly modified by these toxins; in both cardiac (52) and neuronal (47) tissue the tendency, after long (> 1 sec) depolarizations, of the channels to dwell in a state from which they can only be slowly made activatable (i.e., a "slow inactivated state"), is furthered by TTX and STX (Figure 7). The underlying mechanism for this gating modulation is completely unknown.

Certain occluders also discriminate among Na^+ channels from neuronal and skeletal muscle. But in this case the blocking ligands are small peptides, the μ-conotoxins from the mollusc *Conus geographus*. This molecule binds tightly to muscle Na^+ channels, effectively reducing Na^+ current (53; see Figure 6A), and also can displace bound

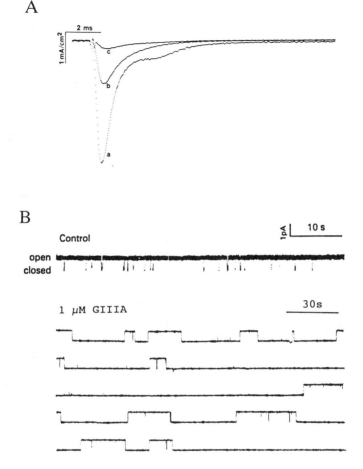

Figure 6. Blockade of Na$^+$ channels by μ-conotoxin (GIIIA) or by saxi-toxin (STX) (*46*).   A. Macroscopic currents in frog skeletal muscle recorded in control solution (a) and 10–15 min (b) and c) 25–30 min (c) after external perfusion with GIIIA (2 μM). B. Single channel currents from skeletal muscle vesicles which are fused with planar lipid bilayers and modified by batrachotoxin, ensuring long open times under control conditions. The downward deflecting closing events are caused by binding of GIIIA or STX, and the duration of the closed (blocked) state is inversely proportional to the particular toxin's dissociation rate. Note the difference in time scales. (Reproduced with permission from Ref. 46. Copyright 1986 The New York Academy of Sciences).

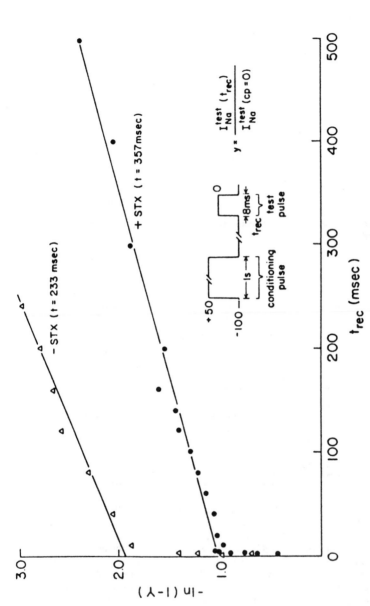

Figure 7. Slow inactivation of Na⁺ channels is potentiated by STX. The graph shows the time required for the recovery of Na⁺ channels to an activatable state after a long (1 sec, +50 mV) inactivating depolarization. When tested by a brief test pulse, control currents (Δ) recovered in a fast (τ = 233 msec) phase. Addition of STX (○, 2 nM, which approximately halved the currents with no inactivating pulse) approximately doubled the fraction of currents recovering in the slow phase and also increased the time constant of slow recovery. The fast recovery rate was unaffected. (Reproduced with permission from Ref. 47. Copyright 1986 The New York Academy of Sciences).

radiolabelled STX, showing that it acts at the same site (54, 55). Reversible blockade of single muscle channels resembles that by STX (Figure 6B); although the potency of μ-conotoxins is lower, the dissociation rate is also smaller. In neuronal and cardiac Na^+ channels, the blockade of Na^+ currents and the inhibition of STX binding are far weaker, thus μ-conotoxins discriminate between channels that bind TTX and STX about equally.

The results from such occluders show that Na^+ channels differ in structure between their internal and external surfaces, have binding sites for a variety of monovalent and divalent cations, and are pharmacologically different despite very similar physiological actions in nerve, muscle, and mammalian cardiac cells.

It is noteworthy that two other groups of small peptides from the same organism's venom affect other ion channels. The ω-conotoxins block certain Ca^{2+} channels, and α-conotoxins block acetylcholine-activated ion channels as found at the neuromuscular junction (see Figure 1; see ref. 56). *Conus* venoms contain a variety of toxins that affect Na^+ channel gating (40, 57) and others that modify K^+ channels, as well as a high concentration of phospholipase A_2 (58). As with other offensive (predatory) venoms, the mixture of active components has a synergism that exploits the prey's normal physiology to effect a rapid, convincing paralysis.

The Na^+ Channel Is a Modulated Receptor

In all of the examples cited above, changes in the states of the channel are affected by toxin action. Reciprocally, the binding and actions of the toxins are changed by voltage-driven gating among different channel conformations. Such a dynamic interaction has been termed the "modulated receptor" model, and was originally proposed as an explanation for effects of local anesthetics on Na^+ channels (59). For the toxin classes reviewed here, stabilizers have selective affinity for resting over inactivated channels, occluders seem to increase the tendency of channels to enter the "slow inactivated" state, and most activators bind more tightly and certainly more rapidly to an activated state of the Na^+ channel. An exception are the brevetoxins, which, like BTX or veratridine, alter channel kinetics to favor the open state, but whose onset of action is not accelerated by channel activation.

A second aspect of the modulated receptor is the interaction among toxins of different classes. We previously noted that activators (including brevetoxin), were potentiated by stabilizers. In addition, the binding of brevetoxin is slightly enhanced by occluders (60), and the blocking action of occluders (STX and TTX) becomes voltage-dependent in activator (BTX)-modified channels (61), whereas it normally exhibits no voltage-dependent action (62). Thus, the Na^+ channel has multiple, separate, yet interactive binding sites for several types of toxins (Figure 8). Taken together with the broad effects of membrane potential on individual toxin actions, it seems probable that structural changes ripple through extensive domains of the channel during normal physiological gating.

Toxin-Induced Permeabilities

Certain marine toxins induce "new" ion permeabilities, rather than modifying existing ion channels. Palytoxin (PTX), a large organic molecule, for example, irreversibly increases the cation permeability of many cell membranes, to both Na^+ and other metal and organic cations (63–65; see chapter by Ohizumi in this volume). The mode of action may involve the Na^+/K^+ pump, converting it to a passive channel (or carrier-exchanger), for the depolarizing effect of PTX can be inhibited by preincubation of the tissue with ouabain (strophanthin), a glycoside that selectively inhibits the Na^+/K^+ pump (65). Alternatively, some other protein or carbohydrate on the

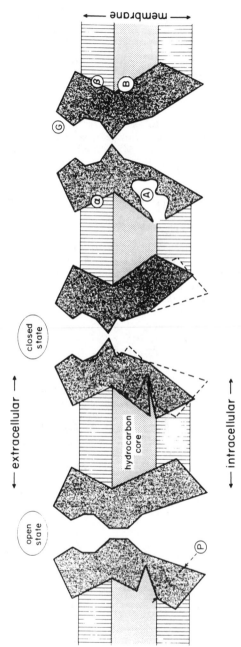

Figure 8. A schematic for the toxin binding sites on the voltage-gated Na^+ channel. Toxin-free open and closed conformations are drawn at the left and center. Separate sites are depicted within the membrane for activators such as BTX, VTD (A), and brevetoxin (B); these are coupled to each other and to the α-peptide toxin site (α), which is kinetically linked to the β-peptide toxin site (β; see ref. 20). Near the outer opening of the pore is a site (G) for STX and TTX which is affected by binding at site A and which can modify inactivation gating.

cell's surface may act as a common anchor for both drugs, accounting for their antagonism, independent of the Na^+/K^+ pump.

A similar phenomenology attends the actions of maitotoxin, another large organic molecule that induces a Ca^{2+} selective membrane permeability (*see* chapter by Ohizumi & Kobayashi in this volume). None of the known inhibitors of Ca^{2+} channels, including Co^{2+}, Cd^{2+}, dihydropyridines, and verapamil or diltiazem affect the maitotoxin-induced increase in Ca^{2+} permeability. To date, an association between maitotoxin and an existing Ca^{2+} pump or exchange protein has not been demonstrated.

Indirect Effects of Toxins

It is worth noting that toxins do not necessarily act directly upon the channel protein to modify its function. As mentioned earlier, it has become increasingly clear that the gating processes of many ion channels are intimately associated with activation of receptors by neurotransmitter or hormone. The degree of this interaction ranges from complete control of the gating process (66) to just a modulatory role. These receptors are linked to a wide variety of molecular cascades which provide a rich repertoire for modulation of ion channel function. Apart from the cyclic-AMP cascade which has received wide attention (67), several other second messenger pathways play significant roles in channel modulation. These include the products of phosphatidylinositol metabolism; inositol 1,4,5-triphosphate, which through its release of intracellular stores of Ca^{2+}, activates a variety of calcium-dependent ion channels (68,69); and diacylglycerol, the activator of protein kinase C, which has been shown to both augment and inhibit the gating of K^+ and Ca^{2+} channels (70,71). Direct coupling of neurotransmitter receptors to K^+ and Ca^{2+} channels via GTP-binding proteins (G-proteins) also occurs (66). More recently, the arachidonic acid metabolic cascade has been implicated in the activation of a K^+ channel in aplysia neurons by the neuropeptide, FMRFamide (72).

By interfering with any one of the many phases associated with these second messenger pathways, toxins may alter channel gating. For example, the blue green algal toxins, aplysiatoxin, and lyngbyatoxin bind to and activate protein kinase C in a manner similar to phorbol esters (73). They also stimulate arachidonic acid metabolism (74). The coral toxin, palytoxin, also stimulates arachidonic acid breakdown albeit by an unknown mechanism (74) and affects other biochemical activities of the cell (*see* chapters by Fujiki et al., Wattenberg et al., and Levine et al., this volume).

Several toxins, in addition to acting upon ion channels appear to effect second messenger processes. Maitotoxin, whose primary action has been proposed to be activation of a Ca^{2+} channel (75) also stimulates phosphotidylinositol turnover (76). Although this latter action does not appear to be a consequence of its channel activating ability, it is unclear whether the converse is true. However, it is interesting to note that manoalide, a sesterterpenoid from sponge which inhibits Ca^{2+} channels in a number of cell types (77), also inhibits norepinephrine-stimulated IP_3 production (78). Since manoalide is known to be an irreversible inhibitor of phospholipase C, a key enzyme in the inositol trisphospate/diacylglycerol second messenger pathway, it is tempting to associate this action with its ability to inhibit Ca^{2+} channels.

Conclusion

Marine toxins modify the functions of many different types of ion channels in animal cell membranes. These channels may be important for maintaining the cell's resting potential, for generating electrical membrane signals, such as impulses, and for controlling hormonally triggered or metabolic responses. Thus toxins may depolarize membranes, leading to a (sometimes transient) increase in cellular activities, or they may inhibit activities. Toxin actions may be direct, equivalent to binding at a channel, or indirect, resulting from modifications mediated by second messengers, G-proteins, cAMP, etc. At present we can describe the biochemistry and biophysics of several toxin:channel interactions, and we have phenomenological reports of the pathophysiology of intoxication. What we lack is an understanding of the roles of ion channels in the normal physiology of cells and tissues. Knowledge of these processes are essential for linking the two aspects of our knowledge, and using such knowledge to design effective therapies against poisoning by marine toxins.

Acknowledgements

We thank Dr. Daniel Chernoff for providing the simulation program used in Figure 2. Sherwood Hall provided enthusiastic commentary, beyond the editorial call of duty, and Rachel Abrams and Mary Gioiosa were helpful and patient in the facile preparation of the manuscript.

Literature Cited

1. Hille, B. *Ionic Channels of Excitable Membranes*; Sinauer Associates: Sunderland, MA, USA, 1984.
2. Rubin, R.P. *Calcium and the Secretory Process*; Plenum Press: New York, 1974.
3. Fox, A.P.; Nowycky, M.C.; Tsien, R.W. *J. Physiol.* **1987**, *394*, 149–172.
4. Almers, W.; Stirling, C. *J. Membrane Biol.* **1984**, *77*, 169–186.
5. Connor, J.A. *Federation Proceedings* **1978**, 37 (*8*), 2139–2145.
6. Hermann, A.; Kartung, K. *Cell Calcium* **1983**, *4*, 387–05.
7. Goldin, S.M.; Moczydlowski, E.G.; Papazian, D.M. *Ann. Rev. Neurosci.* **1983**, *6*, 419–46.
8. Kaczmarek, L.K. *TINS* **1987**, *10*, 30–34.
9. Fabiato, A.; Fabiato, F. *J. Physiol. London* **1975**, 249 469–495.
10. Meyer, T.; Holowka, D.; Stryer, L. *Science* **1988**, *240*, 653–56.
11. Pouyéssgur; Jacques, Y.; Lazdunski, M. *Nature* **1980**, *286*, (*5769*), 162–164.
12. Dubinsky, J.M.; Oxford, G.S. *J. Gen. Physiol.* **1984**, *83*, 309–339.
13. Lewis, D.H.; Weight, F.F.; Luini, A. Proc. Natl. Acad. Sci. USA **1986**, *83*, 9035–9039.
14. Pappone, P.A.; Lucero, M.T. *J. Gen. Physiol.* **1988**, *91*, 817–833.
15. DeCoursey, T.E.; Chandy, K.G.; Gupta, S.; Cahalan, M.D. *J. Neuroimmunology* **1985**, *10*, 71–95.
16. Hille, B. *J. Gen. Physiol.* **1972**, *59*, 637–58.
17. Hille, B. Biophys. J. **1975**, *15*, 164a.
18. Oxford, G.S.; Yeh, J.Z. *J. Gen. Physiol.* **1985**, *85*, 583–02.
19. Henderson, R.; Ritchie, J.M.; Strichartz, G.R. *Proc. Nat. Acad. Sci. USA* **1974**, *71*, 3936–3940.
20. Strichartz, G.; Rando, T.A.; Wang, G.K. *Ann. Rev. Neurosci.* **1987**, 10, 237–67.
21. Miller, J.A.; Agnew, W.S.; Levinson S.R. *Biochemistry* **1983**, *22*, 462–470.
22. Levinson S.R.; Duch, D.S.; Urban B.W.; Recio-Pinto, E. In *Tetrodotoxin, Saxitoxin and the Molecular Biology of the Sodium Channel*; Levinson, S.R., Kao, C.Y., Eds.; New York Acad. Sci.: New York, N.Y., 1986; pp 162–178.
23. Armstrong, C.M.; Bezanilla,F. *J. Gen. Physiol.* **1974**, *63*, 533–52.
24. Catterall, W.A. *Proc. Natl. Acad. Sci. USA* **1975**, *72*, 1782–1786.
25. Catterall, W.A. *Ann. Rev. Pharmacol. Toxicol.* **1980**, *20*, 15–43.
26. Lazdunski, M.; Balerna, M.; Barhanin, J.; Chicheportiche, R.; Fosset, M.; Frelin, C.; Jacques, Y.; Lombet, A.; Pouyssegur, J.; Renaud, J.F.; Romey, G.; Schweitz, H.; Vincent, J.P. *Neurotransmitters and their Receptors*; Littauer, U.Z., Dudai, Y., Silman, I., Teichberg, V.I., Vogel, Z., Eds.; John Wiley and Sons Ltd., 1980.

27. Ulbricht, W. *Pflugers Archiv, European Journal of Physiology* 1972, *336*, 201–212.
28. Sutro, J.B. *J. Gen. Physiol.* 1986, *87*, 1–24.
29. Leibowitz, M.D.; Sutro, J.B.; Hille, B. *J. Gen. Physiol.* 1986, *87*, 25–46.
30. Rando, T.A. *J. Gen. Physiol.* 1989, *93*, 43–65.
31. Huang, J.M.C.; Wu, C.H.; Baden, D.G. *J. Pharmacol. Exp. Therap.* 1984, *229*, 615–621.
32. Lombet, A.; Bidard, J.N.; Lazdunski, M. *FEBS Letters* 1987, *219*, 355–359.
33. Wu, C.H.; Huang, J.M.C.; Vogel, S.M.; Luke, V.S.; Atchison, W.D.; Narahashi, T. *Toxicon* 1985, *23*, 481–87.
34. Catterall, W.A.; Gainer, M. *Toxicon* 1985, *23*, 497–504.
35. Strichartz, G.R.; Crill, E.A.; Rando, T.A.; Blizzard, T.; Qin, G.W.; Lee M.S.; Nakanishi, K. *Biophys. J.* 1987, *51*, 193a.
36. Khodorov, B.I. *Prog. Biophys. Molec. Biol.* 1985, *45*, 57–148.
37. Rando, T.A.; Wang, G.K.; Strichartz, G.R. *Mol. Pharmacol.* 1986, 29, 467–477.
38. Garber, S.S.; Miller, C. *J. Gen. Physiol.* 1987, 89, 459–480.
39. Ulbricht, W.; Schmidtmayer, J. *J. Physiol. Paris* 1981, *77*, 1103–1111.
40. Gonoi, T.; Ohizumi, Y.; Kobayashi, J.; Nakamura, H.; Catterall, W.A. *Molecular Pharmacology* 1987, *32*, 691–698.
41. Wang, G.K.; Strichartz, G. *J. Gen. Physiol.* 1985, *86*, 739–762.
42. Catterall, W. *J. Biol. Chem.* 1976, *251*, 5528–5536.
43. Mozhayeva, G.N.; Naumov, A.P.; Nosyreva, E.D.; Grishin, E.V. *Biochim. Biophys. Acta* 1980, *597*, 587–602.
44. Warashina, A.; Fugita, S.; Satake, M. *Pflgers Arch.* 1981, *391*, 273–276.
45. Strichartz, G.R.; Wang, G.K. *J. Gen. Physiol.* 1986, *88*, 413–435.
46. Moczydlowski, E.; Uehara, A.; Guo, X.; Heiny, J. In *Tetrodotoxin, Saxitoxin and the Molecular Biology of the Sodium Channel*; Kao, C.Y., Levinson, S.R., Eds.; New York Acad. Sci.: New York, 1986; pp. 269–292.
47. Strichartz, G.; Rando, T.; Hall, S.; Gitschier, J.; Hall, L.; Magnani, B.; Hansen Bay, C. In *Tetrodotoxin, Saxitoxin and the Molecular Biology of the Sodium Channel*; Kao, C.Y., Levinson, S.R., Eds.; New York Acad. Sci.: New York, 1986; pp 96–112.
48. Kao, C.Y.; Nishiyama, A. *J. Physiol. London* 1965, *180*, 50–66.
49. Hille, B. *J. Gen. Physiol.* 1968, *51*, 199–219.
50. Kao, C.Y. *Toxicon* 1983, *21* (Suppl. 3), 211–219.
51. Cohen, C.J.; Bean, B.P.; Colatsky, T.J.; Tsien, R.W. *J. Gen. Physiol.* 1981, *78*, 383–411.
52. Burnashev, N.; Sokolova, S.N.; Khodorov, B.I. *Gen. Physiol. Biophys.* 1984, *3*, 507–509.
53. Cruz, L.J.; Gray, W.R.; Olivera, B.M.; Zeikus, R.P.; Kerr, L.; Yoshikami, D.; Moczydlowski, E. *J. Biol. Chem.* 1985, *260*, 9280–9288.
54. Moczydlowski, E.; Olivera, B.M.; Gray, W.R.; Strichartz, G.R. *Proc. Natl. Acad. Sci USA* 1986, *83*, 5321–25.
55. Yanagawa, V.; Abe, T.; Satake, M. *Neurosci. Lett.* 1986, *64*, 7–12.
56. Olivera, B.M.; Gray, W.R.; Zeikus, R.; McIntosh, J.M.; Varga, J.; River, J.; Santos, V.D.; Cruz, L.J. *Science* 1985, *230*, 1338–1343.
57. Strichartz, G.; Wang, G.K.; Schmidt, J.S.; Hahin, R.; Shapiro, B.I. *Fed. Proc.* 1980, *39*, 2428.
58. Chesnut, T.J.; Carpenter, D.O.; Strichartz, G.R. *Toxicon* 1987, 25, 267–278.
59. Hondeghem, L.M.; Katzung, B.G. *Ann. Rev. Pharmacol. Toxicol.* 1984, *24*, 387–423.
60. Poli, M.A.; Mende, T.J.; Baden, D.G. *Mol. Pharmacol.* 1986, *30*, 129–135.
61. Krueger, B.K.; Worley, J.F.; French, R.J. *Nature* 1983, *303*, 172–175.
62. Rando, T.A.; Strichartz, G.R. *Biophys. J.* 1986, *49*, 785–794.

63. Chhatwal, G.S.; Hessler, H.J.; Habermann, E. *Naunyn-Schmiedeloergs Arch. Pharmacol.* **1983**, *323*, 261–268.
64. Maramatsu, I.; Nishio, M.; Kigoshi, S.; Uemura, D. Br. *J. Pharmacol.* **1988**, *93*, 811–816.
65. Castle, N.A.; Strichartz, G.R. *Toxicon* **1988**, *26*, 941–51
66. Brown, A.M.; Birnbaumer, L. Am. *J. Physiol.* **1988**, *254*, H401–H410.
67. Levitan, I.B. *J. Membrane Biol.* **1985**, *87*, 177–190.
68. Berridge, M.J. *Ann. Rev. Biochem.* **1987**, *56*, 159–93.
69. Blatz, A. L.; Magleby, K. L. *Trends in Neurosci.* **1987**, *10*, 463–467.
70. Kaczmarek, L.K. *Trends in Neurosci.* **1987**, *10*, 30–34.
71. Levitan, I.B. *Ann. Rev. Neurosci.* **1988**, *11*, 119–36.
72. Belardetti, F.; Siegelbaum, S.A. *Trends in Neurosci.* **1988**, 11 232–238.
73. Fujiki, H.; Suganuma, M.; Tahira, T.; Yoshioka, A.; Nakayasu, M.; Endo, Y.; Shudo, K.; Takayarva, S.; Moore, R.E.; Sugimura, T. In *Cellular Interactions by Environmental Tumor Promoters*; Fujiki, H., Hecker, E., Moore, R.E., Sugimura, T., Weinstein, I.B., Eds.; Japan Sci. Soc. Press, Utrecht, Tokyo/VNV Science Press: Tokyo, 1984; pp 37–45.
74. Levine, L; Fujiki, H. *Carcinogenesis* **1985**, *6*, 1631–1634.
75. Kobayashi, M.; Ochi, R.; Ohizumi, Y. Br. *J. Pharm.* **1987**, *92*, 665–671.
76. Berta, P.; Sladeczek, F.; Derancourt, J.; Durand, M.; Travo, P.; Haiech, J. *FEBS Lett.* **1986**, *197*, 349.
77. Wheeler, L.A.; Sachs, G.; DeVries, G.; Goodrum, D.; Wolderussie, E.; Mwallem, S. *J. Biol. Chem.* **1987**, *262*, 6531–6538.
78. Bennett, C.F.; Mong, S.; Wu, H.-L.W.; Clark, M.A.; Wheeler, L.A.; Crooke, S.T. *Molecular Pharmacol.* **1987**, *32*, 587–593.

RECEIVED May 4, 1989

Chapter 2

Biosynthesis of Red Tide Toxins

Yuzuru Shimizu, Sandeep Gupta, and Hong-Nong Chou

Department of Pharmacognosy and Environmental Health Sciences, College
of Pharmacy, University of Rhode Island, Kingston, RI 02881–0809

The biosynthesis of two major classes of red tide toxins, saxitoxin analogs and brevetoxins, have been studied. It was shown that saxitoxin
is biosynthesized from arginine, acetate, and methionine methyl group.
Brevetoxins were shown to be unique polyketides, which are probably
biosynthesized from dicarboxylic acids. Some details of the biosynthetic mechanism have been elucidated.

The elucidation of the biosynthetic mechanism of red tide toxins is an important
step to understand the toxigenesis of the deleterious red tide organisms. Questions
have been raised about what triggers the production of the toxins in the organisms,
where the regulating step is, or how the organisms acquire the ability to produce
the toxins. More than a decade ago when we started to study the biosynthesis of
saxitoxin analogs and brevetoxins, very little was known about the molecular origins
of the unique classes of compounds. We now have the knowledge of all the
building blocks of saxitoxin molecule, and some details of the biosynthetic process.

Biosynthesis of Saxitoxin Analogs

Saxitoxin was first isolated from Alaskan butter clam *Saxidomus giganteus*, but is
actually produced by the dinoflagellate, *Gonyaulax* spp *(1,2)*. More than a dozen
related toxins have been isolated by several groups including ours *(3)*. All of the
toxins have the same unique tricyclic perhydropurine skeleton, and there were various speculations about the origin of the perhydropurine ring (I).

In the early stage of our investigation, we obtained experimental results which
indicated that the major portion of the ring system is derived from arginine as
shown in Scheme 1 *(4)*. However, our repeated attempts to incorporate [1-
^{13}C]arginine into the toxin molecule did not result in any specific labeling. In
Scheme 1, two C_1 units or one C_2 unit would become C-5 and C-13 of the toxin
molecule. Therefore, as an alternate approach, we fed [1,2-$^{13}C_2$] acetate to a
toxin-producing strain of the blue-green alga, *Aphanizomenon flos-aquae (5)*.

Two intact units of acetate were incorporated into neosaxitoxin as evidenced by
the appearance of two sets of AB type signals in the ^{13}C-NMR spectrum. The
orientation of the incorporated acetate units was determined by feeding [2-
^{13}C]acetate as shown in Scheme 2. This result clearly excludes the original pathway
in which C-5 should come from C-1 of arginine.

To accommodate this new finding and the previous results, we considered a
new pathway (Scheme 3), in which acetate or its derivative condenses with arginine
followed by decarboxylation. Such Claisen-type condensation on alpha-amino acid
has some precedent in biochemical systems *(6)*. To prove this hypothesis, we synthesized [2-^{13}C, 2-^{15}N]arginine and ornithine and fed to *A. flos-aquae (5)*.

0097–6156/90/0418–0021$06.00/0

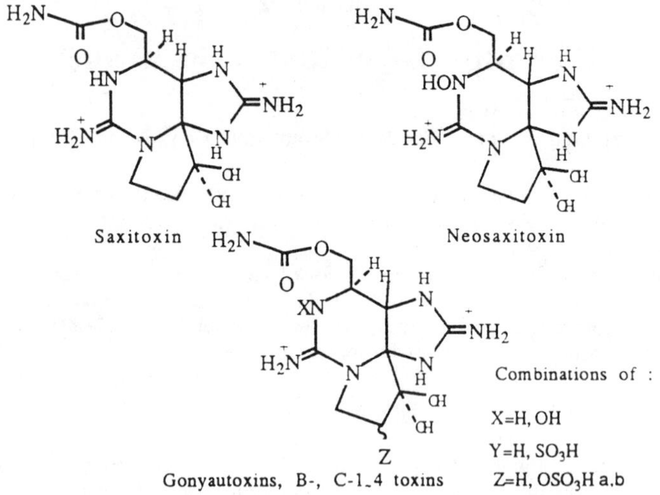
Saxitoxin
Neosaxitoxin
Combinations of :
X=H, OH
Y=H, SO_3H
Z=H, OSO_3H a,b
Gonyautoxins, B-, C-1-4 toxins

I

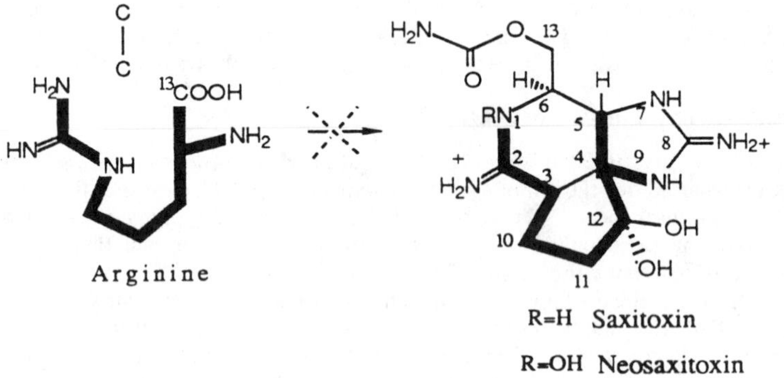
Arginine
R=H Saxitoxin
R=OH Neosaxitoxin

Scheme 1

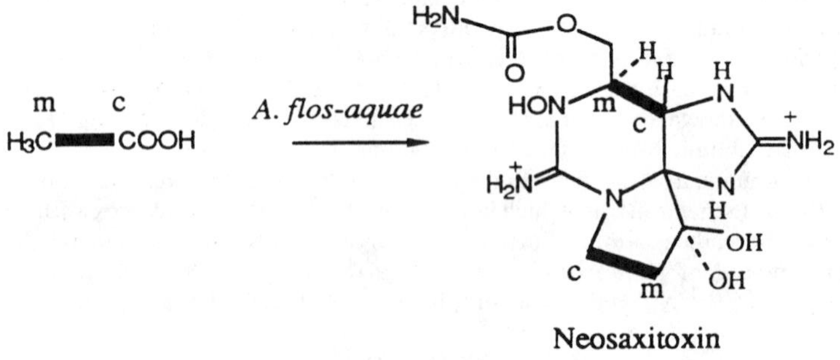
A. flos-aquae
Neosaxitoxin

Scheme 2

The ^{13}C-NMR spectrum of neosaxitoxin obtained from the feeding experiment showed an enhanced signal for C-4, which was split into a doublet by the spin-spin coupling with the neighboring ^{15}N (J=9.3 Hz) (Scheme 4). The result clearly indicated that the connectivity C-2 - N-2 of arginine was incorporated intact into the toxin molecule, supporting the pathway in Scheme 3.

The next question was the origin of the side-chain carbon C-13. We first fed formate as a general precursor of C_1 units without success. The failure made us consider CO_2 as a possible source. The idea seemed attractive, since CO_2 can be fixed as malonyl CoA, which could undergo condensation with arginine (Scheme 5). Vigorous efforts were made to effect the selective enrichment of C-13 by pulse-feeding of $^{13}CO_2$, but only some random incorporation was observed after prolonged incubation. Meanwhile, β-alanine isolated from the feeding experiments showed enrichment in C-3, suggesting the pulse feeding was operating. [3-^{13}C]-3-Hydroxypropionate was also prepared and was fed as a closer putative intermediate in such a pathway. Again no significant incorporation was observed. The negative results made us reexamine C_1 precursors.

Despite our earlier failure in formate feeding experiments, [3-^{13}C]serine, [1,2-^{13}C]glycine, and [Me-^{13}C]methionine were found to enrich C-13 in neosaxitoxin effectively (7). The best incorporation was observed with methionine, indicating it is the direct precursor via S-adenosylmethionine. Glycine C-2 and serine C-3 must have been incorporated through tetrahydrofolate system as methyl donors in methionine biosynthesis.

In order to obtain more insights into the biosynthetic mechanism, [Me-^{13}C-Me-d_3] double-labeled methionine was fed to the organism. The ^{13}C-NMR spectrum of resulting neosaxitoxin showed a clean triplet for C-13 beside the natural abundance singlet. The result indicated that only one deuterium was left on the methylene carbon.

In another experiment, [1,2-^{13}C2-2-d_3] double-labeled acetate was fed. First we observed a complete loss of deuterium atoms. In a short incubation, however, we obtained neosaxitoxin partially retaining a deuterium atom (40% equivalent of incorporated acetate molecule). The location of the deuterium atom was on C-5, which was originally the carboxyl carbon of acetate, suggesting that it migrated from the adjacent methyl-derived carbon C-6.

The result is not totally surprising, because hydride ion shifts are known in many methylations. Thus, it was proposed that the methyl carbinol is formed by the sequence: methylation of a double bond - hydride shift - formation of terminal methylene - epoxidation - opening of the epoxide to aldehyde - reduction to carbinol (Scheme 6). The pathway can explain well the loss of two original hydrogens in methionine methyl group.

Biosynthesis of Brevetoxins

Brevetoxins are the toxic principles of the Florida red tide organism, *Gymnodinium breve* (*Ptychodiscus brevis*). The compounds are potent ichthyotoxins. Several toxins have been isolated from the organisms and their structures elucidated. They can be divided into two types of toxins having different ring systems, represented by brevetoxin A and brevetoxin B (II). Brevetoxin B was the first brought to a pure state, and its structure was determined by X-ray crystallography (8). Brevetoxin A is the most toxic principle in the organism, and its structure was also determined by X-ray crystallography (9). Both compounds have unique all-*trans* polyether ring systems, which made us believe that the compounds are products of successive opening of *trans*-epoxides. The precursors of the epoxides should be polyketidal *trans* polyenes.

We started feeding experiments using ^{13}C-labeled acetate long before the establishment of the toxin structures with a hope that the analysis of the ^{13}C-NMR

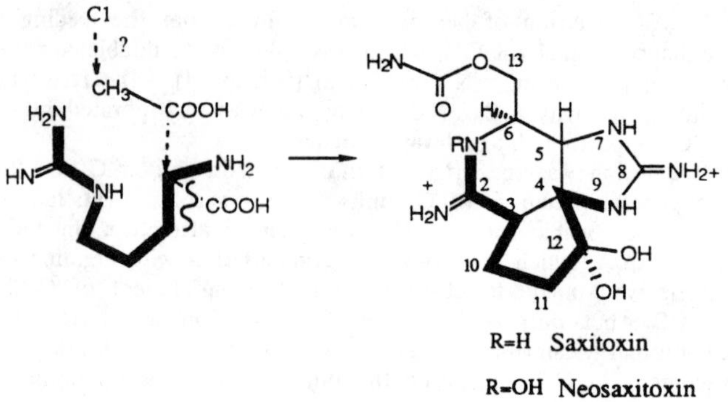

R=H   Saxitoxin

R=OH   Neosaxitoxin

Scheme 3

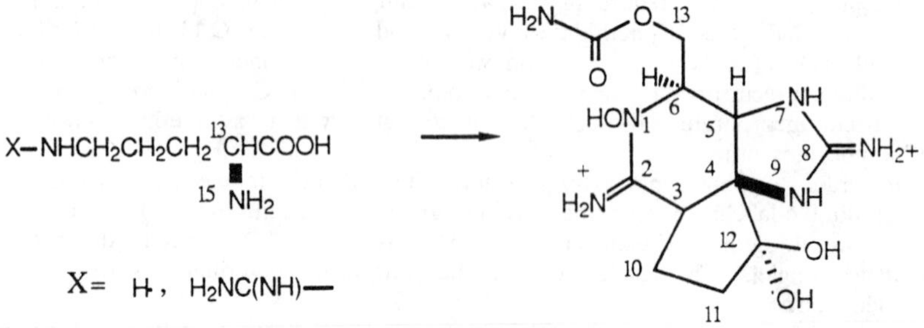

Scheme 4

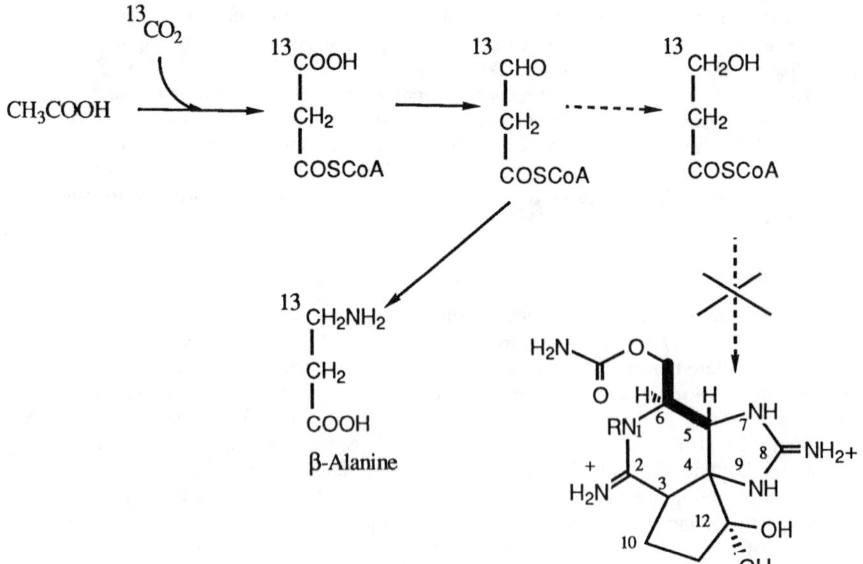

Scheme 5

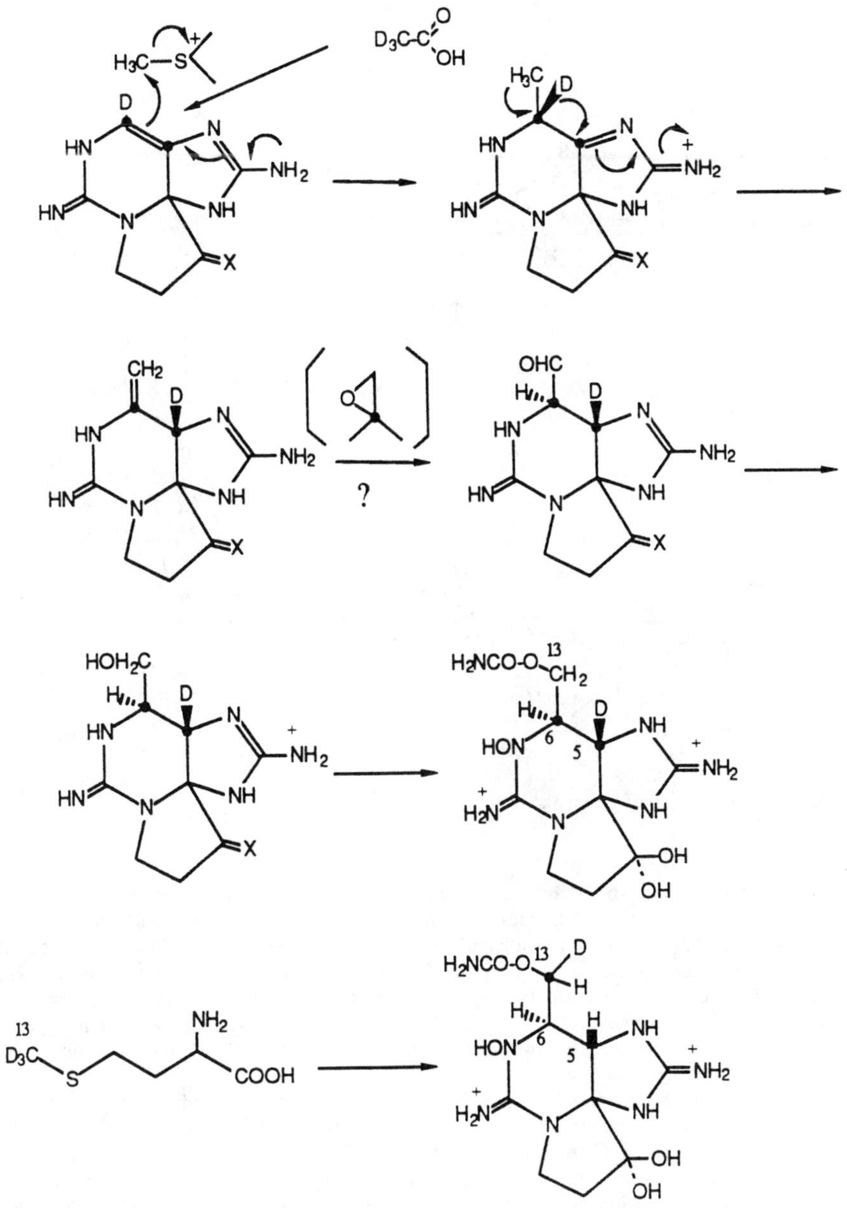

D_3C-C(=O)OH
H_3C—S
D
HN
HN
N
N
NH_2
NH
X
H_3C
D
NH_2
CH_2
?
OHC
HOH$_2$C
H_2NCO-O-CH_2
13
HON
H_2N
6
5
OH
D_3C
S
COOH
H

Scheme 6

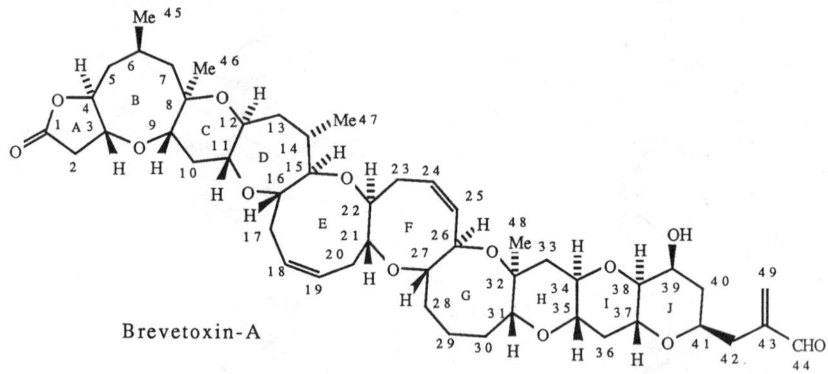

Brevetoxin-A

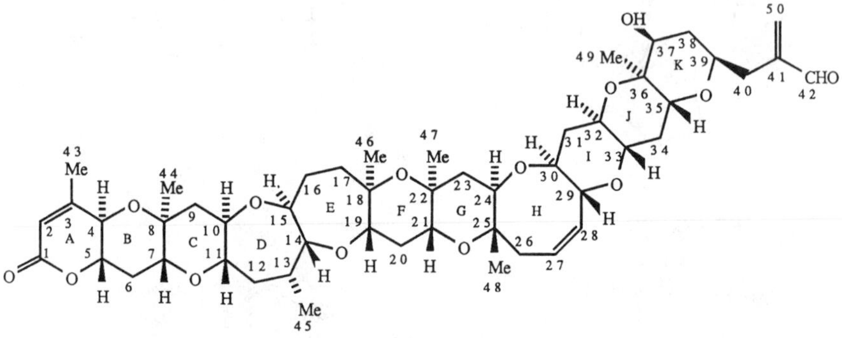

Brevetoxin-B

II

spectra might help the structure elucidation. It was assumed then that the compounds were normal polyketides. In fact, the elucidated structures have all the appearance of the ordinary polyketides with some methyl branches, which could come from methionine or propionate. However, the labeling pattern of brevetoxin B obtained from the acetate feedings was totally inexplicable in terms of the known polyketide biosynthesis (10). Similar observations were made by Nakanishi's group (11). We now believe that we have evidence to explain the apparent randomization.

When the organism was exposed to a high concentration of $[2-^{13}C]$acetate for a brief period, certain rows of carbons (mostly three carbons in a row) in brevetoxin B had C-C spin-spin couplings (10) (Scheme 7A). These couplings clearly indicate rather "fresh" condensations between actate methyl-derived carbons. Similar spin-spin couplings were not observed when $[1-^{13}C]$acetate was fed under the same condition.

The most plausible explanation of this very unusual labeling pattern is to assume the involvement of succinate, which is synthesized from acetate as shown in Scheme 8a.

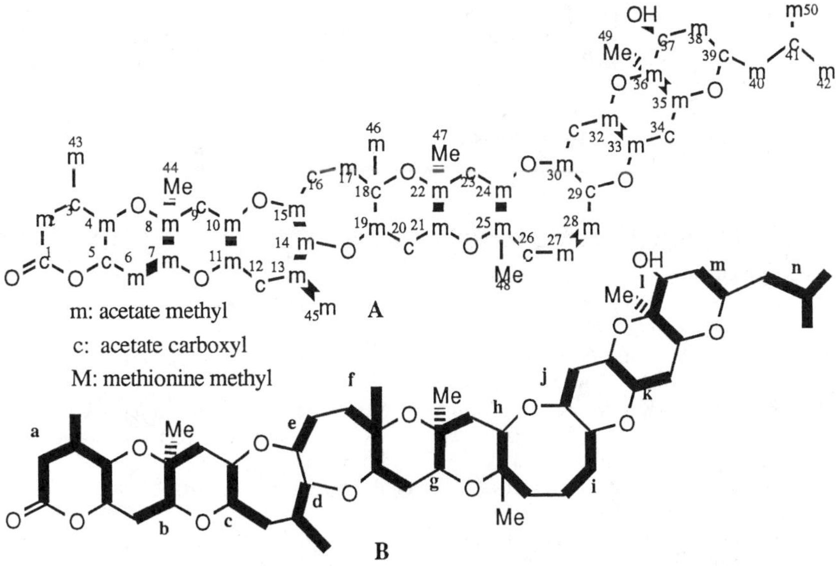
m: acetate methyl
c: acetate carboxyl
M: methionine methyl
A
B

Scheme 7

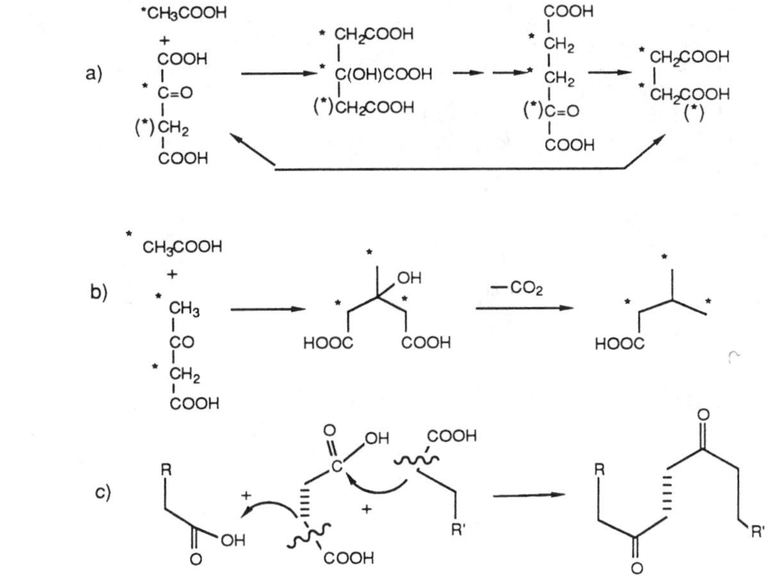
a)
b)
c)
$-CO_2$

Scheme 8

Three other portions of the molecule which do not fit the polyketide combination pattern (Scheme 7B a,f,n) can be also explained by assuming they are derived from a dicarboxylic acid such as 3-hydroxy-3-methylglutaric acid (Scheme 8b). If that is the case, brevetoxins are a new type of mixed polyketides which are formed by condensation on both ends of dicarboxylic acids (Scheme 8c). In order to prove further the hypothesis, feeding experiments with such putative precursors as succinate, acetoacetate, and propionate are in progress.

Literature Cited

1. Schantz, E. J.; Mold, J. D.; Stanger, D. W.; Shave, J.; Riel, F. J.; Bowden, J. P.; Lynch, J. M.; Wyler, R. S.; Riegel, B. R.; Sommer, H. *J. Am. Chem. Soc.* **1957**, *79*, 5230.
2. Schantz, E. J.; Lynch, J. M.; Vayada, G.; Matsumoto, K.; Rapoport, H. *Biochemistry* **1966**, *5*, 1191.
3. Shimizu, Y. In *Progress in the Chemistry of Organic Natural Products;* Springer-Verlag: New York, 1984; Vol 45, p 235.
4. Shimizu, Y.; Kobayashi, M.; Genenah, A.; Ichihara, N. In *Seafood Toxins -* ACS Symposium Series 262; Ragelis, E. R., Ed.; American Chemical Society: Washington, DC, 1984; p 151.
5. Shimizu, Y.; Norte, M.; Hori, A.; Genenah, A.; Kobayashi, M. *J. Am. Chem. Soc.* **1984**, *106*, 6433.
6. Ohuchi, S.; Okuyama, A.; Nakagawa, H.; Aoyagi, T.; Umezawa, H. *J. Antibiot.* **1984**, *37*, 518
7. Shimizu, Y.; Gupta, S.; Norte, M.; Hori, A.; Genenah, A. In *Toxic Dinoflagellates*; Anderson, D. M.; White, A.; Baden, G. D., Eds.; Elsevier: New York, 1985; p 271.
8. Lin, Y. Y.; Risk, M. M.; Ray, S. M.; Van Engen, D.; Clardy, J.; Golik, J.; James, C.; Nakanishi, K. *J. Am. Chem. Soc.* **1981**, *103*, 6773.
9. Shimizu, Y.; Chou, H-N.; Bando, H.; Van Dyne, G.; Clardy, J. *J. Am. Chem. Soc.* **1986**, *108*, 514.
10. Chou, H-N.; Shimizu, Y. *J. Am. Chem. Soc.* **1987**, *109*, 2184.
11. Lee, M. S.; Repeta, D. J.; Nakanishi, K.; Zagorski, M. G. *J. Am. Chem. Soc.* **1986**, *108*, 7855.

RECEIVED May 22, 1989

Chapter 3

The Saxitoxins

Sources, Chemistry, and Pharmacology

Sherwood Hall[1], Gary Strichartz[2], E. Moczydlowski[3], A. Ravindran[3,5], and P. B. Reichardt[4]

[1]U.S. Food and Drug Administration, 200 C Street, SW, Washington, DC 20204
[2]Anesthesia Research Laboratories, Harvard Medical School, Brigham and Women's Hospital, 75 Francis Street, Boston, MA 02115
[3]Yale University, School of Medicine, 333 Cedar Street, New Haven, CT 06510
[4]Department of Chemistry, University of Alaska, Fairbanks, AK 99701

Saxitoxin and its derivatives, referred to collectively as saxitoxins, are found in marine dinoflagellates and various other organisms. Due to their high potency, they are a public health concern and may have a significant influence on ecological relationships. The saxitoxins in general block sodium current through most known classes of voltage-activated sodium channels by a direct binding reaction. The molecular basis for this interaction, which at present remains obscure, is of great interest because of the selectivity and high affinity of the binding.

The saxitoxins were originally recognized and are still most frequently encountered as the compounds responsible for the human ailment named paralytic shellfish poisoning (PSP). They are accumulated from dinoflagellates by filter-feeders and thenceforth passed through the food web to humans. Since the original work that identified the dinoflagellate *Gonyaulax catenella* as a source and led eventually to the elucidation of the structure of the parent compound, saxitoxin, other studies have revealed a large family of natural saxitoxin derivatives and a variety of apparent source organisms. At present, new saxitoxin derivatives are still being encountered and recent work has raised interesting questions about the ultimate metabolic source of the toxins.

This chapter briefly reviews the present understanding of the chemistry, origin, and distribution of the saxitoxins and methods for their detection. The second section of this chapter discusses studies on their pharmacology directed toward an understanding of the molecular basis for their strong, highly selective interaction with the sodium channel binding site.

[5] Current address: Department of Neuroscience, Johns Hopkins University School of Medicine, 725 Wolfe Street, Baltimore, MD 21205

0097–6156/90/0418–0029$10.25/0
© 1990 American Chemical Society

Structures and Key Properties

The structure of saxitoxin (1,2) is shown in Figure 1, accompanied by 11 derivatives formed by N-1-hydroxyl, 11-hydroxysulfate, and 21-sulfo substitutions (3–13). Structural relationships among these compounds are summarized in Figure 2. The purpose of these diagrams is to summarize structural relationships, not to suggest synthetic or biosynthetic transformations.

Saxitoxin contains two guanidinium groups. In Figure 1 these are shown protonated, with positive charge fully delocalized over each. Whether this electron distribution properly describes the molecule, or whether bond and charge fixation occur to some extent, has yet to be shown. Titrations of saxitoxin (1; 14) reveal two proton dissociations at pK_a 8.22 and 11.28. The lower value, which is unusually low for a guanidinium deprotonation, is assigned to the guanidinium centered on on C-8. This group will be deprotonated in a very small fraction of saxitoxin molecules at physiological pH. Titrations of neosaxitoxin (7; 3) reveal a dissociable proton with a pK_a 6.75, due to the presence of the N-1-hydroxyl but possibly involving the entire C-2 guanidinium system. The relative extent to which the protons on N-13 and the N-1-hydroxyl participate in this dissociation has not yet been established, nor have the dissociation constants of N-1-hydroxyl derivatives other than neosaxitoxin. However, it is clear from their chromatographic behavior (9,10) that the other derivatives have a proton dissociation at about the same pK_a. Dissociation of this proton reduces net charge by one unit; at physiological pH, the dissociated form should predominate. The 11-hydroxysulfate and 21-sulfo groups each reduce net charge by one unit. Net charge, which appears to have a significant influence on both the chromatographic behavior and the pharmacology of the saxitoxins, is summarized in Figure 3. Under acidic isocratic conditions [0.1 M aqueous acetic acid 9,10)] these twelve toxins elute from Bio-Gel P2 (a polyacrylamide gel) as three groups in descending order of net charge: group A, compounds 1 and 7, charge +2; group B, compounds 2, 3, 5, 8, 9, and 11, charge +1; group C, compounds 4, 6, 10, and 12, no net charge. These compounds are found in extracts from various toxic dinoflagellates and cyanobacteria (see below).

Two further structural variations are found in nature – a carbonyl group at C-11, and the decarbamoyl compounds (15). Enzymatic decarbamoylation of the saxitoxins by certain shellfish has been demonstrated (16,17) and decarbamoylsaxitoxins 13–18 have been either clearly demonstrated or strongly suggested to occur in shellfish (16,18). Compound 19, 11-ketosaxitoxin, has been reported as a minor component of the saxitoxins found in extracts of toxic scallops (19,20) from the Bay of Fundy. The compounds that may result from these variations are summarized in Figure 4. A full complement of 24 toxins, summarized in Figure 5, could therefore result from all possible combinations of these 5 modes of variation. Whether all such combinations will in fact be encountered, or are even chemically compatible, remains to be seen. Other structural modifications, such as reduction at C-12, have been performed in vitro but have not yet been reported from biological samples.

The conformation of saxitoxin C4, compound 12, is shown in Figure 6 as it occurs in the crystal lattice (10,21). It is important to recognize that, contrary to the impression given by the conventional 2-dimensional drawing of the saxitoxins, the 5-membered ring containing the C-8 guanidinium is essentially perpendicular to the plane roughly defined by the other two rings. The orientation of substituents at carbons 10, 11, and 12 is described relative to this plane. In Figure 6, α substituents (like oxygen-15) are located below the plane, on the same side as the C-8 guanidinium, and β substituents (like oxygen-14) are located above the plane, on the same side as the C-17 side chain.

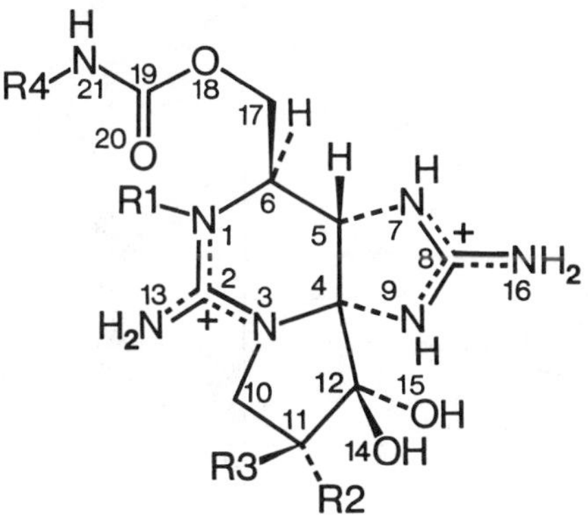

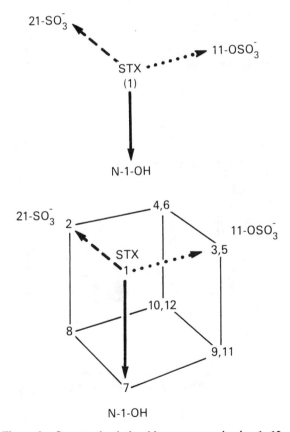

	R1	R2	R3	R4	
1	H	H	H	H	STX
2	H	H	H	SO_3^-	B1
3	H	OSO_3^-	H	H	GTX2
4	H	OSO_3^-	H	SO_3^-	C1
5	H	H	OSO_3^-	H	GTX3
6	H	H	OSO_3^-	SO_3^-	C2
7	OH	H	H	H	NEO
8	OH	H	H	SO_3^-	B2
9	OH	OSO_3^-	H	H	GTX1
10	OH	OSO_3^-	H	SO_3^-	C3
11	OH	H	OSO_3^-	H	GTX4
12	OH	H	OSO_3^-	SO_3^-	C4

Figure 1.   Structures of the twelve saxitoxins found in *Alexandrium* sp.

Figure 2.   Structural relationships among saxitoxins 1–12.

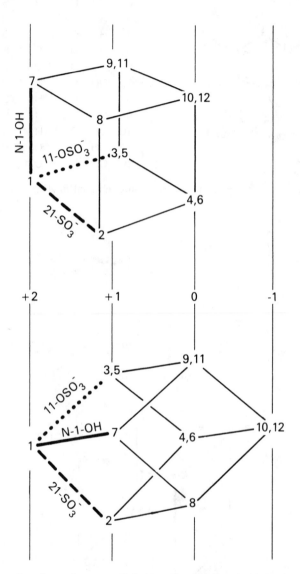

Figure 3. Approximate net charge of saxitoxins **1–12** under acidic conditions (upper) and physiological conditions (lower).

Decarbamoyl Saxitoxins

	R1	R2	R3
13	H	H	H
14	H	OSO$_3^-$	H
15	H	H	OSO$_3^-$
16	OH	H	H
17	OH	OSO$_3^-$	H
18	OH	H	OSO$_3^-$

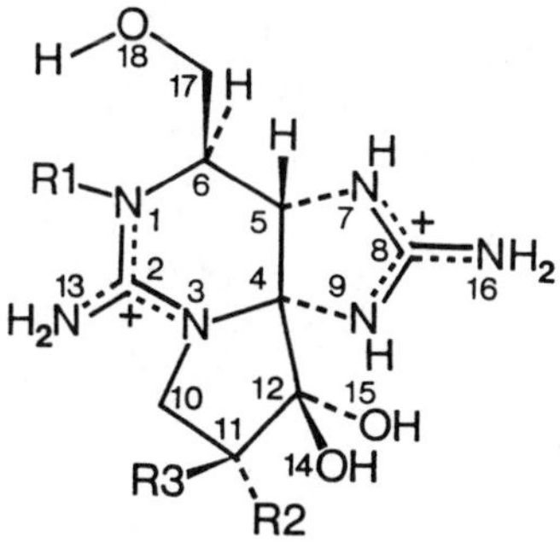

11-Keto Saxitoxins

	R1	R5
19	H	CONH$_2$
20	H	CONHSO$_3^-$
21	H	H
22	OH	CONH$_2$
23	OH	CONHSO$_3^-$
24	OH	H

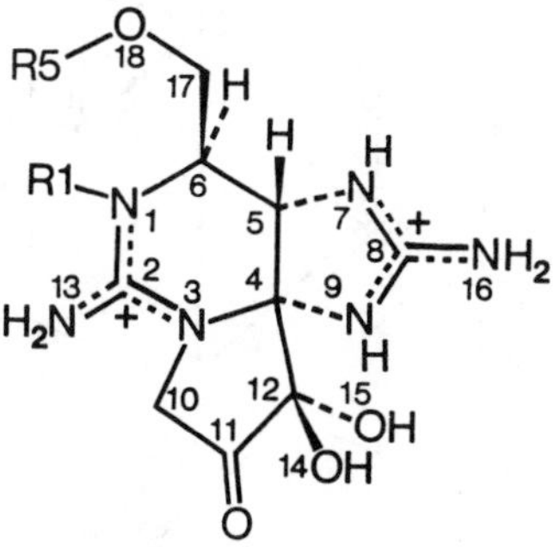

Figure 4. Known and implied structures for other natural saxitoxin derivatives. The structure of decarbamoylsaxitoxin, 13, is well established. Compounds 14, 15, 16, and 19 have been isolated and characterized to varying degrees. Compounds 17, 18, and 20–24 have not yet been reported, but are anticipated as combinations of known variations.

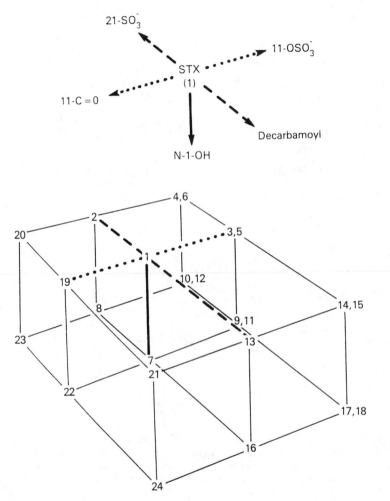

Figure 5. Summary of structural relationships among known modes of variation and the saxitoxins that are known or anticipated.

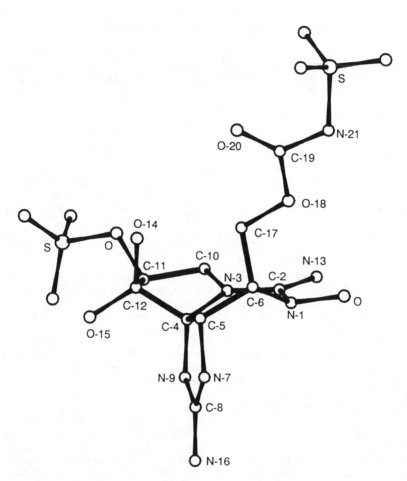

Figure 6. Conformation of compound **12**, toxin C4, in the crystal lattice. Data from diffraction studies by S. D. Darling (*13*); computer graphics by T. Chambers.

The functional group shown at C-12 is a hydrated ketone, a carbonyl group with which water has formed a reversible covalent bond, resulting in a gemdiol, $C(OH)_2$. Reduction of the C-12 ketone leads to two epimeric mono-ols, with a single hydroxyl oriented either α or β (22,23). Despite its tendency to hydrate, C-12 retains the properties of a ketone, particularly the ability to enolize, which renders protons at C-11 readily exchangeable and permits epimerization of the 11-hydroxysulfate. While the exchange of the protons at C-11 complicates the acquisition of NMR spectra in D_2O solutions, reaction conditions can be manipulated to permit exchange labelling with tritium at these positions to produce tritiated saxitoxin with a high specific activity. The development of this technique by Ritchie et al. (24) made it practical to do pharmacological binding experiments with tritium-labelled saxitoxin. Because of the ease with which the 11-hydroxysulfate group epimerizes under most conditions of purification or storage, it is difficult to produce and maintain pure preparations of the various 11-hydroxysulfate saxitoxins.

Saxitoxins with the C-12 ketone/gemdiol function intact are susceptible to oxidation at elevated pH, probably due to a shift in the ketone/gemdiol equilibrium, increasing carbonyl character. They are relatively stable to atmospheric oxygen at low pH, and can be manipulated without significant loss at elevated pH if oxygen is excluded. Oxidation leads to rupture of the C-12/C-4 bond and re-orientation of the remaining two rings to a planar, aromatic configuration, forming a fluorescent purine derivative (25,26). This transformation has been the key to efficient chemical detection of the toxins for both preparative and analytical purposes. It is greatly attenuated in the dihydrosaxitoxins and related compounds where C-12 has been reduced.

Origins and Occurrence

Source Organisms. The organisms that are known to contain the saxitoxins and are generally recognized as primary sources include three morphologically distinct genera of dinoflagellates and one species of blue-green alga.

The link between shellfish toxicity and dinoflagellates was first established through the efforts of Sommer et al. in studies following an outbreak of PSP in the San Francisco Bay area in 1927 (27–29). The toxic dinoflagellate they discovered was assigned to the genus *Gonyaulax* and named *G. catenella* (30). It is useful to recognize that, while the group had suspected from the outset that the toxicity of shellfish might be acquired from plankton, results from their early attempts to prove this were quite misleading and it was only due to their persistence, over several years, that the actual cause was eventually found.

Subsequent work in various regions has established that toxigenic gonyaulacoid dinoflagellates of similar morphology are a source of toxicity in many areas (10,31,32). These organisms have usually been assigned to the genus *Gonyaulax*, in which they formed a distinct subgroup with relatively minor morphological distinctions among species (33–34). Recent taxonomic revisions have variously reassigned the toxigenic gonyaulacoid dinoflagellates to the genera *Protogonyaulax* (35), *Gesnerium* (36), and, finally, *Alexandrium* (37). Thankfully, there appears to be a consensus developing toward the validity and acceptance of the last assignment, which will be used here. Species distinctions remain less clear. It is of paramount importance to stipulate, wherever possible, the source or strain of material under discussion. Members of the genus are similar in gross morphology. All have thecae, with plates that lack obvious detail. Cells of clone EC06, now assigned to *Alexandrium* and generally tamarensis-like in morphology, are shown in Figure 7A. Details of the thecal plates are more evident in empty thecae, such as that of clone HG01, shown in Figure 7B.

While this situation has been developing with the morphologically similar toxigenic gonyaulacoid dinoflagellates, other, clearly distinct dinoflagellates have been recognized as sources of the saxitoxins. *Pyrodinium bahamense* var. *compressa* (*38*), a thecate dinoflagellate with distinctly detailed plates, was first recognized to be toxic by MacLean in his investigation of an outbreak of PSP in Papua New Guinea (*39*). It has since caused several other outbreaks in the West Pacific (*40,41*), and an outbreak on the Pacific coast of Guatemala (*42*) that caused 26 deaths out of 187 confirmed cases. Toxicity has not yet been reported in association with blooms of *Pyrodinium bahamense* var. *bahamense* from the locations in the tropical Atlantic where the organism was originally described.

Gymnodinium catenatum, first recognized as a cause of toxicity in an outbreak in Mazatlan (*43,44*), lacks a theca but otherwise bears some resemblance to the catenate *Alexandrium* species. It has now been recognized as a source of PSP in Spain (*45*), Portugal (*46*), Venezuela (A. La Barbera-Sanchez, personal communication), Tasmania (*47*), and Japan (*48*).

The freshwater cyanophyte *Aphanazomenon flos-aquae* (*see* Chapter 6, by Carmichael), long suspected to contain saxitoxin-like compounds, has been shown to contain saxitoxin and neosaxitoxin, and has been an important tool in the elucidation of saxitoxin biosynthesis (*49,50; see* Chapter 2, by Shimizu). This research clearly demonstrates that the saxitoxins can be produced by a procaryote; thus, the suggestion that bacteria may produce the saxitoxins becomes more plausible.

The pioneering work of Sommer et al. (*28*) and the large body of data accumulated since clearly show that the saxitoxins are produced through the growth of dinoflagellates, accumulate in filter-feeders such as bivalves, and may be further distributed through trophic transfer. However, the saxitoxins are also found in organisms (particularly certain species of Xanthid crabs from the West Pacific) where a dietary food source is not evident and the sporadic distribution among individual specimens is difficult to reconcile with a dietary source (*51–53*). Also, there have been persistent suggestions that the saxitoxins in dinoflagellates are produced by endosymbiont bacteria (*54*). Accounts have recently appeared describing bacteria, isolated from toxic dinoflagellates, that produce the saxitoxins, although the levels reported so far are quite low (*55*). Thus, while dinoflagellates are clearly the proximate source of the saxitoxins that accumulate to high levels in filter-feeders, they may not be the sole or ultimate source of the saxitoxins in the marine environment. (*See* Chapter 5, by Tamplin, for further discussion.)

The composition of the saxitoxins, the relative proportion of the various derivatives, found in *Alexandrium* differs from strain to strain and appears to be a useful, fairly conservative characteristic for distinguishing among strains and among the populations from which they are obtained. Early studies (*9,10*) of toxin composition in extracts of cultured isolates from the Northeast Pacific (Figure 8) revealed dramatic differences among strains from several different locations (Table I), while few if any morphological criteria were found that could reliably discriminate among them. Successive toxin analyses of each strain gave similar results, indicating that toxin composition is a relatively conservative characteristic. Total toxicity was found to vary substantially with culture conditions and with the stage in the batch culture growth cycle, but changes in composition accompanying those of total toxicity were relatively small. In contrast to the differences among strains from different locations, strains from the same area tended to have the same composition, suggesting that there are indigenous, regional populations of *Alexandrium* distinguishable on the basis of toxin composition despite their morphological similarity. Other investigations (*56–60*) have largely supported this concept although, as would be expected, more detailed studies have detected minor variations within regions and have succeeded in inducing variations of toxin composition within a strain through the

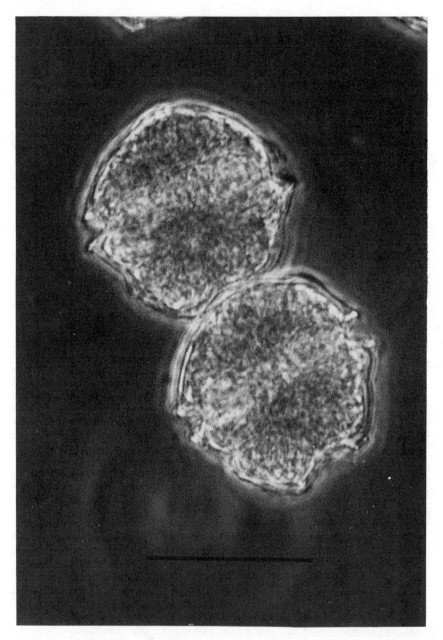

Figure 7A. The toxigenic dinoflagellate *Alexandrium* sp. from Southeast Alaska. Two vegetative cells of clone EC06, from Elfin Cove. Bar is 30 μm. Photomicrography by R. A. Horner.

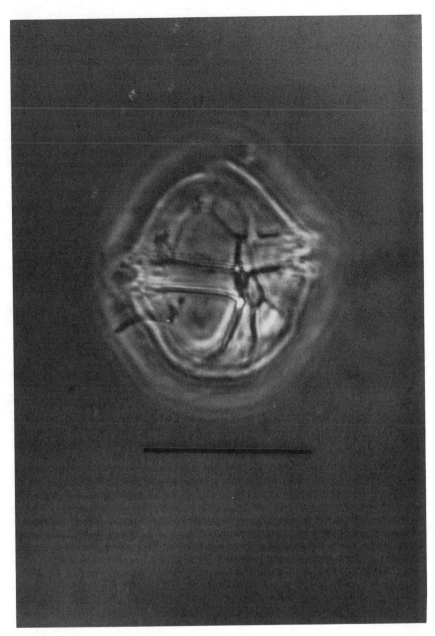

Figure 7B. The toxigenic dinoflagellate *Alexandrium* sp. from Southeast Alaska. An empty theca of clone HG01, from Haines. Bar is 30 μm. Photomicrography by R. A. Horner.

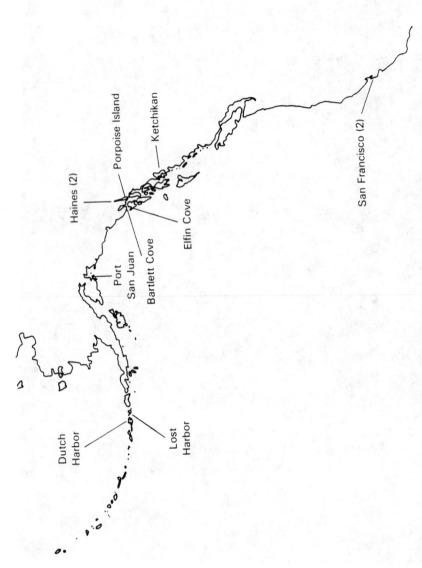

Figure 8. Origins of *Alexandrium* cultures used for the analyses of toxin composition summarized in Table I. Data from Ref. *10*.

Table I. Toxin Composition of *Alexandrium* from the Northeast Pacific

Location	Strain	Toxin Composition[a]						
		1	2	3/5	4/6	7	8	9/11
Dutch Harbor	DH07			?	+	+		+
Lost Harbor	LH01			?	+	+		+
Port San Juan	PW06	+				+		
Elfin Cove	EC06		+		+	+	+	+
Bartlett Cove	BC07		+		+	+	+	+
Porpoise Island	PI07		+		+	+	+	+
Haines	HG01		+	+	+	+	+	+
	HG27		+	+	+	+	+	+
Ketchikan	KN03		+		+	+	+	?
San Francisco	SF01	+			+	+		
	SF05	+			+	+		

SOURCE: Data from Ref. 10.
[a]Compound number per Figure 1.

manipulation of physiological parameters. It nevertheless continues to appear that toxin composition is a useful and relatively conservative characteristic of an isolate and of the population from which the isolate was obtained.

As discussed above, there is the possibility that toxigenesis in *Alexandrium* is not intrinsic but due to symbionts. Whichever proves to be the case, the observed patterns of toxin composition, whether they are for the dinoflagellate itself or the isolated assemblage of dinoflagellate and symbiont, are a basis for recognizing and distinguishing the regional populations.

In Table I the 11-hydroxysulfates are entered as epimeric pairs (3,5; 4,6; 9,11) because they are seldom encountered singly; an extract found to contain one will generally be found to contain some amount of its epimer. This phenonenon is due at least in part and perhaps entirely to the ease with which the 11-hydroxysulfate group spontaneously epimerizes. Despite the observed predominance of the 11α epimers (3, 4, 9) in equilibrated solutions in vitro, the ratio of 11β epimers (5, 6, 11) to the corresponding 11α epimers is consistently found to be higher in fresh dinoflagellate extracts and lower in older extracts, suggesting that the 11β epimers are the principal or only forms synthesized in the organism and that the 11α forms arise through epimerization. Saxitoxins C3 and C4 (10,12) are not included in the table because, although they were obtained from strain PI07 and detected in extracts of it under optimal conditions, they were not reliably detected at the sensitivity of the method used for the survey.

The Saxitoxins in Animals. Filter-feeders accumulate the saxitoxins from the toxigenic dinoflagellates in the phytoplankton they consume and pass the toxins on to the predators that consume them in turn. In general, the result at each step will depend on the sensitivity of the consumer to the toxins and the extent to which it retains the toxins and modifies them. Thus a filter-feeder or subsequent carnivore that is relatively insensitive to the saxitoxins and retains them efficiently can accumulate high levels and cause illness or death when consumed by a predator that is sensitive. This interaction seems likely to affect predator/prey relationships, serving as a deterrent to the consumption of efficient accumulators by sensitive predators. In contrast, the potential for intoxication of predators may also be increased by sensitivity to the toxins among potential prey organisms that are normally able to move to avoid capture, if intoxication of the prey makes it more likely to be caught.

The various species of bivalve molluscs differ in the way they deal with and respond to the saxitoxins. Mussels, for instance, appear in general to accumulate much higher levels of PSP than oysters under similar circumstances; such a relationship was indeed found in well controlled studies at a test site in Puget Sound (*61*). Subsequent laboratory feeding studies showed that mussels readily consumed concentrations of *Alexandrium* equal to or greater than those that caused oysters to cease pumping and close up (*62*). Electrophysiological investigations (*63,64*) of isolated nerves from Atlantic coast bivalves demonstrated that those from oysters were sensitive to the toxins, while those from mussels were relatively insensitive.

Sensitivity of bivalves to the saxitoxins appears to be more general than has been recognized. Shellfish that contain levels of the saxitoxins that make them dangerous for human consumption can seldom be distinguished visibly from those that are safe. Nevertheless, detailed study has shown (*64*; R. Mann and S. Hall, unpublished results) that *Alexandrium* cells have significant effects on bivalve responses, presumably due to the contained saxitoxins.

The saxitoxins accumulated from the plankton by molluscan filter-feeders can be passed on to other animals. Gastropods such as whelks, which prey on bivalves, have been found to be toxic, apparently from consuming toxic bivalves (*65*). Such phenomena occur with organisms other than bivalves and their predators. White has shown that zooplankters can accumulate the saxitoxins (66). Finfish are in general sensitive to the saxitoxins (67) and may be killed by consuming toxic zooplankters (68). It has recently been found, however, that the livers of mackerel from the Atlantic coast of the United States can contain moderate amounts of toxin that resembles saxitoxin. Studies on two market samples of mackerel are described in the pharmacology section. The origin of this toxin is not yet clear.

The composition of toxins in the primary consumer starts out resembling that of the dinoflagellate population consumed and then undergoes diagenesis due to spontaneous transformations, metabolism, and selective retention of the toxins. The primary spontaneous transformation is 11-hydroxysulfate epimerization which, as mentioned above, proceeds from the predominantly 11β composition found in dinoflagellates to the predominantly 11α composition of equilibrated mixtures. Metabolic removal of N-1-hydroxyl and 11-hydroxysulfate groups (69) and either carbamate or sulfamate sidechains (16,17) has been demonstrated in some shellfish species. Selective retention of particular toxins has not been demonstrated but seems likely in view of the different binding properties demonstrated for the various saxitoxins in chromatography and in pharmacological studies.

Saxitoxin composition therefore originates as that of the local dinoflagellate population and evolves with time depending on the characteristics of the filter-feeders and subsequent consumers that accumulate and transform the toxins.

Detection Methods

Some Formal Considerations. As with most natural toxins, detection methods for the saxitoxins are an essential prerequisite for most studies of them, as well as for monitoring programs to ensure the safety of food products that may contain them. Furthermore, the degree of success of such efforts is dependant on the characteristics of the detection method used. Detection of the saxitoxins is particularly challenging because of the large number of different but related compounds that must be dealt with, the low levels that must be detected, and their chemical characteristics. Given these factors it is useful to dwell briefly on some underlying principles.

Detection methods may be considered to fall into two classes which, for the purposes of this discussion, will be designated assays and analyses. Assays are those methods which provide a single response to the sample, which may contain several compounds to which the assay is sensitive. Analyses are those methods in which the compounds of interest in the sample are resolved, and a response obtained separately for each.

It is generally necessary to multiply the response obtained from a detection method by a response factor to convert the response into a useful value. For instance, the response of a fluorescence detector would be multiplied by an appropriate factor (f_c) to obtain the concentration of the particular toxin present, or by a different factor (f_t) to calculate the toxicity. Since the specific toxicities of the various toxins – the ratios of toxicity to concentration – vary over a broad range, the f_c and f_t for a given toxin will generally be different, often greatly so. Furthermore, the f_c may vary for the different toxins, and the f_t may also vary. In an analysis, multiplication by the appropriate factor is straightforward because the various components of interest are resolved and the response for each can be multiplied by the appropriate response factor, f_c or f_t, for each toxin. Assays, however, present a dilemma. Because the components are not resolved and only one response is obtained, only one response factor can be used. The potential accuracy of an assay is therefore limited principally by the range of response factors to the components that may be present. An ideal assay for toxicity would have the same toxicity/response factor, f_t, for each of the saxitoxins, while an ideal assay for toxin concentration would have the same concentration/response factor, f_c, for each. Because the specific toxicities of the saxitoxins differ, a single assay cannot be optimized for both. Few currently available assays are satisfactory for either.

Assays for the Saxitoxins. The original bioassay for the saxitoxins, based on i.p. injection into laboratory mice (70,71), is still in use and will be discussed below. Alternatives have been sought for many years. Bioassays employing house flies (72) and other insects have been described, although they have yet to be shown to be as precise or efficient. Given the phylogenetic distance between mammals and insects, they may be less accurate as predictors of human oral potency, due to likely differences in the binding properties of their sodium channel saxitoxin binding sites.

Immunoassays for the saxitoxins have been developed and are currently being studied (73,74). Most antibodies have been raised to conjugates of carrier protein linked to saxitoxin through C-12, and have response factors to the various toxins that differ greatly on both molar and toxicity bases, responding particularly poorly to saxitoxins bearing the N-1-hydroxyl substituent (74). Immunoassays based on such antibodies therefore fail as general assays for either concentration or toxicity of the saxitoxins. Immunoassays remain tempting, nevertheless, because their initial stochiometric response (binding of the analyte and antibody) can often be chemically amplified (usually by a covalently linked enzyme which catalyzes a secondary reaction) to the extent that traces of analyte bring about easily visible color changes. In favorable cases they can therefore form the basis for true field assays which can be performed under primitive conditions with minimal equipment. Such an assay for the saxitoxins, with response factors designed to be suitable for determining concentration, could eventually be developed either by using a toxin–protein conjugate that induces an antibody with suitable response factors (f_c uniform for the saxitoxins), or by producing several different antibodies with differing responses and concocting a mixture of these which gives the desired uniform net response. An immunoassay with response factors designed to be suitable for determining toxicity to humans could in principle also be developed using either approach.

Of the various chemical assays that have been developed for the saxitoxins (75,76), that described by Bates and Rapoport, based on the oxidation of saxitoxin to a fluorescent derivative, has proved to be the most useful. Other assay methods have been developed from it (77–79). The Bates and Rapoport method is virtually insensitive to the N-1-hydroxyl saxitoxins as originally described and so, like the presently available immunoassays, fails as a general assay for either concentration or toxicity. However, it is quite sensitive for those toxins it does detect and has been the basis for other useful methods.

That only limited success has been attained so far in developing assays for the saxitoxins would seem to suggest that they are difficult to detect. Yet they are potent toxins only because they interact with the sodium channel binding site in a strong, highly selective manner, suggesting that the interaction with the binding site may itself be the most useful basis for detection of the saxitoxins. Although it may in the future be possible to develop synthetic analogs of the binding site that mimic its selective properties, this interaction is at present exploited by using the native binding site, both in bioassays and with in vitro methods that use preparations of nerve or muscle tissue that contain the binding site. In Chapter 5, Tamplin discusses a novel assay in which the presence of saxitoxin and other sodium channel blockers is detected by their ability to protect cultured neuroblastoma cells, presumably by blocking sodium channels, from the effects of two other toxins that permit an otherwise lethal sodium influx.

Various electrophysiological techniques developed for research on excitable tissue preparations can be used to determine toxin potency, but require specialized equipment and expertise that are not generally available. In contrast, the membrane binding techniques, originally developed for pharmacological research and based on suspensions of mammalian tissue, are simple, well adapted to large numbers of samples, sensitive, precise, and accurate as a measure of mammalian toxicity. The accuracy of such assays as a predictor of mammalian toxicity follows by definition, insofar as toxicity is principally determined by binding to sodium channels (*see* below) and the population of sodium channels used in the assay is similar in binding properties to those in the population at risk. The application of a binding assay for routine detection of saxitoxin and tetrodotoxin in clinical samples has been described by Davio (*80*).

Some Chemical Considerations Relevant to the Mouse Bioassay. Net toxicity, determined by mouse bioassay, has served as a traditional measure of toxin quantity and, despite the development of HPLC and other detection methods for the saxitoxins, continues to be used. In this assay, as in most others, the molar specific potencies of the various saxitoxins differ; thus, net toxicity of a toxin sample with an undefined mixture of the saxitoxins can provide only a rough approximation of the net molar concentration. Still, to the extent that limits can be placed on variation in toxin composition, the mouse assay can in principle provide useful data on trends in net toxin concentration. However, the somewhat protean chemistry of the saxitoxins makes it difficult to define conditions under which the composition of a mixture of toxins will remain constant; thus, attaining a reproducible level of mouse bioassay toxicity is difficult. It is therefore useful to review briefly some of the chemical factors that should be considered when employing the mouse bioassay for the saxitoxins or when interpreting results. Similar concepts will apply to other assays.

Two factors will be discussed here: 11-hydroxysulfate epimerization and sulfamate hydrolysis. Other transformations, mediated by shellfish or bacteria, are possible and the saxitoxins are also subject to gross decomposition, resulting in nontoxic products. These factors will complicate the picture slightly, but not alter the basic point.

Epimerization of the 11-hydroxysulfate group occurs without assistance and is difficult to arrest entirely. Although, for reasons yet to be understood, the 11-β-hydroxysulfates appear to predominate in dinoflagellates to the virtual exclusion of 11-α-hydroxysulfates, epimerization proceeds once the toxins have been liberated either through extraction or through digestion in filter-feeders. Epimer ratios thus evolve from predominantly β to predominantly α, and under some circumstances may serve as a clock indicating the time since toxins were accumulated by a filter-feeder. The equilibrium epimer ratios of the various 11-hydroxysulfates are not well established but, on the basis of preliminary data (S. Hall, unpublished data), $\alpha:\beta = 4:1$ appears satisfactory for this discussion.

The susceptibility of the sulfamates to hydrolysis is intermediate with respect to procedures commonly used for extraction and manipulation of extracts. Quantitative hydrolysis of the pure sulfamate toxins can be accomplished (*9*) by heating at $100\,^{\circ}$C for 5 min in the presence of not less than 0.1 M free acid (pH 1 or below). Milder conditions appear insufficient (*10*). Figure 9 summarizes results from two separate experiments in which samples of nontoxic clam flesh, enriched with constant amounts of saxitoxin C1 (**4**), were acidified to differing final concentrations of HCl and heated for 5 min at $100\,^{\circ}$C. The difference between 0.1 M HCl, which would be sufficient for hydrolysis of the pure toxin, and the HCl concentration required to attain plateau toxicity, probably reflects the buffer capacity of

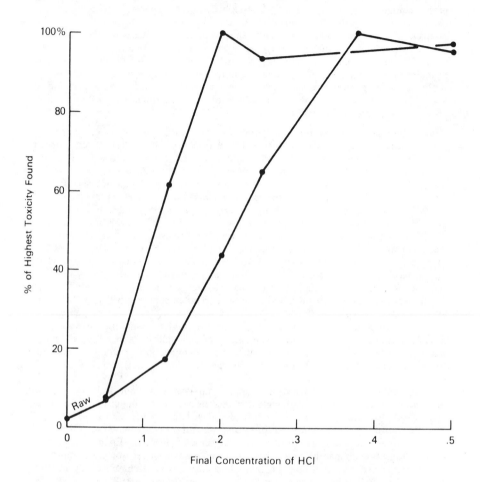

Figure 9. Observed increases in mouse assay toxicity of test mixtures of shellfish meat and toxin C1 (**4**), hydrolyzed with varying concentrations of HCl acid. Two series of experiments are shown. The initial concentration of toxin C1 was uniform for all samples in a series. Toxicity is expressed on the vertical axis as percentage of the maximum toxicity attained for that series.

the sample. The acid strength specified for the standard AOAC mouse bioassay (*81*), a 1:1 mixture of 0.1 M HCl and sample (corresponding to 0.05 M HCl added in the final mixture and intended to give approximately pH 3), is quite insufficient to ensure complete sulfamate hydrolysis. However, trace hydrolysis is difficult to avoid and, due to the substantially higher potency of the hydrolysis products, makes it difficult to obtain reliable values for the potency of samples containing the sulfamates.

Figure 10 summarizes these considerations for a hypothetical sample containing 1 μmol of saxitoxin C2 (**6**), which is both a sulfamate and an 11-β-hydroxysulfate. The sulfamates are frequently the predominant saxitoxins found in dinoflagellates (*10,21*), and 11-hydroxysulfates (*7*) are found almost entirely as the 11-β epimer in fresh dinoflagellate extracts (*10*). Toxin C2 itself is the principal component of the saxitoxins in dinoflagellates from many regions.

At the start, the sample containing 1 μmol of **6** has a toxicity of 230 mouse units. Epimerization to a mixture of C1 (**4**) and **6** will reduce the toxicity of the sample. At a ratio of 4:1 α:β, the decrease will be about 11-fold, to 21 mouse units. Hydrolysis of the original sample to the corresponding carbamate, GTX3 (**5**), would increase the toxicity of the sample 6-fold, to 1400 mouse units. Subsequent epimerization of the hydrolysate, **5**, to a 4:1 mixture of the carbamates GTX2 (**3**) and **5** (or hydrolysis of the epimeric sulfamates **4** and **6** to the same mixture of carbamates) would then change the toxicity to 1,120 mouse units.

The observed toxicity and composition of the hypothetical sample which started as pure **6** may lie anywhere within the indicated bounds as these interconversions occur. The toxicity will decrease with epimerization, along the bottom line in Figure 10, until it reaches an equilibrium ratio. It can be increased at any point by hydrolysis, toward the toxicity shown on the upper line. The highest toxicity will be attained if the hydrolysis is performed on **6** prior to epimerization, resulting in **5**. Hydrolysis of the epimeric **4** and **6**, or epimerization of the hydrolysate, will ultimately result in a stable, equilibrated mixture of the epimers **3** and **5**, with a slightly lower toxicity. The hypothetical sample discussed here, containing pure **6** at the start, was chosen because it is the limiting case among the known toxins. Observed changes will generally be smaller due to the presence of other saxitoxins, but may well approach those described here since many samples have been encountered in which the group C toxins predominate.

It is important that the employment of the mouse bioassay and the interpretation of its results take these factors into consideration to ensure that the data produced are as reliable as possible and that the eventual conclusions are well founded.

Analyses for the Saxitoxins. Early methods for analysis of the saxitoxins evolved from those used for toxin isolation and purification. The principal landmarks in the development of preparative separation techniques for the saxitoxins were: 1) the employment of carboxylate cation exchange resins by Schantz et al. (*82*); 2) the use of the polyacrylamide gel Bio-Gel P2 by Buckley and by Shimizu (*5,78*); and 3) the development by Buckley of an effective TLC system, including a new solvent mixture and a new visualization technique (*83*). The solvent mixture, designated by Buckley as "E", remains the best for general resolution of the saxitoxins. The visualization method, oxidation of the saxitoxins on silica gel TLC plates to fluorescent degradation products with hydrogen peroxide and heat, is an adaptation of the Bates and Rapoport fluorescence assay for saxitoxin in solution. Curiously, while peroxide oxidation in solution provides little or no response for the N-1-hydroxy saxitoxins, peroxide spray on TLC plates is a sensitive test for all saxitoxin derivatives with the C-12 gemdiol intact.

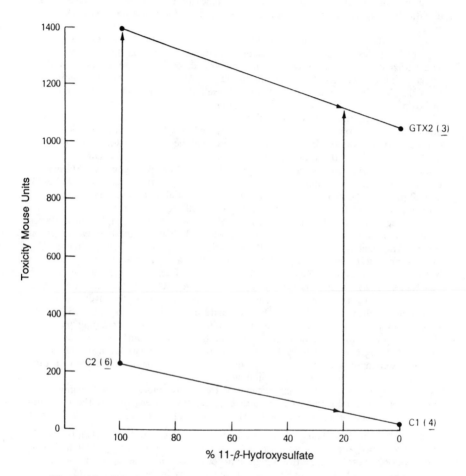

Figure 10. Calculated changes in the mouse bioassay toxicity of a sample initially containing 1 μmol of toxin C2. The horizontal axis represents the percentage of 11-β-hydroxysulfate. The lower line indicates the toxicity of the sample with the 21-sulfo group intact, but with varying degrees of epimerization. The upper line indicates the toxicity of the corresponding carbamates, formed by hydrolysis of the 21-sulfo group.

Carboxylate cation exchange column separation, followed by TLC of collected fractions, has been employed for the analyses of shellfish and dinoflagellate extracts for the saxitoxins (*5,84,85*). This process can provide useful results for the group A and B toxins. Unfortunately, the group C toxins are not retained on carboxylate resins and therefore co-elute with the bulk of the nontoxic matrix, generally precluding their detection by TLC. However, by isocratic column chromatography on Bio-Gel P2 followed by TLC of collected fractions (*9,10*), it is possible to separate all of the saxitoxins from the bulk of the matrix and provide analytical resolution of the twelve toxins found in *Alexandrium* sp. Unfortunately, the method, although simple and reliable, is slow, difficult to quantitate, and requires a relatively large amount of material.

HPLC is clearly the approach most suited to quantitative analysis of the saxitoxins, but the development of HPLC methods was complicated by the absence of a chromophore in the saxitoxins and by their highly polar nature. The most significant work in this area has been that of Sullivan (*see* Chapter 4), who developed a detection system involving post-column oxidation of the column effluent with periodate followed by fluorescence detection (*86*). Separation of the eight group A and B toxins is provided by a binary gradient of two solvent mixtures that include pH buffers and alkyl sulfonate ion pair reagents. The group C toxins can be determined indirectly by hydrolysis to their corresponding carbamates and analysis under the same conditions, or directly by changing to a different mobile phase. The technique developed by Sullivan has proved quite powerful, but requires careful management of operating parameters to produce reliable results. Other investigators have developed slightly different systems (*87–89*) in which separate runs are conducted under isocratic conditions appropriate for each group of toxins, either in sequence or concurrently in parallel columns. Since this approach avoids the need for pumping reproducible gradients, it is less expensive and likely to be both more rugged and reliable, but involves a more complex procedure and intricate plumbing. Also, the more strongly retained toxins (such as group A, during an analysis for group B) will accumulate on the column and bleed off very slowly. For the toxin mixtures typically encountered in nature, the column must be flushed frequently with a stronger eluant to remove the other accumulated toxins to avoid an increase in baseline as they bleed off.

The method described later in this chapter for pharmacological investigation of the saxitoxins using single sodium channels (*90,91*) also bears mention as an entirely different sort of analytical method. While it is by no means practical for general use, it is quite specific for sodium channel blockers, offers a moderate degree of resolution among them, is quantitative, requires no sample cleanup, and for the majority of the saxitoxins is the most sensitive detection method available. Unfortunately, it is not well suited to complex mixtures and requires expensive, dedicated equipment, and about two days effort by a qualified investigator to obtain a result.

In summary, analyses for the saxitoxins provide information that, in principle, is clear and readily interpreted while the design and interpretation of assays for the saxitoxins is complicated by the multiplicity of the toxins, their differing properties, and the variations in toxin composition that will be encountered. Assays, however, can in principle be very simple to perform while the presently available analyses for the saxitoxins require fairly complex equipment.

Pharmacological Studies of the Saxitoxins

The saxitoxins function by binding to a site on the extracellular surface of the voltage-activated sodium channel, interrupting the passive inward flux of sodium ions that would normally occur through the channel while it is in a conducting state. (For further discussion, *see* Chapter 1, by Strichartz.) Although pharmacological studies of saxitoxin and tetrodotoxin have played a major role in the development of the concept of the sodium channel and membrane channels in general, both the mechanism by which the saxitoxins bind and that by which a bound toxin molecule interrupts sodium ion conductance remain obscure. This section will outline studies directed toward the still distant goal of understanding the molecular basis of the strong, selective interaction between the saxitoxins and their binding site on the sodium channel. These studies are principally investigations of structure/activity relationships among several of the saxitoxins.

To clarify the discussion of structure/activity relationships, this section will attempt to illuminate the practical distinction between the equilibrium dissociation constant, K_d, and the rate constants, k_{on} and k_{off} and their relationship, particularly:

- Differences in the equilibrium dissociation constant, K_d, for the binding of the various saxitoxins to the sodium channel binding site largely determine the differences in the potencies of the toxins in whole animal assays and in tissue preparations.

- Structural modifications of the toxins influence the rate constants k_{on} and k_{off} and as a result indirectly affect the equilibrium dissociation constant K_d. However, since a structural alteration can affect both k_{on} and k_{off} and affect both differently, there is not a simple correlation between changes in structure and changes in the K_d or the phenomena that derive from it.

Given this distinction, it is evident that the relationships between structure and activity that are most useful for the elucidation of binding mechanisms are not those between structure and the K_d, in which the effects are obscured, but those between structure and the separate kinetic parameters, k_{on} and k_{off}, which combine to determine the K_d. Useful trends therefore become more apparent when the K_d can be dissected into its component kinetic parameters.

In addition to variations in toxin structure, the nature of the binding site and the medium in which binding occurs both influence the observed behavior of the toxins and can be systematically varied to study the toxin/binding site interaction. These factors will not be discussed at length here, but should be remembered as complications in comparing different data sets.

Pharmacological Methods and Results. The data upon which the following discussion is based were accumulated using three techniques: mouse bioassay, displacement of radiolabelled saxitoxin from rabbit brain membranes, and blockage of sodium conductance through rat sarcolemmal sodium channels incorporated into planar lipid bilayers. The results are summarized in Figures 11 and 12.

The mouse bioassay for PSP, described in its original form by Sommer in 1937 (*29*), involves i.p. injection of a test solution, typically 1 mL, into a mouse weighing 17–23 g, and observing the time from injection to death. From the death time and mouse weight, the number of mouse units is obtained by reference to a standard table; 1 mouse unit is defined as the amount of toxin that will kill a 20-g mouse in 15 min (*71*). The sensitivity of the mouse population used is calibrated using reference standard saxitoxin (*70*). In practice, the concentration of the test solution is adjusted to result in death times of approximately 6 min. Once the correct dilution has been established, 5 mice will generally provide a result differing by less than 20% from the true value at the 95% confidence level. The use of this method for the various saxitoxins and indeterminate mixtures of them would appear

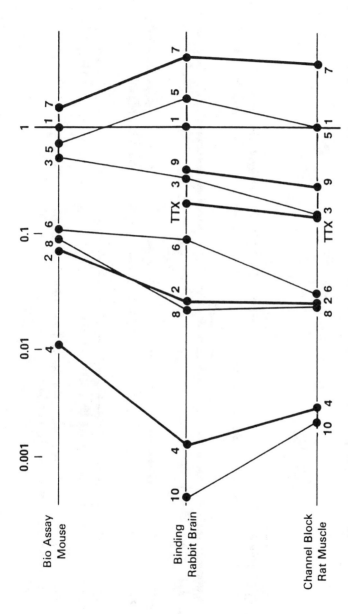

Figure 11. Comparative potencies of several saxitoxins and tetrodotoxin in the mouse bioassay, displacement of radiolabelled saxitoxin from rabbit brain, and block of single batrachotoxin-treated rat sarcolemmal sodium channels incorporated into planar lipid bilayers. The horizontal axis is the log of the potency relative to saxitoxin. Compound numbering corresponds to that in Figure 1. Data from Ref. *10, 94, 95*.

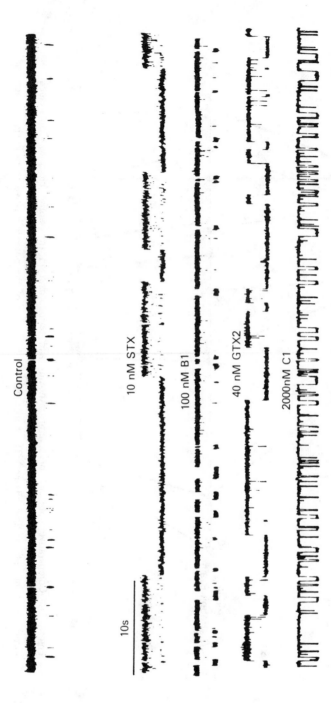

Figure 12. Records of current through single batrachotoxin-treated rat sarcolemmal sodium channels incorporated into planar lipid bilayers, before and after treatment with saxitoxin and derivatives. Data from Ref. 91.

to involve a conceptual flaw, implicitly requiring that the shapes of the dose-response curves for the various derivatives be the same. However, the assay as employed is essentially a titration of the sample to a 6-min death time, so that divergence in the slopes of the dose response curves is of little consequence. The mouse i.p. potencies of several of the saxitoxins (*10*) are summarized in Figure 11. The values shown are the ratios of potency to that of saxitoxin, plotted on a logarithmic scale. Note that the sulfamates **2**, **4**, **6**, and **8** are substantially less potent than the carbamates **1**, **3**, **5**, and **7**, and that toxins **2**, **6**, and **8** are virtually equipotent, despite their differences in structure and net charge (*see* Figure 3).

Binding assays for the saxitoxins were conducted with homogenized rabbit brain and saxitoxin exchange-labelled with tritium at C-11 (*92, 93*). If the various saxitoxins were available with suitably intense radiolabels, then the equilibrium dissociation constant, K_d, could be measured directly for each. Since only saxitoxin is currently available with the necessary label, the binding experiments instead measure the ability of a compound to compete with radiolabelled saxitoxin for the binding site. The value obtained, K_i, corresponds to the equilibrium dissociation constant, K_d, that would be observed for the compound if it were measured directly. Affinity is defined for this assay as the reciprocal of K_i. The affinities of several of the saxitoxins (*94*) are summarized in Figure 11, expressed relative to saxitoxin and plotted on a logarithmic scale.

Single channel studies were performed according to Moczydlowski et al. (*90,91*), measuring the current through single sodium channels that had been incorporated into planar lipid bilayers and treated with batrachotoxin to shift their activation potential into an accessible range. The sample toxins were added and blocking events recorded at several different transmembrane potentials. Current records obtained before and after the addition of several of the saxitoxins are shown in Figure 12. Open and blocked times for each potential were then measured and evaluated statistically to obtain the k_{on} and k_{off} for each potential. Values for k_{on} and k_{off} at zero transmembrane potential were then interpolated from these results, and the equilibrium dissociation constant K_d was determined from the ratio of the kinetic constants, k_{off}/k_{on}. The data shown here are from Guo et al. (*95*). The values shown in Figure 11 are affinities relative to saxitoxin, the affinity in this case being defined as the reciprocal of the computed K_d.

The naturally occurring saxitoxins used in these studies were isolated from cultured *Alexandrium* sp. (*10*). The carbamates **3**, **5**, and **9** were produced by hydrolysis of the corresponding sulfamates **4**, **6**, and **10**. Although the solutions of **4**, **6**, and **10** were prepared from crystalline material, single-channel experiments on these occasionally revealed blocking events with long dwell times; trace hydrolysis of the sulfamates to the corresponding carbamates causes these events. The contaminating carbamates were removed by passing the toxin solutions through small carboxylate cation exchange columns.

To prepare α-saxitoxinol, saxitoxin was reduced with sodium borohydride and purified by chromatography on IRP-64 and Bio-Gel P2 (*14,22,23*). Current records from single-channel experiments using this material are shown in Figure 13. Note the absence of blocking events with long dwell times that would indicate the presence of residual saxitoxin.

Mackerel liver extracts were prepared from commercial samples of whole, fresh mackerel purchased at fish markets in the Washington, D.C., area and claimed by the vendors to be from Boston and New Jersey. The extracts were tested using single channel techniques to provide qualitative data on the toxins present. Traces from these experiments are shown in Figure 14. Preliminary evaluation of the dwell times of the blocking events suggests a single class of blocking events similar to those of saxitoxin, clearly distinct from other known saxitoxins except GTX3 (**5**).

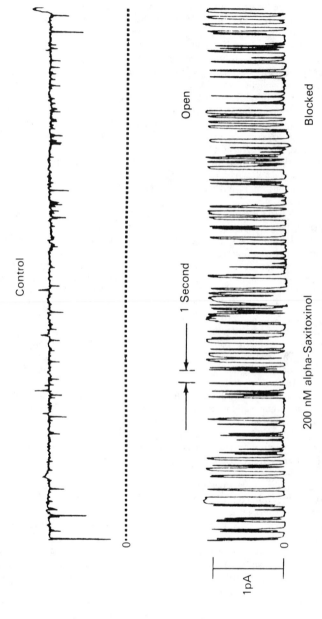

Figure 13. Records of current through a single batrachotoxin-treated rat sarcolemmal sodium channel before and after treatment with 200 nM α-saxitoxinol.

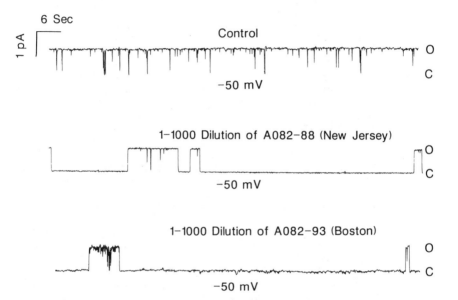

Figure 14. Records of current through single batrachotoxin-treated rat sarcolemmal sodium channels before and after treatment with extracts from mackerel livers.

GTX3 can be ruled out because, due to the conditions of sample preparation, it would have been accompanied by its epimer, GTX2 (3). The blocking kinetics of GTX2 are sufficiently different from those observed that it would have been easily detected if present.

Discussion. *Correlation Among the Three Data Sets.* A general correspondence is evident among the relative potencies of the various derivatives observed in the three systems (Figure 11). That the mouse bioassay results agree with the other more direct measures of toxin binding suggests that binding affinity is a significant determinant of the potency of the saxitoxins administered to mice. Some differences would be expected because the population of sodium channels in the mouse that determines its response is not well defined and the pharmacodynamics of the individual toxins in the whole animal would be expected to be influenced by their differing chemical properties, particularly charge. In view of these factors, the degree of correlation is surprising.

The correspondence between the relative affinities from binding experiments and those calculated from single channel data confirms the validity of the interpretation of the bilayer results and indicates that the employment of batrachotoxin to open the channels does not significantly perturb the the saxitoxin binding site interactions studied here. The observed differences may be due to the different classes of sodium channels employed, since the saxitoxin binding sites of sodium channels from muscle and brain can be distinguished on the basis of their very different interactions with a naturally occuring peptide from *Conus* venom, μ-conotoxin GIIIA (95).

Single Channel Data. Figure 12 shows data from single channel experiments. The upper trace is a control record showing the current through a single channel that has been treated with batrachotoxin so that it remains open. The measured current, due to the flux of sodium ions, is interrupted by brief, spontaneous closures but is otherwise constant. The next trace shows the blocking events due to single molecules of saxitoxin, present at the concentration of 10 nM, as they bind to and depart from the binding site. Note that the duration of the blocking events is variable and tends to be relatively long. The arrival at and departure from the binding site are unrelated, random events beyond the natural constraint that one must precede the other. The duration of an individual open state is thus indeterminate, governed only by the probability that a saxitoxin molecule will arrive and bind successfully. However, statistical summary of a large number of open times provides a mean value which, coupled with the concentration of the toxin, characterizes the tendency of the toxin to bind: the on rate, k_{on}, under the conditions of the experiment (transmembrane potential, temperature, pH, etc.). Similarly, the duration of an individual blocked state is indeterminate, governed only by the probability that the bound toxin molecule will depart. Statistical summary of a large number of blocked times provides a mean value which, independent of toxin concentration in the medium, characterizes the tendency of the toxin to remain bound to the site, the mean dwell time. This value is the reciprocal of the off rate, k_{off}, again under the conditions of the experiment. It is important to note that, while the mean open time is directly affected by the concentration of the blocker, the mean dwell time is independent of blocker concentration. Procedures for the analysis of single channel data and the underlying concepts are presented in detail in Moczydlowski et al. (90,91).

The subsequent records in Figure 12 are for saxitoxin derivatives bearing a 21-sulfo group (B1, **2**), an 11-α-hydroxysulfate (GTX2, **3**), and both substituents (C1, **4**). Note the difference in the dwell times of the blocking events, in particular the great reduction in dwell time associated with the presence of the 21-sulfo group.

It is important to distinguish between the kind of information being obtained from single channel methods and others where one measures the collective response of a large number of channels to the applied sample. In single channel data, the duration of the blocked state, which apparently corresponds to the dwell time of the toxin molecule at the binding site, contains information about the toxin molecule that is bound. Although the durations of individual events are indeterminate, their distribution can be characterized. A population of blocking events for a single channel that does not conform to the distribution expected for a pure compound therefore indicates that the sample applied is a mixture. In cases where the mean dwell times of two toxins differ greatly, simple inspection of the recording can reveal trace contamination of one with the other. Such information cannot generally be extracted from the "collective response" data provided by binding experiments and most electrophysiological techniques, in which one observes the net effect of a large number of blocking events at a large number of binding sites. In each of the following two cases, collective response data obtained using mouse bioassay, membrane binding, and a variety of electrophysiological techniques were unable to unequivocally define the potency of the subject toxins because their observed activity could have been due to contamination by traces of more potent toxins at concentrations below the practical limits of analytical detection.

When saxitoxin C1 (**4**) was first isolated and tested by mouse bioassay (*9,10*), it was found to have a very low potency and the assay results were quite variable. Subsequent study with electrophysiological and binding techniques gave similar results. The observed potency was low enough that it could be accounted for by amounts of its epimer, C2 (**6**), or hydrolysis product, GTX2 (**3**), below the limits of detection by chemical methods. Due to the ease with which these transformations occur, it was not originally clear whether C1 was intrinsically toxic (*10*). However, the population of blocking events observed for a clean preparation of **4**, such as that shown in Figure 12, is clearly that of a pure compound with a short dwell time, distinct from the populations observed for **6** or **3**. This observation confirms that, although the potency of C1 is greatly attenuated, it does have intrinsic activity. Had the toxicity been due only to traces of the other compounds, the observed record would have been a very low frequency of long blocking events. Recordings from some preparations of **4** included distinctly different, long blocking events as would be expected if the preparation were contaminated with the hydrolysate **3** or epimer **6**. The long blocking events were eliminated by treatment of the preparation with carboxylate cation exchange resin, indicating that the contaminant was **3** which, unlike **4** or **6**, binds to the resin.

A similar situation exists with the saxitoxins reduced at C-12. Chemical reduction of saxitoxin to the two epimeric saxitoxinols gives preparations found to have very low toxicity by a variety of collective response methods (*23,96,97*). However, the saxitoxinols can be re-oxidized under very mild conditions to yield saxitoxin (*23*) which implies that, regardless of the extent to which a saxitoxinol preparation is purified, it might develop trace impurities of saxitoxin during subsequent handling. It has therefore remained unclear whether the observed potency of the saxitoxinols was intrinsic, or due to contaminating saxitoxin at concentrations below the limits of chemical detection. Figure 13 shows traces for single channel experiments with α-saxitoxinol. Note the population of short blocking events, quite distinct from those characteristic of saxitoxin. These experiments provide the first unambiguous evidence that α-saxitoxinol is itself active on sodium channels. When similar studies on β-saxitoxinol have been completed, it will be possible to address the nature of the interaction between the binding site and the functional group at C-12, which is clearly a key component in toxin binding.

In an entirely different vein, single channel methods provide an exceptionally sensitive and reliable technique for detecting the saxitoxins and related compounds and, under favorable circumstances, providing information on their identity. Extracts can be added to the test solution with minimal preparation, eliminating concerns about contamination during the sort of sample cleanup required for other techniques. The fact that discrete blocking events are observed establishes the presence of a reversible sodium channel blocker, and thus one of very few classes of compounds. The mean dwell time of blocking events provides qualitative information on the identity of the toxin. Figure 14 shows traces for experiments with extracts from mackerel livers; similar extracts had given a positive test for PSP using the mouse bioassay. Preliminary analysis of the populations of blocking events suggests a single toxin with a mean dwell time approximately that of saxitoxin. The origin of this toxin is not yet clear, but it seems likely to have been accumulated at a low rate, insufficient to cause the death of the fish, probably by the consumption of zooplankters that had accumulated the saxitoxins, ultimately from *Alexandrium*. If this is the case, the observed composition is interesting since saxitoxin is not known to be the sole or major component in the mixtures of saxitoxins found along the Atlantic coast.

Relationship Between Kinetic and Equilibrium Parameters. From the single channel data, the k_{on} and mean dwell time can be determined for each toxin, under the conditions of the experiment. A plot can then be constructed, using k_{on} and mean dwell time as the axes, in which each toxin is assigned a locus on the basis of these two parameters. Figure 15 is such a plot, using a logarithmic scale for each axis. The construction of this kinetic/equilibrium (K/E) plot will be discussed in detail in a future publication. For the present, it is sufficient to note that it is a convenient property of a log/log plot that the distance along the diagonal is proportional to the ratio of the distances along the axes. Thus, since $K_d = k_{off}/k_{on}$, with k_{on} increasing vertically and dwell time ($1/k_{off}$) increasing to the right, distances along a diagonal going upward to the right correspond to increasing affinity ($1/K_d$). Perpendiculars to this diagonal are "isoaffinity" lines, shown here at decade intervals of K_d. The plotted position of each toxin thus corresponds vertically to its k_{on}, horizontally to its dwell time ($1/k_{off}$), and diagonally to its affinity ($1/K_d$). Note the correlation between the geometric projections on the diagonal at the lower right of Figure 15 and the calculated affinities plotted along the bottom line in Figure 11. The K/E plot can be viewed as representing a dissection of the equilibrium dissociation constant K_d (the diagonal) into its component kinetic parameters k_{on} (vertical) and k_{off} (horizontal).

In the K/E plot, effects of structure on activity that are obscure in the equilibrium data become evident when the K_d is resolved into kinetic parameters. Neosaxitoxin (7) differs from saxitoxin (1) by the presence of a hydroxyl group at N-1. In the mouse bioassay (Figure 11), 7 is slightly more potent than 1; in the rabbit brain radiolabel displacement assay the difference is greater. These results would suggest that the N-1-hydroxyl substituent does something to enhance potency, but comparison of the results in Figure 11 for other pairs of toxins, with (3, 2, 4) and without (9, 8, 10) the N-1-hydroxyl substituent, seems not to support this hypothesis. The binding assay potency of 9 is only slightly greater than that of 3, while that of 8 is actually less than 2, and that of 10 is a good deal less than 4. The effects become clear in Figure 15 where in each case the N-1-hydroxyl toxin (7, 9, 8, 10) has both a lower k_{on} and a longer mean dwell time than the corresponding toxin without the N-1-hydroxyl (1, 3, 2, 4). The relative amounts of change in each parameter differ slightly so that the net result in the K_d differs in both magnitude and sense. In contrast, the changes in k_{on} and k_{off} are similar in magnitude, themselves larger than the resulting change in K_d, and uniform in sense.

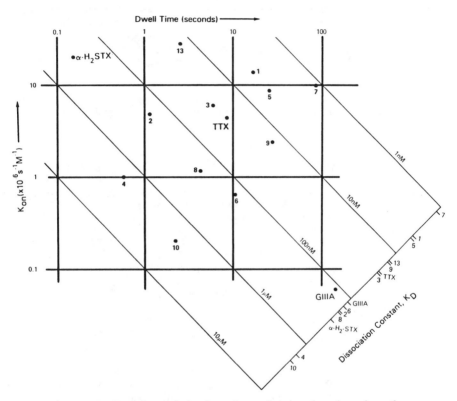

Figure 15.   Data from single channel experiments, plotted to show the relationship between kinetic and equilibrium parameters for several of the saxitoxins, tetrodotoxin, and *Conus geographus* toxin GIIIA. Compound numbering corresponds to that in Figure 1. The vertical axis is k_{on} and the horizontal axis is dwell time, the reciprocal of k_{off}. The dissociation constant, the ratio of k_{off}/k_{on}, therefore corresponds to distance along the diagonal. Data primarily from Ref. *95*.

Similarly, from the comparison in Figure 11 of the equilibrium data for the 11-β-hydroxysulfates 5 and 6 with the corresponding 11-H compounds 1 and 2, the effect of the 11-β-hydroxysulfate substituent is not clear. In Figure 15 it is apparent that the substituent in both cases reduces the k_{on} and increases the mean dwell time. In Figure 15 note also the several toxins with equilibrium dissociation constants clustered near the 100 nM isoaffinity line which might, on the basis of K_d alone, be considered to have similar properties. The cluster includes α-saxitoxinol; saxitoxins 2, 8, and 6; and an entirely different kind of compound, the cone shell peptide μ-conotoxin GIIIA (see Chapter 20). Once the K_d is resolved into its kinetic components the differences among these toxins become quite clear, with α-saxitoxinol having a fast k_{on} and a very short mean dwell time, while μ-conotoxin GIIIA, which binds at the binding site, has both the slowest k_{on} and longest mean dwell time of the toxins shown.

Structure/Activity Relationships. Once the equilibrium data are resolved into their kinetic components, trends are readily apparent, although these have yet to be integrated satisfactorily into a structural hypothesis for the toxin/binding site interaction. This is due in part to two complicating factors. The first is a need for adequate definition of the structure of the toxin as it interacts with the binding site. The x-ray crystal structures that have been determined (1,2,10,11,13) for several of the saxitoxins provide a precise definition of the toxin molecule in the crystal lattice (see Chapter 10). The conformation of the toxin in solution may differ and cannot be determined easily, although it can be defined to some extent by detailed NMR studies (see Chapter 22). The greater problem is that, for such a high affinity binding as this, the binding energy may be large with respect to energy differences among conformations, such that the conformation at the binding site may resemble neither that in the crystal lattice nor that which predominates in solution. The need is to define those toxin conformations that are likely to be energetically accessible under the conditions at the binding site. The answers will likely be derived from computations which use the x-ray structures and NMR data as bases. The second complicating factor is that, although a single modification in structure may have a single major effect – eliminating a hydrogen bond by deleting one group or steric hindrance by adding another – any modification of the toxin molecule in principle affects the behavior of the entire molecule. While modification of a single site has the net observable effects of increasing or decreasing the k_{on} and k_{off}, these observable effects are actually the sums of a multitude of interactions which individually make positive or negative contributions to each rate. The analysis of molecular interactions in terms of a finite set of parameters is in itself an approximation. Accepting that approximation, one would ideally have experimental means to measure separately a large number of parameters relating to fit and bonding. In practice, relatively little of this sort of information is obtainable.

Despite these uncertainties, some of the observable trends clearly suggest effects. In Figure 16, a plot similar to Figure 15, the approximate net charge (see Figure 3) is shown for each of the saxitoxins. Note the general correlation between on rate, k_{on}, and net charge. Removal of the carbamate side chain from 1 to form decarbamoylsaxitoxin 13 reduces the mean dwell time but increases k_{on}. This removal may deprive the toxin molecule of a favorable binding interaction, while simplifying and thus accelerating orientation of the toxin molecule for binding. The significance of this portion of the toxin molecule is further suggested by the great decrease in mean dwell time seen for the addition of the 21-sulfo group: compare 1 and 2; 3 and 4; 5 and 6; 7 and 8; 9 and 10. In contrast, the presence of the N-1-hydroxyl group is consistently associated with an increase in mean dwell time: compare 1 and 7; 2 and 8; 3 and 9; 4 and 10. This effect might be due to

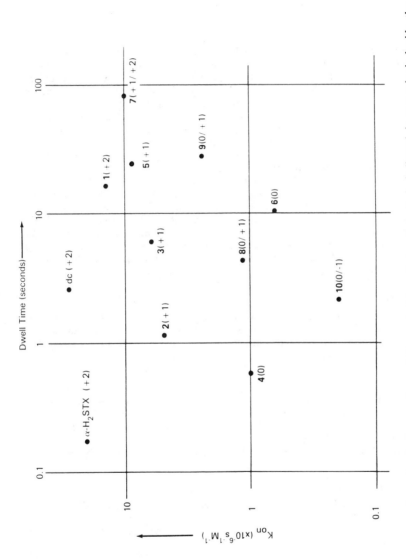

Figure 16. The effect of net charge. Approximate net charge for each toxin is indicated in parenthesis beside the compound number. Note the strong correlation between net charge and k_{on}. Net charge does not appear to have a significant effect on dwell time.

a favorable interaction in the same portion of the molecule, or to the influence, demonstrated in vitro, of the N-1-hydroxyl substituent on the behavior of C-12 (97). The greatly reduced dwell time of α-saxitoxinol emphasizes the importance of the C-12 functionality. More extensive discussion of these effects, their significance, and their interpretation will be presented as more data become available.

Summary

The saxitoxins are a large family of related compounds with differing properties, derived from a single parent compound through relatively few modes of variation. The natural saxitoxins are found in three genera of marine dinoflagellates and at least one procaryote. The composition of toxins found in *Alexandrium*, one of these genera, tends to be uniform within regions but quite different among them, suggesting that the organism occurs as discrete regional populations. Animals vary in their sensitivity to the saxitoxins, some accumulating the toxins to high levels and altering the composition of the mixture through a variety of processes. Detection of the saxitoxins is complicated by their chemistry, variety, the variations in composition, and the low levels that must be detected.

The saxitoxins are reversible blockers of voltage-activated sodium channels. The differences in structure of the various saxitoxins alter the rates at which they bind to and depart from the binding site on the sodium channel. The ratio of these rates determines the equilibrium dissociation constant which, in turn, largely controls their relative toxicity in vivo. Studies of their activity, directed primarily at elucidating the interaction of the toxin molecule with the binding site, also provide insights useful in understanding the pharmacology of other toxins and bimolecular interactions in general.

Acknowledgements

We wish to express our appreciation to J. Daly, NIH, for generously supplying the batrachotoxin essential for the bilayer experiments; to S. D. Darling, University of Akron, for his work on the crystal structures of C2 and C4; and to T. Chambers, FDA, for molecular graphics.

Literature Cited

1. Schantz, E. J.; Ghazarossian, V. E.; Schnoes, H. K.; Strong, F. M.; Springer, J. P.; Pezzanite, J. O.; Clardy, J. *J. Am. Chem. Soc.* **1975**, *97*(5), 1238-9.
2. Bordner, J.; Thiessen, W. E.; Bates, H. A.; Rapoport, H. *J. Am. Chem. Soc.* **1975**, *97*(21), 6008-12.
3. Shimizu, Y.; Hsu, C.-P.; Fallon, W. E.; Oshima, Y.; Miura, I.; Nakanishi, K. *J. Am. Chem. Soc.* **1978**, *100*, 6791-6793.
4. Shimizu, Y.; Fallon, W. E.; Wekell, J. C.; Gerber, D. J.; Gauglitz, E. J., Jr. *J. Agric. Food Chem.* **1978**, *26*, 878-881.
5. Shimizu, Y.; Alam, M.; Oshima, Y.; Fallon, W. E. *Biochem. Biophys. Res. Commun.* **1975**, *66*, 731-737.
6. Boyer, G. L.; Schantz, E. J.; Schnoes, H. K. *J. Chem. Soc., Chem. Commun.* **1978**, (20), 889-90.
7. Boyer, G. L. Ph.D. Thesis, University of Wisconsin, 1980.
8. Wichmann, C. F.; Boyer, G. L.; Divan, C. L.; Schantz, E. J.; Schnoes, H. K. *Tetrahedron Lett.* **1981**, *22*(21), 1941-4.
9. Hall, S.; Reichardt, P. B.; Neve, R. A. *Biochem. Biophys. Res. Comm.* **1980**, *97*, 649-653.
10. Hall, S. Ph.D. Thesis, University of Alaska, 1982.

11. Wichmann, C. F.; Niemczura, W. P.; Schnoes, H. K.; Hall, S.; Reichardt, P. B.; Darling, S. D. *J. Am. Chem. Soc.* **1981**, *103*(23), 6977-8.
12. Koehn, F. E.; Hall, S.; Wichmann, C. F.; Schnoes, H. K.; Reichardt, P. B. *Tetrahedron Lett.* **1982**, *23*(22), 2247-8.
13. Hall, S.; Darling, S. D.; Boyer, G. L.; Reichardt, P. B.; Liu, H. W. *Tetrahedron Lett.* **1984**, *25*(33), 3537-8.
14. Rogers, R. S.; Rapoport, H. *J. Am. Chem. Soc.* **1980**, *102*(24), 7335-9.
15. Ghazarossian, V. E.; Schantz, E. J.; Schnoes, H. K.; Strong, F. M. *Biochem. Biophys. Res. Commun.* **1976**, *68*(3), 776-80.
16. Sullivan, J. J.; Iwaoka, W. T.; Liston, J. *Biochem. Biophys. Res. Commun.* **1983**, *114*(2), 465-72.
17. Sullivan, J. J. Ph.D. Thesis, University of Washington, 1982.
18. Oshima, Y.; Kotaki, Y.; Harada, T.; Yasumoto, T. In *Seafood Toxins*; Ragelis, E. P., Ed.; ACS Symposium Series No. 262; American Chemical Society: Washington, DC, 1984; pp. 161-170.
19. Wichmann, C. F. Ph.D. Thesis, University of Wisconsin, 1981.
20. Koehn, F. E. Ph.D. Thesis, University of Wisconsin, 1983.
21. Hall, S.; Reichardt, P. B. In *Seafood Toxins*; Ragelis, E. P., Ed.; ACS Symposium Series No. 262; American Chemical Society: Washington, DC, 1984; 113-123.
22. Shimizu, Y.; Hsu, C.-P.; Genenah, A. *J. Am. Chem. Soc.* **1981**, *103*, 605-609.
23. Koehn, F. E.; Ghazarossian, V. E.; Schantz, E. J.; Schnoes, H. K.; Strong, F. M. *Bioorg. Chem.* **1981**, *10*(4), 412-428.
24. Ritchie, J. M.; Rogart, R. B.; Strichartz, G. R. *J. Physiol. (London)* **1976**, *261*, 477-494.
25. Wong, J. L.; Brown, M. S.; Matsumoto, K.; Oesterlin, R.; Rapoport, H. *J. Amer. Chem. Soc.* **1971**, *93*(18), 4633-4.
26. Bates, H. A.; Rapoport, H. *J. Agric. Food Chem.* **1975**, *23*(2), 237-9.
27. Meyer, K. F.; Sommer, H.; Schoenholz, P. *J. Prevent. Med.* **1928**, *2*, 365-394.
28. Sommer, H.; Whedon, W. F.; Kofoid, C. A.; Stohler, R. *Arch. Pathol.* **1937**, *24*, 537-559.
29. Sommer, H.; Meyer, K. F. *Arch. Pathol.* **1937**, *24*, 560-598.
30. Whedon, W. F.; Kofoid, C. A. *Univ. Calif. Publ. Zool.* **1936**, *41*, 25-34.
31. Needler, A. B. *J. Fish. Res. Board Can.* **1949**, *7*, 490-504.
32. Neal, R. A. Ph.D. Thesis, University of Washington, 1967.
33. Steidinger, K. A. *Phycologia* **1971**, *10*, 183-187.
34. Taylor, F. J. R. *Environ. Letters* **1975**, *9*, 103-119.
35. Taylor, F. J. R. In *Toxic Dinoflagellate Blooms*; Taylor. D. L.; Selinger, H. H., Eds.; Developments in Marine Biology 1, Elsevier/North Holland: New York, 1979; 47-56.
36. Loeblich, A. R.; Loeblich, L. A., III In *Toxic Dinoflagellate Blooms*; Taylor, D. L.; Selinger, H. H., Eds.; Developments in Marine Biology 1, Elsevier/North Holland: New York, 1979; 41-46.
37. Balech, E. In *Toxic Dinoflagellates*; Anderson, D. M.; White, A. W.; Baden, D. G., Eds.; Elsevier: New York, 1985; pp. 33-38.
38. Steidinger, K. A.; Tester, L. S.; Taylor, F. J. R. *Phycologia* **1980**, *19*, 329-337.
39. MacLean, J. L. *Limnol Oceanogr.* **1977**, *22*, 234-254.
40. Harada, T.; Oshima, Y.; Yasumoto, T. *Agric. Biol. Chem.* **1982**, *46*(7), 1861-4.
41. Gacutan, R. Q.; Tabbu, M. Y.; Aujero, E. J.; Icatlo, F. *J. Marine Biol.* **1985**, *87*, 223-227.
42. Rosales-Loessener, R.; de Porras, E.; Dix, M. W. In *Red Tides: Biology, Environmental Science, and Toxinology*; Okaichi, Y.; Anderson, D. M.; Nemoto, T., Eds.; Elsevier: New York, 1989; pp. 113-116.

43. Morey-Gaines, G. *Phycologia* **1982**, *21*, 154-163.
44. Mee, L. D.; Espinosa, M.; Diaz, G. *Mar. Envir. Res.* **1986**, *19*, 77-92.
45. Fraga, S.; Sanchez, F. J. In *Toxic Dinoflagellates*; Anderson, D. M.; White, A. W.; Baden, D. G., Eds.; Elsevier: New York, 1985; 51-54.
46. Franca, S.; Almeida, J. F. In *Red Tides: Biology, Environmental Science, and Toxinology*; Okaichi, Y.; Anderson, D. M.; Nemoto, T., Eds.; Elsevier: New York, 1989; pp. 93-96.
47. Hallegraeff, G. M.; Stanley, S. O.; Bolch, C. J.; Blackburn, S. I. In *Red Tides: Biology, Environmental Science, and Toxinology*; Okaichi, Y.; Anderson, D. M.; Nemoto, T., Eds.; Elsevier: New York, 1989; pp. 77-80.
48. Ikeda, T.; Matsuno, S.; Sato, S.; Ogata, T.; Kodama, M.; Fukuyo, Y.; Takayama, H. In *Red Tides: Biology, Environmental Science, and Toxinology*; Okaichi, Y.; Anderson, D. M.; Nemoto, T., Eds.; Elsevier: New York, 1989; pp. 411-414.
49. Shimizu, Y. *Pure Appl. Chem.* **1986**, *58*(2), 257-62.
50. Shimizu, Y.; Norte, M.; Hori, A.; Genenah, A.; Kobayashi, M. *J. Am. Chem. Soc.* **1984**, *106*(21), 6433-4.
51. Konosu, S.; Inoue, A.; Noguchi, T.; Hashimoto, Y. *Bull. Japan. Soc. Sci. Fish.* **1969**, *35*, 88-92.
52. Noguchi, T.; Koyama, K.; Uzu, A.; Hashimoto, K. *Nippon Suisan Gakkaishi* **1983**, *49*(12), 1883-6.
53. Yasumura, D.; Oshima, Y.; Yasumoto, T.; Alcala, A. C.; Alcala, L. C. *Agric. Biol. Chem.* **1986**, *50*, 593-598.
54. Silva, E. S. *Notas E Estudos Inst. Biol. Marit.* **1962**, No. 26, plates I-X.
55. Kodama, M.; Ogata, T.; Sato, S. *Agric. Biol. Chem.* **1988**, *52*, 1075-1077.
56. Alam, M. I.; Hsu, C. P.; Shimizu, Y. *J. Phycol.* **1979**, *15*(1), 106-10.
57. Oshima, Y.; Hayakawa, T.; Hashimoto, M.; Kotaki, Y.; Yasumoto, T. *Bull. Japan. Soc. Sci. Fish.* **1982**, *48*, 851-854.
58. Boyer, G. L.; Sullivan, J. J.; Andersen, R. J.; Taylor, F. J. R.; Harrison, P. J.; Cembella, A. D. *Mar. Biol. (Berlin)* **1986**, *93*(3), 361-9.
59. Cembella, A. D.; Sullivan, J. J.; Boyer, G. L.; Taylor, F. J. R.; Anderson, R. J. *Biochem. Systematics Ecol.* **1987**, *15*, 171-186.
60. Sribhibhadh, A. Ph.D. Thesis, University of Washington, 1963.
61. DuPuy, J. L. Ph.D. Thesis, University of Washington, 1968.
62. Twarog, B. M.; Hidaka, T.; Yamaguchi, H. *Toxicon* **1972**, *10*(3), 273-8.
63. Twarog, B. M.; Yamaguchi, H. *Proc. 1st Int. Conf. Toxic Dinoflagellate Blooms*; LoCicero, V. R., Ed.; Mass. Sci. Technol. Found., 1975; pp. 381-93.
64. Shumway, S. E.; Cucci, T. L. *Aquatic Toxicol.* **1987**, *10*, 9-27.
65. Caddy, J. F.; Chandler, R. A. *Proc. Nat. Shellfish Assoc.* **1968**, *58*, 46-50.
66. White, A. W. In *Toxic Dinoflagellate Blooms*; Taylor, D. L.; Seliger, H. H., Eds.; Developments in Marine Biology 1; Elsevier North Holland: New York, 1979; pp. 381-384.
67. White, A. W. *Mar. Biol.* **1981**, *65*, 255-260.
68. White, A. W. *Limnol. Oceanogr.* **1981**, *26*, 103-109.
69. Shimizu, Y.; Yoshioka, M. *Science* **1981**, *212*, 547-549.
70. Schantz, E. J.; McFarren, E. F.; Schaeffer, M. L.; Lewis, K. H. *J. Assoc. Off. Agric. Chem.* **1958**, *41*, 160-168.
71. McFarren, E. F. *J. Assoc. Off. Agric. Chem.* **1959**, *42*, 263-271.
72. Ross, M. R.; Siger, A.; Abbott, B. C. In *Toxic Dinoflagellates*; Anderson, D. M.; White, A. W.; Baden, D. G., Eds.; Elsevier: New York, 1985; 433-8.
73. Chu, F. S.; Fan, T. S. L. *J. Assoc. Off. Anal. Chem.* **1985**, *68*(1), 13-16.
74. Yang, G. C.; Imagire, S. J.; Yasaei, P.; Ragelis, E. P.; Park, D. L.; Page, S. W.; Carlson, R. W.; Guire, P. W. *Bull. Environ. Contam. Toxicol.* **1987**, *39*, 264-271.

75. Gershey, R. M.; Neve, R. A.; Musgrave, D. L.; Reichardt, P. B. *J. Fish. Res. Board Can.* **1977**, *34*(4), 559-63.

76. McFarren, E. F. *J. Assoc. Off. Anal. Chem.* **1960**, *43*, 544-547.

77. Shoptaugh, N. H.; Ikawa, M.; Foxall, T. L.; Sasner, J. J., Jr. *Environ. Sci. Res* **1981**, *20*, 427-35.

78. Buckley, L. J.; Ikawa, M.; Sasner, J. J., Jr. In *Proc. 1st Intl. Conf. Toxic Dinoflagellate Blooms*; LoCicero, V. R., Ed.; Mass. Sci. Technol. Found., 1975; 423-31.

79. Jonas-Davies, J.; Sullivan, J. J.; Kentala, L. L.; Liston, J.; Iwaoka, W. T.; Wu, L. J. Food Sci. **1984**, *49*, 1506.

80. Davio, S. R.; Fontelo, P. A. *Anal. Biochem.* **1984**, *141*(1), 199-204.

81. Horwitz, W. *Official Methods of Analysis of the Association of Official Analytical Chemists, 12th Edition*; AOAC: Washington, D. C., 1975, pp 319-321.

82. Schantz, E. J.; Mold, J. D.; Stranger, D. W.; Shavel, J.; Riel, F.; Bowden, J. P.; Lynch, J. M.; Wyler, R. S.; Riegel, B.; Sommer, H. *J. Am. Chem. Soc.* **1957**, *79*, 5230-5235.

83. Buckley, L. J.; Ikawa, M.; Sasner, J. J., Jr. *J. Agric. Food Chem.* **1976**, *24*(1), 107-11.

84. Yasumoto, T.; Oshima, Y.; Konta, T. *Bull. Japan Soc. Sci. Fish.* **1981**, *47*, 957-959.

85. Koyama, K.; Noguchi, T.; Ueda, Y.; Hashimoto, K. *Nippon Suisan Gakkaishi* **1981**, *47*(7), 965.

86. Sullivan, J. J.; Iwaoka, W. T. *J. Assoc. Off. Anal. Chem.* **1983**, *66*, 297-303.

87. Oshima, Y.; Machida, M.; Sasaki, K.; Tamaoki, Y.; Yasumoto, T. *Agric. Biol. Chem.* **1984**, *48*(7), pp 1707-11.

88. Oshima, Y.; Hasegawa, M.; Yasumoto, T.; Hallegraeff, G.; Blackburn, S. *Toxicon* **1987**, *25*(10), 1105-11.

89. Nagashima, Y.; Maruyama, J.; Noguchi, T.; Hashimoto, K. *Nippon Suisan Gakkaishi* **1987**, *53*, 819-823.

90. Moczydlowski, E.; Garber, S. S.; Miller, C. *J. Gen. Physiol.* **1984**, *84*(5), 665-686.

91. Moczydlowski, E.; Hall, S.; Garber, S. S.; Strichartz, G. S.; Miller, C. *J. Gen. Physiol.* **1984**, *84*(5), 687-704.

92. Strichartz, G. R.; Hansen Bay, C. M. *J. Gen.Physiol.* **1981**, *77*(2), 205-21.

93. Ritchie, J. M.; Rogart, R. B.; Strichartz, G. *J. Physiol. (London)* **1976**, *258*(2), 99P-100P.

94. Strichartz, G.; Rando, T.; Hall, S.; Gitschier, J.; Hall, L.; Magnani, B.; Bay, C. H. *Ann. N. Y. Acad. Sci.* **1986**, *479*, 96-112.

95. Guo, X.; Uehara, A.; Ravindran, A.; Bryant, S. H.; Hall, S.; Moczydlowski, E. *Biochemistry* **1987**, *26*(24), 7546-56.

96. Kao, C. Y.; Kao, P. N.; James-Kracke, M. R.; Koehn, F. E.; Wichmann, C. F.; Schnoes, H. K. *Toxicon* **1985**, *23*(4), 647-55.

97. Strichartz, G. *J. Gen. Physiol.* **1984**, *84*(2), 281-305.

RECEIVED October 4, 1989

Chapter 4

High-Performance Liquid Chromatographic Method Applied to Paralytic Shellfish Poisoning Research

John J. Sullivan[1]

U.S. Food and Drug Administration, 22201 Twenty-Third Drive, SE, Bothell, WA 98021

The use of high performance liquid chromatography (HPLC) for the study of paralytic shellfish poisoning (PSP) has facilitated a greater understanding of the biochemistry and chemistry of the toxins involved. HPLC enables the determination of the type and quantity of the PSP toxins present in biological samples. An overview of the HPLC method is presented that outlines the conditions for both separation and detection of the PSP toxins. Examples of the use of the HPLC method in toxin research are reviewed, including its use in the determination of the enzymatic conversion of the toxins and studies on the movement of the toxins up the marine food chain.

Advances in our understanding of the biochemistry and chemistry of natural products are necessarily tied to the availability of analytical methods for their detection. Without the means to rapidly and accurately detect a compound in biological samples, research into its biogenesis and metabolic pathways is severely limited. This has been the case in the marine toxins field and especially for paralytic shellfish poisoning (PSP) where the development of rapid and accurate analytical methods for the toxins involved has paralleled the growth in our understanding of the phenomenon. Analytical methods for PSP range from the somewhat inaccurate and non-quantitative mouse bioassay developed in the 1930's to more efficient and sophisticated methods based upon separation techniques such as TLC, electrophoresis, and HPLC. A historical perspective on these methods is the subject of a recent review (1). In most cases, the identification of the PSP toxins and advances in our knowledge of the biochemistry of PSP were a direct result of the development of new analytical techniques, particularly the HPLC method.

With the development of HPLC, a new dimension was added to the tools available for the study of natural products. HPLC is ideally suited to the analysis of non-volatile, sensitive compounds frequently found in biological systems. Unlike other available separation techniques such as TLC and electrophoresis, HPLC methods provide both qualitative and quantitative data and can be easily automated. The basis for the HPLC method for the PSP toxins was established in the late 1970's when Buckley et al. (2) reported the post-column derivatization of the PSP toxins based on an alkaline oxidation reaction described by Bates and Rapoport (3). Based on this foundation, a series of investigations were conducted to develop a rapid, efficient HPLC method to detect the multiple toxins involved in PSP. Originally, a variety of silica-based, bonded stationary phases were utilized with a low-pressure post-column reaction system (PCRS) (4,5). Later, with improvements in toxin separation mechanisms and the utilization of a high efficiency PCRS, a

---

[1] Current address: Varian Associates, Inc., 2700 Mitchell Drive, Walnut Creek, CA 94599

routine method for separation of the 12 most common PSP toxins was achieved (6,7).

The HPLC method (7) for the PSP toxins has a variety of applications in both research and in public health monitoring programs. A number of advances in our understanding of the biochemistry of PSP are a direct result of this technique. Following is a brief overview of the HPLC method with a couple of examples of its utility in PSP research.

Overview of the HPLC Method

The PSP toxins represent a real challenge to the analytical chemist interested in developing a method for their detection. There are a great variety of closely related toxin structures (Figure 1) and the need exists to determine the level of each individually. They are totally non-volatile and lack any useful UV absorption. These characteristics coupled with the very low levels found in most samples (sub-ppm) eliminates most traditional chromatographic techniques such as GC and HPLC with UV/VIS detection. However, by the conversion of the toxins to fluorescent derivatives (3), the problem of detection of the toxins is solved. It has been found that the fluorescent technique is highly sensitive and specific for PSP toxins and many of the current analytical methods for the toxins utilize fluorescent detection. With the toxin detection problem solved, the development of a useful HPLC method was possible and somewhat straightforward.

PSP Toxin Separations

A variety of HPLC separation strategies have been reported using silica, C-18, and ion exchange columns (8,9,10). Additionally, toxin separations were developed in the author's laboratory utilizing amino and cyano bonded phase HPLC columns (4,5). All of these techniques achieved some degree of toxin separation but either could not separate all of the toxins desired or the columns were found to be some-what unstable under the conditions used. The most efficient and reliable toxin separations that have been reported utilize ion-interaction chromatography on porous polymer columns (6,7). In this separation strategy, the column packing acts as a support for the formation of a dynamic ion-exchange mechanism between the toxin molecules and a counter ion contained in the mobile phase. As in conventional ion-exchange, pH and ionic strength control toxin elution. However, the primary advantage of ion-interaction chromatography is that the number and characteristics of the exchange sites on the column can be varied by adding an organic solvent to the mobile phase and choosing the proper type and concentration of counter ion. This greatly increases the efficiency of the separations, particularly for a series of compounds such as the PSP toxins where net charges for the various toxins are between -1 and +2 at pH ca. 7.0.

The HPLC conditions utilized for the PSP toxin separations (7) are outlined in Table I. Since there is such a range in net charge for the toxins, conditions have not been found in which all of the toxins could be separated in a single HPLC run. Method 2 can be utilized to separate toxins C1, C2, C3, and C4, while Method 1 is used to separate the remainder of the toxins (Figure 2). The separation mechanism operating in Method 1 is by far the most complex, with a gradient being performed over the course of the HPLC run that varies pH, ionic strength, and acetonitrile concentration. It is critical, particularly for this separation, that great care be taken in mobile phase preparation to achieve consistent toxin separations; details of these procedures are outlined elsewhere (7). Nevertheless, with care, these HPLC separations are reproducible and column lifetimes exceeding 1000 sample analyses are common.

R1	R2	R3	Carbamate Toxins	Sulfamate Toxins	Decarbamoyl Toxins
H	H	H	STX	B1	dc–STX
OH	H	H	NEO	B2	dc–NEO
OH	H	OSO_3^-	GTX I	C3	dc–GTX I
H	H	OSO_3^-	GTX II	C1	dc–GTX II
H	OSO_3^-	H	GTX III	C2	dc–GTX III
OH	OSO_3^-	H	GTX IV	C4	dc–GTX IV

R4:

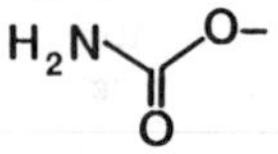

H_2N group — carbamate

R4: H

O_3^-S — sulfamate

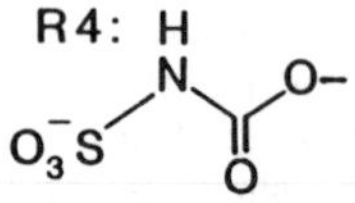

R4:

HO–

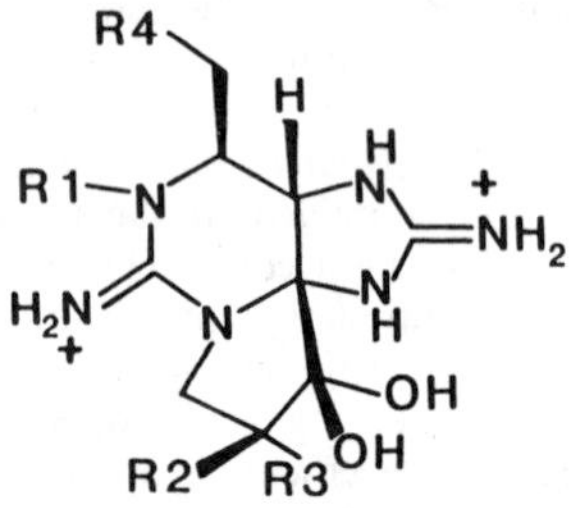

Figure 1. Structures of the PSP toxins. Several of the decarbamoyl toxins have not been reported but are postulated to occur in nature based on presence of the others.

Table I. HPLC Conditions for Separation of the PSP Toxins

Post Column Reaction System (PCRS):
Oxidant - 0.5 mL/min of 5 mM periodic acid in 100 mM phosphate buffer, pH 7.80
Acid - 0.3 mL/min of 0.75 M nitric acid
Reaction Coil - 1 mL volume held at 90°C
Fluorescence detector at 340 nm excitation, 400 nm emission

Method 1:
HPLC column - Hamilton PRP-1, 15 cm x 4.1 mm, 10 μm packing
Flow Rate - 1.3 mL/min, Column Temperature - 35°C
Mobile Phase A: Water with 1.5 mM each hexane and heptane sulfonate, and 1.5 mM ammonium phosphate (as PO_4), pH 6.70
Mobile Phase B: 25% Acetonitrile with 1.5 mM each hexane and heptane sulfonate and 6.25 mM ammonium phosphate, pH 7.00
Gradient:

Time (min)	A(%)	B(%)
0	100	0
4	100	0
11	70	30
17	10	90
17.5	0	100

Method 2:
(Same PCRS conditions, column, and flow rate as Method 1)
Mobile Phase A: 3 mM tetrabutylammonium phosphate, 15 mM ammonium phosphate, pH 7.50
Mobile Phase B: 3 mM tetrabutylammonium phosphate, 30 mM ammonium phosphate, pH 6.00
Gradient:

Time (min)	A(%)	B(%)
0	100	0
2.5	100	0
6	80	20
7	0	100

Detection of the Toxins

Detection of the PSP toxins has proven to be one of the largest hurdles in the development of analytical methods. The traditional means, and still in wide use today, is determination of mouse death times for a 1 mL injection of the test solution. There are a variety of drawbacks to utilization of this technique in routine analytical methods, that have prompted the search for replacements. In 1975 Bates and Rapoport (3) reported the development of a fluorescence technique that has proven to be highly selective for the PSP toxins, and very sensitive for many of them. This detection technique has formed the basis for analytical methods involving TLC (11), electrophoresis (12), column chromatography (13), autoanalyzers (14), and HPLC (5,6,7).

The application of the fluorescence derivatization technique in an HPLC method involves utilization of a post column reaction system (PCRS) as shown in Figure 3 to carry out the wet chemistry involved. The reaction is a 2-step process with oxidation of the toxins by periodate at pH 7.8 followed by acidification with nitric acid. Among the factors that influence toxin detection in the PCRS are periodate concentration, oxidation pH, oxidation temperature, reaction time, and final pH. By far, the most important of these factors is oxidation pH and, unfortunately, there is not one set of reaction conditions that is optimum for all of the PSP toxins. The reaction conditions outlined in Table I, while not optimized for any particular toxin, were developed to allow for adequate detection of all of the toxins involved. Care must be exercised in setting up an HPLC for the PSP toxins to duplicate the conditions as closely as possible to those specified in order to achieve consistent adequate detection limits.

Research Applications of HPLC

By utilizing the HPLC method, it is possible to determine the level of each individual toxin in sample solutions. This provides a "toxin profile" that can be very useful in PSP toxin research studies. The ability to examine relative changes in toxin concentration and profile has greatly facilitated studies relating to toxin production by dinoflagellates, metabolism of toxins in shellfish, and movement of toxins up the food chain. Since the HPLC method is easily automated and requires only very small sample sizes (< 1 g tissue), it has clear advantages over other analytical procedures for the toxins in many research situations. Two examples of the utilization of HPLC for the study of the PSP toxins follow.

Metabolism of the PSP Toxins in Clams

One of the first applications of the HPLC method was the investigation of differences in toxin profiles between shellfish species from various localities (4). It became apparent immediately that there were vast differences in these toxin profiles even among shellfish from the same beach. There were subtle differences between the various shellfish species, and butter clams had a completely different suite of toxins than the other clams and mussels. It was presumed that all of the shellfish fed on the same dinoflagellate population, so there must have been other factors influencing toxin profiles such as differences in toxin uptake, release, or metabolism. These presumptions were strengthened when toxin profiles in the littleneck clam (*Prototheca Staminea*) were examined. It was found that, in this species, none of the toxin peaks in the HPLC chromatogram had retention times that matched the normal PSP toxins. It was evident that some alteration in toxin structure had occurred that was unique in this particular shellfish species.

Based on this observation, a series of investigations were conducted to

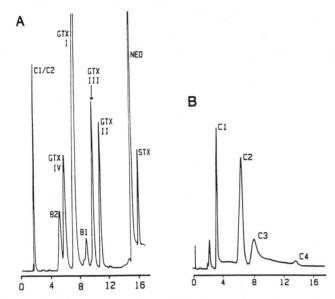

Figure 2. Chromatogram of the PSP toxin standards. Separation conditions from Table I: (A) Method 1; (B) Method 2.

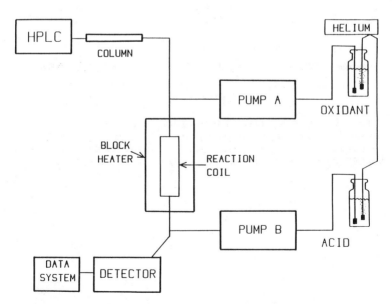

Figure 3. Flow diagram of the PCRS utilized in the HPLC separation of the PSP toxins.

characterize the toxins in littleneck clams (*4, 15*). These studies involved determining the toxin profiles in various clam tissues, incubating known PSP toxins with tissue preparations, and conducting chemical stability and modification studies on the unknown toxins. All of these investigations utilized HPLC separations to characterize the changes taking place in toxin structure. Figure 4 illustrates some of these studies which involved conversion of the known toxins to their unknown form and then back to the known form again.

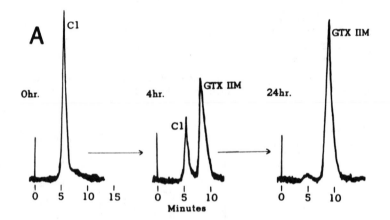

Figure 4A. Chromatogram illustrating the studies conducted on the enzymatic conversion of the PSP toxins to decarbamoyl metabolites (appended with an "M" in these figures). Conversion of C1 to dcGTXII (GTX IIM) in a homogenate of littleneck clam tissue after 4 and 48 hr.

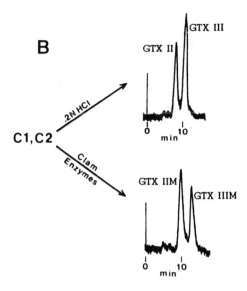

Figure 4B. Chromatogram illustrating the studies conducted on the enzymatic conversion of the PSP toxins to decarbamoyl metabolites (appended with an "M" in these figures). Conversion of a mixture of C1 and C2 by the clam enzymes was chromatographically distinct from conversion to GTX II and GTX III with weak acid.

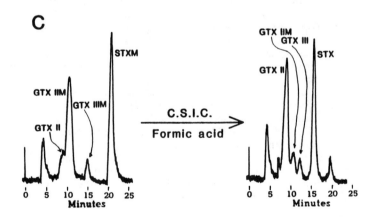

Figure 4C. Chromatogram illustrating the studies conducted on the enzymatic conversion of the PSP toxins to decarbamoyl metabolites (appended with an "M" in these figures). HPLC was used to confirm that the toxin metabolites (labeled "M") were the decarbamoyl form through conversion back to the carbamate form with chlorosulfonyl isocyanate. (*See* Ref. 15 for HPLC conditions.)

Results of these investigations indicated that there were enzymes present in the littleneck clam that facilitated decarbamoylation of either the carbamate or sulfocarbamoyl toxins to their decarbamoyl form (*see* Figure 5). These enzymes may be unique to this particular species of shellfish since, although the decarbamoyl toxins have been found in small quantities in other species, no shellfish species examined to date contains the predominance of the decarbamoyl toxins found in the littleneck clams.

These studies would not have been possible without the availability of the HPLC method that enabled both qualitative and quantitative determination of the toxins in extremely small amounts of biological samples.

Tracing Toxins Up the Food Chain

In order to be a hazard to humans, or to marine vertebrates, the PSP toxins have to be transferred up the food chain from the producing dinoflagellates. By gaining insights into this transport, the nature and extent of the impact of the toxins on sensitive organisms can be determined. Utilization of HPLC in studies such as these has the advantage of providing information on both toxin composition and concentration. A number of investigations have utilized HPLC to study the transport of the toxins up the food chain. Ideally, in a study such as this, toxin analyses need to be performed on both the causative dinoflagellates and the accumulating organisms. This is often very difficult to accomplish due to the transient nature of most dinoflagellate blooms. Consequently, many investigations have relied on culturing dinoflagellates from the same locality in which the bloom occurred along with collecting the accumulating organism from the vicinity. A third experimental design is to conduct the entire study in the laboratory, feeding cultured dinoflagellates to the accumulating organisms. All of these approaches have been utilized along with HPLC detection of the toxins.

One of the first uses of the HPLC method was in the study of a *Protogonyaulax catenella* bloom in Quartermaster Harbor, Washington (*4*). This incident provided some of the first evidence that toxin metabolism was occurring in littleneck clams; chromatograms from these studies are illustrated in Figure 6. Jonas-Davies et al. (*16*) studied a variety of organisms during a bloom several years later in the same bay and determined that the toxins were accumulated in a variety of intertidal organisms besides bivalve molluscs.

HPLC was used to determine that *Gymnodinium catenatum* was the causative organism in PSP outbreaks in Tasmania (*17*) and Spain (Sullivan, unpublished). *G. catenatum* produces a suite of toxins predominated by the low-toxicity sulfamate toxins, C1 to C4, and while shellfish feeding on *G. catenatum* contain predominantly these toxins, they also contain some carbamate and decarbamoyl toxins. Currently, it is not known if the carbamate and decarbamoyl toxins are present due to chemical conversion from the sulfocarbamoyl form or due to selective retention of trace quantities already present in the dinoflagellates. The HPLC played a key role in these studies by providing a rapid, quantitative means to differentiate the various toxins present.

HPLC has also been utilized in more complex food chain transfers. It has been known for some time that the toxins can be responsible for fish kills in the Bay of Fundy (*18*). The vectors for these fish kills are zooplankton that feed on toxic dinoflagellates. In two related studies (*19*; Sullivan, unpublished), HPLC was utilized to investigate the transport of toxins from dinoflagellates to zooplankton and then to fish. The HPLC method is ideally suited for this since only very small sample sizes (ca. 100,000 dinoflagellate cells) are required.

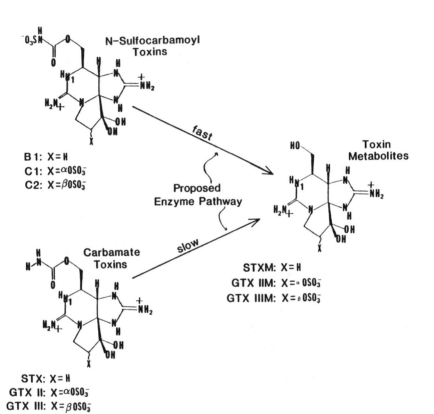

Figure 5.   Enzymatic transformation of the PSP toxins in the littleneck clams.   HPLC was used extensively to determine the presence and characteristics of these conversions.   (Reproduced with permission from Ref. 15.   Copyright 1983 Academic Press, Inc.).

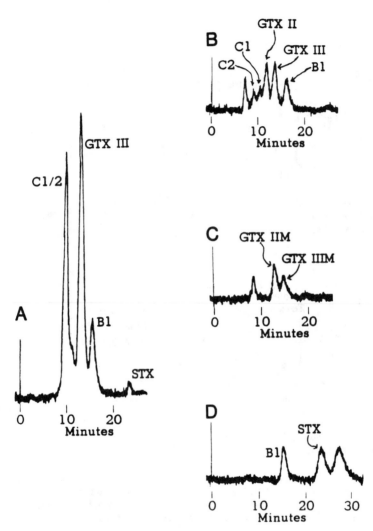

Figure 6. HPLC separation of the toxins present in various organisms during a dinoflagellate bloom in Quartermaster Harbor, Washington: (A) dinoflagellates, (B) mussels, (C) littleneck clams, (D) butter clams. The toxin metabolites ("M") were later found to be the decarbamoyl form. (*See* Ref. 4 for HPLC conditions.)

Conclusion

HPLC is a key tool in the study of PSP. Although this technique has only been utilized for approximately 5 years in PSP research, a number of important discoveries can be linked directly to it. It is unlikely that these studies would have been possible without the availability of HPLC. As HPLC becomes more widely available to researchers in the toxin field, progress will continue to be made in elucidating the nature of PSP, particularly in an understanding of mammalian uptake, distribution, and metabolism, an area that, to date, has not been investigated.

Acknowledgment

I would like to thank Sherwood Hall for the supply of toxin standards used in these studies.

Literature Cited

1. Sullivan, J. J.; Wekell, M. M.; Hall, S. In *Handook of Natural Toxins, Vol. 4, Marine Toxins and Venoms*; Tu, A. T., Ed.: Marcel Dekker, Inc., 1988, in press.
2. Buckley, L. J.; Oshima, Y.; Shimizu, Y. *Anal. Biochem.* **1978**, 85, 157–64.
3. Bates, H. A.; Rapoport, H. J. *J. Agric. Food Chem.* **1975**, *23*, 237–39.
4. Sullivan, J. J. Ph.D. Thesis, Univ. of Washington, Seattle, 1982.
5. Sullivan, J. J.; Iwaoka, W. T. *J. Assoc. Offic. Anal. Chem.* **1983**, 66, 297–303.
6. Sullivan, J. J.; Wekell, M. M. In *Seafood Toxins*; Ragelis, E. P., Ed.; ACS Symposium Series No. 262; American Chemical Society: Washington, DC, 1984; pp 197–205.
7. Sullivan, J. J.; Wekell, M. M. In *Seafood Quality Determination*; Kramer, D. E.; Liston, J., Eds.; Univ. of Alaska, Sea Grant College Program, Anchorage; 1987; pp 357–71.
8. Boyer, G. L. Ph.D. Thesis, Univ. of Wisconsin, Madison, 1980.
9. Robinson, K. A. *Biochem. Biophys. Acta* **1982**, *687*, 315–20.
10. Onoue, Y.; Noguchi, T.; Nagashima, Y.; Hashimoto, K.; Kanoh, S.; Ito, M.; Tsukada, K. *J. Chrom* **1983**, *257*, 373–79.
11. Hall, S.; Reichardt, P. B.; Neve, R. A. *Biochem. Biophys. Res. Comm.* **1980**, *97*, 649–53.
12. Oshima, Y.; Fallon, W. E.; Shimizu, Y.; Noguchi, T.; Hashimoto, Y. *Bull. Jap. Soc. Sci. Fish.* **1976**, *42*, 851–56.
13. Ikawa, M.; Wegener, K.; Foxall, T. L.; Sasner, J. J.; Noguchi, T.; Hashimoto, K. *J. Agric. Food Chem.* **1982**, *30*, 526–28.
14. Jonas-Davies, J.; Sullivan, J. J.; Kentala, L. L.; Liston, J.; Iwaoka, W. T.; Wu, L. J. *J. Food Sci.* **1984**, *49*, 1506–09.
15. Sullivan, J. J.; Iwaoka, W. T.; Liston, J. *J. Biochem. Biphys. Res. Comm.* **1983**, *114*, 465–72.
16. Jonas-Davies; J.; Liston, J. In *Toxic Dinoflagellates*; Anderson, D. M.; White, A. W.; Baden, D.G., Eds:, Elsevier Science Publishing: New York, 1985; pp 467–72.
17. Oshima, Y.; Hasegawa, M.; Yasumoto, T.; Hallegraeff, G.; Blackburn, S. *Toxicon* **1987**, *25*, 1105–11.
18. White, A. W. *J. Fish. Res. Board Canada* **1977**, *34*, 2421–24.
19. Boyer, G. L.; Sullivan, J. J.; LeBlanc, M.; Anderson, R. J. In *Toxic Dinoflagellates*; Anderson, D. M.; White, A. W.; Baden, D. G., Eds.; Elsevier Science Publishing: New York, 1985; pp 407–12.

RECEIVED July 10, 1989

Chapter 5

A Bacterial Source of Tetrodotoxins and Saxitoxins

Mark L. Tamplin[1]

Center of Marine Biotechnology, Maryland Biotechnology Institute,
University of Maryland, Baltimore, MD 21202

Tetrodotoxin (TTX) and saxitoxin (STX) are potent sodium channel
blockers that are found in phylogenetically diverse species of
marine life. The wide distribution of TTX and STX has resulted
in speculation that bacteria are the source of these toxins.
Recently, investigators have reported isolation of marine bacteria,
including *Vibrio*, *Alteromonas*, *Plesiomonas*, and *Pseudomonas*
species, that produce TTX and STX. This chapter details the
methods and results of research to define bacterial sources of TTX
and STX.

The tetrodotoxins (TTXs) and saxitoxins (STXs) have in common the ability to
block sodium channels of excitable membranes (*1–5*). Saxitoxin and tetrodotoxin are some of the most potent non-proteinaceous neurotoxins known and
are responsible for significant human morbidity and mortality (*6, 7*). Although
for many years the biosynthetic origin(s) of TTXs and STXs has not been identified, recent evidence indicates that bacteria may be a source.

Tetrodotoxin poisoning has been recognized for more than two thousand
years. Japanese historical records show that the consumption of certain species
of pufferfish (*Tetraodon* spp.) resulted in paralytic intoxication (*8*). This problem continues in modern times in various Asian countries, especially Japan,
where pufferfish are still regarded as a delicacy. Clinical symptoms of TTX
intoxication include numbness, paralysis, and in some instances death. In fact,
the "zombie" state described in the Voodoo religion has been attributed to TTX
in potions derived from pufferfish (*9*).

Numerous species of edible shellfish and finfish are known to contain
natural toxins during all or part of their life cycle. TTXs and STXs have been
isolated from species of fish, starfish, crab, octopus, frog, newt, salamander, goby,
gastropod, mollusk, flatworm, annelid, zooplankton, and algae (*10–16*). For
example, the Australian blue-ringed octopus (*Hapalochlaena maculosa*) contains
tetrodotoxin in its posterior salivary gland and can kill an adult human with a
single bite (*17*).

[1] Current address: U.S. Food and Drug Administration, Fishery Research Branch, P.O. Box 158,
Dauphin Island, AL 36528

0097–6156/90/0418–0078$06.00/0
© 1990 American Chemical Society

Concentrations of TTXs in fish and other marine organisms vary among species, individual, and tissues of animals (*18*). Liver and ovary of fish usually contain the highest concentrations of TTXs, although relatively high quantities can be found in other tissues (*18*). Terrestrial animals have also been shown to contain TTXs, including brightly-colored frogs, newts, and their egg clusters (*19, 20*). There are two documented cases of newt-associated TTX human poisonings, one of which resulted in death (*21*).

The STXs have been purified from many marine species, including fish, crabs, annelids, and algae (*4, 16, 22–26*). STX and its derivatives are the toxic agents of paralytic shellfish poisoning (PSP), an illness caused by ingestion of contaminated shellfish, such as mussels and clams, which accumulate STX by feeding on toxic dinoflagellates (e.g. *Alexandrium* spp.) (*27, 28*). Reports indicate that concentrations of STX in dinoflagellates vary between species and clones (*23, 29*).

For many years the source(s) of TTXs and STXs has been a controversial subject. The discovery that TTXs were in tissues of animals that were phylogenetically distinct resulted in speculation that TTXs were bioaccumulated and/or originated from symbiotic microorganisms, such as bacteria. Since 1986, a growing number of reports describe bacteria that produce TTXs (*30–35*). Bacteria have also been proposed as the source of STX in marine dinoflagellates. Sousa y Silva (*36*) provided early suggestions of a bacterial source of STX. Subsequently, STX and neo-STX were isolated from strains of a cyanobacterium, *Aphanizomenon flos-aquae*, a procaryotic organism (*22*). Recently, Kodama (*37*) has reported that a bacterium cultured from *Gonyaulax* (= *Alexandrium*) *tamarensis* produces STX.

This chapter reviews recent experimental evidence of a bacterial source of sodium channel blockers, principly TTXs. These findings support the hypothesis that procaryotic organisms produce TTXs which contaminate oceanic food chains.

Materials and Methods

Organisms. TTX, anhydro-TTX, and STX have been reported from various procaryotic species (Table I). All of these species, with the exception of *Escherichia coli*, are marine organisms.

Extraction of Sodium Channel Blockers. A review of published reports shows that methods for purification of sodium channel blockers from bacterial cultures are similar to techniques for isolation of TTX and STX from pufferfish and dinoflagellates (*30, 31, 38, 39*). Typically, cell pellets of bacterial cultures are extracted with hot 0.1% acetic acid, the resulting supernatant ultra-filtered, lyophilized, and reconstituted in a minimal volume of 0.1% acetic acid. Culture media can also be extracted for TTX by a similar procedure (*31*). Both cell and supernatant extracts are analyzed further by gel filtration chromatography and other biological, chemical, and immunological methods. Few reports describe purification schemes that include extraction of control samples of bacteriological media (e.g., broths and agars) which may be derived from marine plant and animal tissues.

Mouse Bioassay. The mouse is the traditional animal of choice for detecting biological activity due to STX and TTX. Mice receive an intraperitoneal injection of sample and are observed for symptoms of intoxication, i.e., dypsnea, convulsions, and death. This method is effective for detecting biological activity of STX and TTX in numerous samples. For the standard STX assay, one mouse unit is defined as that quantity of STX injected i.p. in 1 ml solution that will

Table I. Procaryotic Organisms that Produce Sodium Channel Blockers

Species	Source	Reference
TTXs		
Vibrio fischeri-like	Xanthid crab (Atergatis floridus)	35, 31
Pseudomonas sp.	Pufferfish (Fugu poecilonotus)	33
Alteromonas sp.	Red calcareous alga (Jania sp.)	53
Vibrio alginolyticus	Pufferfish (Fugu vermicularis vermicularis)	32, 34
Vibrio anguillarum	ATCC, NCMB[a]	34
Vibrio costicola	ATCC, NCMB	34
Vibrio cholerae	ATCC, NCMB	54
Vibrio fischeri	ATCC, NCMB	34
Vibrio harveyi	ATCC, NCMB	34
Vibrio marinus	ATCC, NCMB	34
Vibrio parahaemolyticus	ATCC, NCMB	34
Photobacterium phosphoreum	ATCC, NCMB	34
Aeromonas hydrophila	ATCC, NCMB	34, 54
Aeromonas salmonicida	ATCC, NCMB	34
Plesiomonas shigelloides	ATCC, NCMB	34
Escherichia coli	ATCC, NCMB	34
Alteromonas communis	ATCC, NCMB	34
Alteromonas haloplanktis	ATCC, NCMB	34
Alteromonas nigrifaciens	ATCC, NCMB	34
Alteromonas undina	ATCC, NCMB	34
Alteromonas vaga	ATCC, NCMB	34
STXs		
Vibrio-like sp.	Protogonyaulax tamarensis	37
Aphnizomenon flos-aquae		4, 22

[a] ATCC = American Type Culture Collection;
NCMB = National Collection of Marine Bacteria

kill a 20 g mouse in 15 min (*40*). This amount is close to the detection limit and under standard conditions corresponds to about 200 ng of STX, a concentration of about 500 nM in the injected solution. The corresponding definition for TTX uses similar conditions and a 30 min death time.

Immunological Assays. Carlson et al. (*41*) and Chu et al. (*42*) describe immunological assays for detecting STX, at a level of sensitivity of approximately 1 ng STX/ml (ca. 3 nM). As with most immunological assays, antigenic, not biologic, activity is measured. The potential for sample matrix and non-specific binding to affect antibody–antigen reactions strictly necessitates that both positive and negative controls be run for each sample.

Davio et al. (*43*) report efforts to obtain monoclonal antibodies (mAbs) to STX. Because STX is a small molecule of approximately 300 daltons, well below the size necessary for immunogenicity, a carrier molecule must be conjugated to the hapten (STX). This technique must minimize alterations of the antigenic form. For the anti-STX antibodies tested to date, the ratios of immunoassay response factor to pharmacological potency for various STX derivatives differ substantially, the immunoassay being virtually unresponsive to some of the common natural derivatives (*44*).

Displacement Assay for TTX and STX. Davio and Fontelo (*45*) and Richie et al. (*46*) describe methods for detecting STX by measuring displacement of radiolabelled STX from brain membranes. The sensitivity of this assay is approximately 1 ng STX/ml. TTX can also be detected since STX and TTX share the same biological receptor on the sodium channel.

Blocking Events of Sodium Currents Applied to Rat Sarcolemmal Sodium Channels in Planar Lipid Bilayers. The physiological effects of STX and TTX on the movement of sodium ions through membrane sodium channels can be detected by a method first described by Krueger et al. (*47*). In this procedure, the voltage-dependent gating of sodium channels is monitored in artifical lipid bilayers containing purified sodium channels. Specific gating patterns can be used to identify different sodium channel-blocking toxins.

Tissue Culture Assay. Kogure et al. (*48*) report a novel tissue culture assay for detecting several types of sodium channel blockers. The mouse neuroblastoma cell line ATCC CCL 131 is grown in RPMI 1640 supplemented with 13.5% fetal bovine serum and 100 μg/ml gentamycin, in an atmosphere of 5% CO_2–95% air at 37°C. Ninety-six well plates are seeded with 1 x 10^5 cells in 200 μl of medium containing 1 mM ouabain and 0.075 mM veratridine. Veratridine and ouabain cause neuroblastoma cells to round-up and die. In the presence of sodium channel blockers (e.g., TTXs or STXs), the lethal action of veratridine is obviated and cells retain normal morphology and viability. An important feature of this assay is that a positive test for sodium channel blockers results in normal cell viability. Since bacterial extracts can contain cytotoxic components, this assay offers an advantage over tests that use cell death as an endpoint. The minimum detectable level of TTX is approximately 3 nM, or approximately 1/1000 mouse unit.

Chemical Methods. Chemical purification and structural analyses of TTXs and STXs include thin layer chromatography (TLC), high performance liquid chromatography (HPLC), ion exchange chromatography, size exclusion chromatography, gas chromatography (GC), mass spectroscopy (MS), and nuclear magnetic resonance spectroscopy (NMR) (*33, 39, 49–52*). Several laboratories have designed continuous TTX analyzers that use HPLC, and in many instances, GC-MS instrumentation.

Results

Biological Characterization of Bacterial Sodium Channel Blockers.   A variety of bacteria are reported to produce sodium channel blockers (Table I).   Yasumoto et al. (*30*) and Noguchi et al. (*31*) have isolated bacteria that produce TTX and its less toxic derivative, anhydro-TTX from tissues of alga and toxic crabs, respectively.   In subsequent reports, TTX, anhydro-TTX, and 4-epi-TTX were shown to be produced by bacteria isolated from various toxic marine organisms.   These include a *Vibrio* sp. cultured from the gut of a xanthid crab (*Atergatis floridus* sp.) (*31*), *V. alginolyticus* isolated from the intestines of a pufferfish (*Fugu vermicularus vermicularis*) (*32*), an *Alteromonas* sp. isolated from reef animals (*53*), and a *Pseudomonas* sp. cultured from the surface of pufferfish (*33*).   Simidu et al. (*34*) describe a variety of marine bacteria, including *Vibrio* sp., *Photobacterium* sp., *Aeromonas* sp., *Plesiomonas* sp., and *Alteromonas* sp. that produce TTX and/or anhydro-TTX.

Interestingly, they report that *E. coli*, a normal inhabitant of the mammalian gastrointestinal tract, produces anhydro-TTX.   Unidentified sodium channel blockers have also been detected in cultures of *Vibrio cholerae*, an estuarine bacterium and human enteropathogen (*54*).

Yasumoto et al. (*30*) describe two components of a *Pseudomonas* sp. culture with identical HPLC retention times to TTX and anhydro-TTX.   These fractions produced typical signs of TTX intoxication in mice, with median death times similar to standard TTX and anhydro-TTX.   Noguchi et al. (*32*) demonstrate by HPLC and GC-MS analyses that 7 biotypes of *Vibrio* sp. produced substances with retention times and molecular weights similar to TTX and anhydro-TTX.   However, they observed mouse toxicity in only 1 biotype.   Likewise, Simidu et al. (*34*) report that extracts of *V. alginolyticus* ATCC 17749 cultures displayed TTX-like toxicity in mice.   The latter study shows that a variety of marine bacteria, plus *E. coli*, produced substances that, by HPLC analysis, were identical to TTX and anhydro-TTX.

It is emphasized that some investigations show that bacterial cultures contain TTX-like substances which are not detected by mouse bioassay and are "difficult to detect" by HPLC and GC-MS analyses.   Structural analyses of these substances with other techniques was not reported.

Tamplin et. al. (*54*) observed that *V. cholerae* and *A. hydrophila* cell extracts contained substances with TTX-like biological activity in tissue culture assay, counteracting the lethal effect of veratridine on ouabain-treated mouse neuroblastoma cells.   Concentrations of TTX-like activity ranged from 5 to 100 ng/L of culture when compared to standard TTX.   The same bacterial extracts also displaced radiolabelled STX from rat brain membrane sodium channel receptors and inhibited the compound action potential of frog sciatic nerve.   However, the same extracts did not show TTX-like blocking events of sodium current when applied to rat sarcolemmal sodium channels in planar lipid bilayers.

Chromatographic Characterization of TTXs.   The vast majority of reports have identified TTX and anhydro-TTX in bacterial cultures using HPLC, TLC, and GC-MS.   Yasumoto et al. (*30*) showed that TTX-like substances extracted from a *Pseudomonas* sp. culture could bind to activated charcoal at pH 5.5 and be eluted with 20% ethanol in 1% acetic acid.   In addition, HPLC analysis demonstrated TTX and anhydro-TTX-like fluorophors following strong base treatment.   These compounds migrated on silica gel comparably to TTX and anhydro-TTX.   Furthermore, when analyzed by electron ionization (EI)-MS and fast atom bombardment (FAB)-MS, they yielded chromatograms identical to C_9 bases produced by alkali treatment of TTX and anhydro-TTX. Similar findings have been reported by others (*30–35*).

Discussion

Experimental evidence indicates that many marine bacteria produce TTXs. However, TTX production by some bacteria has not been validated since TTX and anhydro-like TTX are described as "difficult to detect" by using HPLC and GC-MS methods, and show no activity in the mouse bioassay.

If common marine bacteria, such as *Vibrio* sp. and *Pseudomonas* sp., indeed produce TTXs, it might be expected that more animals, particularly those living in aquatic environments, would be toxic. However, apparently only specific animals can concentrate TTX and/or provide a niche for TTX-producing bacteria.

The mechanisms involved in transfer of bacterial TTXs to animal tissues (e.g., liver, skin, intestines, and gonads of fish) are unknown. Tetrodotoxin could originate from bacteria on skin, intestines, or other internal fish tissues. Indeed, evidence indicates that *Vibrio* sp. may be normal flora of fish tissues (*55*). Diet may also be a potential source of toxin, since algae and other animals in the food chain are known to contain TTX and STX (*23, 29, 30*). This has particular relevance since Kogure et al. (*56*) have shown that relatively high concentrations of TTX can be found in marine sediments, potentially affecting benthic-feeding animals. Hypothetically, animals that concentrate TTX may have evolved unique proteins with the binding properties of the TTX binding site, sequestering TTX, and reducing its toxicity for host tissues. Indeed, it has been reported that the minimum lethal dose of TTX is greater for toxic versus non-toxic species of pufferfish (*57*). Furthermore, experiments measuring the fate of intraperitoneally injected, tritiated TTX demonstrate that TTX accumulates in pufferfish tissues (i.e., skin, liver, intestines, muscle) similar to wild pufferfish stocks (*58*).

It has been suggested that TTX and chemically related compounds are part of an anti-predatory mechanism of animals and offspring (*59–61*). Toxic starfish (*Astropecten polyacanthus*) are known to be eaten by the trumpet shell (*Charonia sauliae*) which accumulates large quantities of TTX (*59, 60*). Furthermore, the role of TTX as a defense agent of pufferfish was supported by studies showing that mild handling causes pufferfish skin to release 5–80 mouse units of TTX (*61*). Such evidence would indicate that TTX-producing bacteria have evolved important relationships with marine animals.

The conditions for production of TTX and STX by bacteria are unknown. The low levels of TTX and STX observed in laboratory cultures may indicate that the host environment has not been duplicated. Likely, the composition of culture medium and other physicochemical parameters for TTX and STX production have not yet been defined in vitro. Conversely, bacteria may actually produce only small amounts of TTX and STX in vivo that accumulate in host tissues over long time intervals.

The experimental evidence described above has lead to new hypotheses on the natural source of TTXs and STXs, offering strong support for a bacterial origin. Independent studies indicate that marine and estuarine bacteria residing on or in tissues of marine organisms are potential sources of TTXs and STXs (*30–35*). Future research will likely determine the distribution of bacteria producing sodium channel blockers, as well as factors that influence the production and distribution of these toxins in tissues of marine organisms.

Acknowledgment

The author wishes to thank Dr. K. Kogure and Dr. S. Hall for their helpful discussions in preparing this manuscript.

Literature Cited

1. Kao, C.Y. *Pharmacol. Rev.* **1966**, *18*, 997–1049.
2. Kao, C.Y. In *Tetrodotoxin, Saxitoxin, and the Molecular Biology of the Sodium Channel*; Kao, C.Y.; Levinson, S.R., Eds.; The New York Academy of Sciences: New York, **1986**; p 52–67.
3. Mosher, H.S. In *Tetrodotoxin, Saxitoxin, and the Molecular Biology of the Sodium Channel*; Kao, C.Y.; Levinson, S.R., Eds.; The New York Academy of Sciences: New York, **1986**; p 32–43.
4. Shimizu, Y. In *Tetrodotoxin, Saxitoxin, and the Molecular Biology of the Sodium Channel*; Kao, C.Y.; Levinson, S.R., Eds.; The New York Academy of Sciences: New York, **1986**; p 24–31.
5. Strichartz, G.; Rando, T.; Hall, S.; Gitschier, J.; Hall, L.; Magnani, B.; Bay, C.H. In *Tetrodotoxin, Saxitoxin, and the Molecular Biology of the Sodium Channel*; Kao, C.Y.; Levinson, S.R., Eds.; The New York Academy of Sciences: New York, **1986**; p96–112.
6. Luthy, J. In *Toxic Dinoflagellate Blooms*; Taylor, D.L.; Seliger, H.H. Eds.; Elsevier/North-Holland: New York, Amsterdam and Oxford, **1979**; p 23–28.
7. Valenti, M.; Pasquini, P.; Andreucci, G.; *Vet. Hum. Toxicol.* **1979**, *21*, 107–110.
8. Kaempfer, E. In *The History of Japan, Together with a Description of the Kingdom of Siam*; MacLehouse: Glasgow, **1906**.
9. Davis, W. *Science* **1988**, *240*, 1715–1716.
10. Fuhrman, F.A. In *Tetrodotoxin, Saxitoxin, and the Molecular Biology of the Sodium Channel*; Kao, C.Y.; Levinson, S.R., Eds.; The New York Academy of Sciences: New York, **1986**; p 1–14.
11. Thuesen, E.V.; Kogure, K.; Hashimoto, K.; Nemoto, T. *J. Exp. Mar. Ecol. Biol.* **1989**, in press.
12. Maruyama, J.; Noguchi, T. *Mer (Tokyo) (Bull. Soc. Fr.-Jpn. Oceanogr.)* **1984**, *22*, 299–304.
13. Miyazawa, K.; Jeon, J.K.; Noguchi, T.; Ito, K.; Hashimoto, K. *Toxicon* **1987**, *25*, 975–980.
14. Noguchi, T.; Hashimoto, Y. *Toxicon* **1973**, *11*, 305–307.
15. Noguchi, T.; Uzu, A.; Koyama, K.; Maruyama, J.; Nagashima,Y.; Hashimoto, K. *Bull. Jpn. Soc. Sci. Fish.* **1983**, *49*, 1887–1892.
16. Yasumoto, T.; Oshima, Y.; Kotaki, Y. *Toxicon* **1983**, (Suppl. *3*), 513–516.
17. Sheumack, D.D.; Howden, M.E.H.; Spence, I.; Quinn, R.J. *Science*, **1978**, *199*, 188–189.
18. Hwang, D.; Noguchi, T.; Arakawa, O.; Abe, T.; Hashimoto, K. *Nippon Suisan Gakkaishi* **1988**, *54*, 2001–2008.
19. Buchwald, H.D.; Durham, L.; Fischer, H.G.; Harada, R.; Mosher, H.S.; Kao, C.Y.; Fuhrman, F.A. *Science* **1964**, *143*, 474–475.
20. Kim, Y.H.; Brown, G.B.; Mosher, H.S.; Fuhrman, F.A. *Science* **1975**, *19*, 151–152.
21. Bradley, S.G.; Klika, L.J. *J. Am. Med. Assoc.* **1981**, *246*, 247.
22. Alam, M.; Shimizu, Y.; Ikawa, M.; Sasner, J.J. *J. Environ. Sci. Health* **1978**, *A13*, 493.
23. Hall, S. *Toxins and Toxicity of Protogonyaulax from the Northeast Pacific.* Ph.D. thesis. University of Alaska **1982**, 1–6.

24. Kotaki, Y.; Tajiri, M.; Oshima, Y.; Yasumoto, T. *Bull. Jpn. Soc. Sci. Fish.* **1983,** *49,* 283–286.
25. Nakamura, M.; Oshima, Y.; Yasumoto, T. *Toxicon* **1984,** *22,* 381–385.
26. Shimizu, Y., M. Kobayashi, A. Genenah, and N. Ichihara. In *Seafood Toxins*; Ragelis, E., Ed.; Amer. Chem. Soc.: Washington, D.C., **1984**; p 151–160.
27. Needler, A.B. *J. Fish. Res. Bd. Can.* **1949,** *7,* 490–504.
28. Prakash, A. *J. Fish. Res. Bd. Can.* **1963,** *20,* 983–996.
29. Maranda, L.; Anderson, D.M.; Shimizu, Y. *Estuarine Coastal and Shelf Science* **1985,** *21,* 401–410.
30. Yasumoto, T.; Yasumura, D.; Yotsu, M.; Michishita, T.; Endo, A.; Kotaki, Y. *Agric. Biol. Chem.* **1986,** *50,* 793–795.
31. Noguchi, T.; Jeon, J.; Arakawa, O.; Sugita, H.; Deguchi, Y.; Shida, Y.; Hashimoto, K. *J. Biochem.* **1986,** *99,* 311–314.
32. Noguchi, T.; Hwang, D.; Arakawa, O.; Sugita, H.; Deguchi, Y.; Shida, Y.; Hashimoto, K. *Marine Biology* **1987,** *94,* 625–630.
33. Yotsu, M.; Yamazaki, T.; Meguro, Y.; Endo, A.; Murata, M.; Naoki, H.; Yasumoto, T. *Toxicon* **1987,** *25,* 225–228.
34. Simidu, U.; Noguchi, T.; Hwang, D.; Shida, Y.; Hashimoto, K. *Appl. Environ. Microbiol.***1987,** *53,* 1714–1715.
35. Sugita, H.; Ueda, R.; Noguchi, T.; Arakawa, O.; Hashimoto, K.; Deguchi, Y. *Nippon Suisan Gakkaishi* **1987,** *53,* 1693.
36. Sousa y Silva, E. *Notas E Estudos Insti. Biol. Marit.* **1962,** *26,* 1–24.
37. Kodama, M. *Agric. Biol. Chem.* **1988,** *52,* 1075–1077.
38. Hall, S.; Shimizu, Y. In *Toxic Dinoflagellates*; Anderson, D.M.; White, A.W.; Baden, D.G., Eds.; Elsevier/North-Holland: New York, Amsterdam and Oxford, **1985**; p 545–548.
39. Yasumoto, T.; Michishita, T. *Agric. Biol. Chem.* **1985,** *49,* 3077–3080.
40. *Offical Methods of Analysis of the Association of Official Analytical Chemists. 14th Ed*; Williams, S., Ed.; AOAC: Arlington, Virginia, **1984**; Chapter 18, p 344.
41. Carlson, R.E.; Lever, M.L.; Lee, B.W.; Guire, P.E. In *Seafood Toxins*; Ragelis, E., Ed.; Amer. Chem. Soc.: Washington, D.C., **1984**; p 181–192.
42. Chu, F.S.; Fan, T.S.L. *J. Assoc. Off. Anal. Chem.* **1985,** *68,* 13–16.
43. Davio, S.R.; Hewetson, J.F.; Beheler, J.E. In *Toxic Dinoflagellates*; Anderson, D.M.; White, A.W.; Baden, D.G., Eds.; Elsevier/North-Holland: New York, Amsterdam and Oxford, **1985**; p 343–348.
44. Yang, G.C.; Imagire, S.J.; Yasaei, P.; Ragelis, E.P.; Park, D.L.; Page, S.W.; Carlson, R.E.; Guire, P.E. *Bull. Environ. Contam. Toxicol.* **1987,** *39,* 264–271.
45. Davio, S.R.; Fontelo, P.A. *Anal. Biochem.* **1984,** *141,* 199–204.
46. Richie, J.M.; Rogart, R.B.; Strichartz, G.R. *J. Physiol.* (London) **1976,** *261,* 477–494.
47. Krueger, B.K.; Worley, J.F.; French, R.R.; *Nature* **1983,** *303,* 172–175.
48. Kogure, K.; Tamplin, M.L.; Simidu, U.; Colwell, R.R. *Toxicon* **1988,** *26,* 191–197.
49. Nakamura, M.; Yasumoto, T. *Toxicon,* **1985** *23,* 271–276.
50. Yasumoto, T.; Nakamura, M.; Oshima, Y.; Takahata, J. *Bull. Jpn. Soc. Sci. Fish.* **1982,** *48,* 1481–1483.
51. Nakayama, T.; Terakawa, S. *Anal. Biochem.* **1982,** *126,* 153–155.
52. Kobayashi, Y.; Kubo, H.; Kinoshita; T. *Anal. Biochem.* **1987,** *160,* 392–398.
53. Yasumoto, T.; Nagai, H.; Yasumura, D.; Michishita, T.; Endo,A.; Yotsu, M.; Kotaki, Y. In *Tetrodotoxin, Saxitoxin, and the Molecular Biology of the*

Sodium Channel; Kao, C.Y.; Levinson,S.R., Eds.; The New York Academy of Sciences: New York, **1986**; p 44–51.
54. Tamplin, M.L.; Colwell, R.R.; Hall, S.; Kogure, K.; Strichartz, G. *Lancet* **1987**, *1*, 975.
55. Grimes, D.J.; Brayton, P.; Colwell, R.R.; Gruber, S.H. *Syst. Appl. Microbiol.* **1986**, *6*, 221–226.
56. Kogure, K.; Do, H.K.; Thuesen, E.V.; Nanba, K.; Ohwada, K.; Simidu, U. *Mar. Ecol. Prog. Ser.*, in press.
57. Saito, T.; Noguchi, T.; Harada, T.; Murata, O.; Abe, T.; Hashimoto, K. *Bull. Jpn. Soc. Sci. Fish.* **1985**, *51*, 1371.
58. Watabe, S.; Sato, Y.; Nakaya, M.; Nogawa, N.; Oohashi, K.; Noguchi, T.; Morikawa, N.; Hashimoto, K. *Toxicon* **1987**, *25*, 283–1289.
59. Noguchi, T.; Narita, H.; Maruyama, J.; Hashimoto, K. *Bull. Jpn. Soc. Sci. Fish.* **1982**, *48*, 1173–1177.
60. Noguchi, T.; Jeon, J.K.; Maruyama, J.; Sato, Y.; Saisho, T.; Hashimoto, K. *Bull. Jpn. Soc. Sci. Fish.* **1985**, *51*, 1727–1731.
61. Saito, T.; Noguchi, T.; Harada, T.; Murata, O.; Hashimoto, K. *Bull. Jpn. Soc. Sci. Fish.* **1985**, *51*, 1175–1180.

RECEIVED July 10, 1989

Chapter 6

Natural Toxins from Cyanobacteria (Blue-Green Algae)

Wayne W. Carmichael, Nik A. Mahmood[1], and Edward G. Hyde

Department of Biological Sciences, Wright State University, Dayton, OH 45435

Acute lethal toxicity from cyanobacteria is caused by ingestion of toxic cells or toxins from certain freshwater/brackish water species of *Anabaena*, *Aphanizomenon*, *Microcystis*, *Nodularia*, and *Oscillatoria*. Contact poisonings are reported from marine species of Lyngbya, *Oscillatoria*, and *Schizothrix*. More recently, cytotoxic compounds with antialgal, antifungal, antibacterial, antiprotozoan, and antineoplastic activity have been reported from species of *Scytonema*, *Oscillatoria*, *Hapalosiphon*, and *Plectonema*. ˙Acutely lethal toxins include a related family of hepatotoxic cyclic hepta and pentapeptides, termed microcystins or cyanoginosins, which contain both D and L amino acids plus two novel amino acids. Species and strains of *Anabaena* produce at least two neurotoxins, termed anatoxins; one a depolarizing neuromuscular blocking agent, the other an irreversible anticholinesterase. Strains of *Aphanizomenon flos-aquae* produce saxitoxin and neosaxitoxin, the primary toxins in cases of paralytic shellfish poisoning (PSP). These various toxins have intraperitoneal mouse LD_{50} values of 10–500 μg/kg. Current research indicates that a third group of toxins with contact irritant properties are produced by some freshwater cyanobacteria. Indirect evidence indicates that one or all of these toxins are responsible for certain cases of human gastroenteritis and dermatitis from municipal and recreational water supplies.

Reports of toxic algae in the freshwater environment are almost exclusively due to members of the division Cyanophyta, commonly called blue-green algae or cyanobacteria. Although cyanobacteria are found in almost any environment ranging from hot springs to Antarctic soils, known toxic members are mostly planktonic. Published accounts of field poisonings by cyanobacteria are known since the late 19th century (*98*). These reports describe sickness and death of livestock, pets, and wildlife following ingestion of water containing toxic algae cells or the toxin released by the aging cells. Recent reviews of these poisonings and the toxins of freshwater cyanobacteria are given by Carmichael (*1–3*), Codd and Bell (*4*), and Gorham and Carmichael (*5*).

---

[1] Current address: Springborn Laboratory, Inc., Mammalian Toxicology Division, 553 North Broadway, Spencerville, OH 45887

Toxins produced by cyanobacteria are in two general groups. The first, which is emphasized in this chapter, includes those toxins responsible for acute lethal poisonings. About 12 genera have been implicated in producing these toxins but only *Anabaena, Aphanizomenon, Microcystis, Nodularia,* and *Oscillatoria* have had toxins isolated, and investigated chemically and toxicologically. The second group includes a number of secondary chemicals that are not highly lethal to animals but instead show more selective bioactivity. These bioactive chemicals (Table I) include scytophycins produced by certain strains of *Scytonema pseudohofmanni.* *S. pseudohofmanni* strain BC-1-2, which was isolated from a forested area on the island of Oahu, Hawaii, was found to produce two lipophilic toxins termed scytophycin A (MW 821) and B (MW 819). Scytophycin B is moderately toxic to mice by the intraperitoneal route (LD$_{50}$ = 650 μg/kg). Scytophycins show a very strong cytotoxic activity, however, when tested against cell cultures, such as KB human epidermoid carcinoma and NIH/3T3 mouse fibroblast cell lines (1 and 0.65 ng/mL, respectively). These toxins are also moderately active against intraperitoneally implanted P388 lymphocytic leukemia and Lewis lung carcinoma (6). *S. hofmanni* UTEX 1581 has been shown to produce the chlorine-containing diaryl-lactone called cyanobacterin. This compound has been shown to have anticyanobacterial activity and has been proposed as a possible algicide against cyanobacteria (7,8).

A cytotoxic alkaloid has been isolated from the filamentous species *Hapalosiphon fontinalis* strain V-3-1. This isolate was made from soil samples collected in the Marshall Islands in 1981. This strain produces the lipophilic compound hapalindole A, which has a broad range of antialgal and antimycotic activity (9). *Oscillatoria acutissima* strain B-1, isolated from a freshwater pond in Oahu, was found to produce two novel macrolide compounds termed acutiphycin and 20, 21-didehydroacutiphycin. These macrolides show cytotoxicity (KB and NIH/3T3) and antitumor activity (Murine Lewis lung carcinoma) (10).

Other toxins that show low lethal toxicity to laboratory test animals include lipopolysaccharide endotoxin produced as part of the cell wall by all cyanobacteria (11) and certain toxins of some cyanobacteria suspected of causing contact irritation in recreational water supplies (4,12; Carmichael and Codd, unpublished results).

Economic losses due to water-based diseases of freshwater cyanobacteria toxins have only been reported from the first group and are the result of contact with or consumption of water containing toxin and/or toxic cells. These toxins are all water soluble and temperature stable. They are either released by the cyanobacterial cell or loosely bound so that changes in cell permeability or age allow their release into the environment. Known occurrences of toxic cyanobacteria in water supplies include Canada (four provinces), Europe (12 countries), United States (20 states), USSR (Ukraine), Australia, India, Bangladesh, South Africa, Israel, Japan, New Zealand, Argentina, Chile, Thailand, and the Peoples Republic of China (13–15). The economic impact from toxic freshwater cyanobacteria include the costs incurred from deaths of domestic animals; allergic and gastrointestinal problems after human contact with waterblooms (including lost income from recreational areas); and increased expense for the detection and removal of taste, odor, and toxins. The remainder of this chapter discusses the acute lethal toxins produced by freshwater cyanobacteria which includes hepatotoxins and neurotoxins (Table I).

Neurotoxins

Anatoxins. Neurotoxins produced by filamentous *Anabaena flos-aquae* are called anatoxins (ANTXS) (15). Currently two anatoxins, from different strains of *A. flos-aquae,* have been isolated and at least partially characterized. ANTX-A from strain (single filament isolate) NRC-44-1 is the first toxin from a freshwater

Table I.   Toxins of Freshwater Cyanobacteria

Species, Strain, and Source	Toxin Term	Structure	LD_{50} µg/kg ip, Mouse
Neurotoxins			
A. flos-aquae Strain NRC-44-1 (Canada, Saskatchewan)	Anatoxin-A	Secondary amine alkaloid, MW 165	200
Strain NRC-525-17 (Canada, Saskatchewan)	Anatoxin-A(S)	N-hydroxy guanidine methyl-phosphate ester, MW 252	20
Aph. flos-aquae Strain NH-1 & NH-5 (U.S., New Hampshire)	Aphantoxin (neosaxitoxin) Aphantoxin II (saxitoxin)	Purine alkaloid MW 315 (neoSTX) MW 299 (STX)	10
Hepatotoxins			
A. flos-aquae Strain S-23-g-1 (Canada, Saskatchewan)	Microcystins[a]	Heptapeptides MW 994	50
M. aeruginosa Strain WR-70 (=UV-010) (South Africa, Transvaal)	Cyanoginosins[a]	Heptapeptides MW 909-1044	50
(Waterbloom, Australia, New South Wales)	Cyanoginosin	Heptapeptide MW 1035	50
(Waterbloom, U.S., Wisconsin	Microcystin	Heptapeptide MW 994	50
Strain NRC-1(SS-17) (Canada, Ontario)	Microcystin	Heptapeptide MW 994	50
Strain 7820 (Scotland, Loch Balgaves)	Microcystin	Heptapeptide MW 994	50
(Waterbloom, Norway, Lake Akersvatn)	Microcystin	Heptapeptide MW 994	50
M. aeruginosa Strain M-228 (Japan, Tokyo)	Microcystin	Heptapeptide MW 994 MW 1044	50

[a] *See* text for explanation of terminology.

Continued on next page

Table I. *Continued*

Species, Strain, and Source	Toxin Term	Structure	LD_{50} $\mu g/kg$ ip, Mouse
M. aeruginosa		MW 1039	
M. viridis	Cyanoviridin[a]	Heptapeptide MW 1039	not reported
N. spumigena	Nodularin	Pentapeptide MW 824	30-50
O. agardhii var. isothrix (Waterbloom, Norway, Lake Froylandsvatn)	Microcystins	Heptapeptides MW 1009	300-500
O. agardhii var. (Waterbloom, Norway, Lake Kolbotnvatn)	Microcystins	Heptapeptides MW 1023	500-1000
Cytotoxins			
S. pseudohofmanni Strain BC-1-2 (U.S., Hawaii)	Scytophycin	Methylformamide A & B A = MW 821; B = MW 819	650 (scytophycin B)
S. hofmanni Strain UTEX-1581 (U.S., Texas)	Cyanobacterin	Chlorinated diaryllactone	not reported
H. fontinalis Strain V-3-1 (Marshall Islands)	Hapalindole A	Substituted indole alkaloid	not reported
T. byssoidea Strain H-6-2 (U.S., Hawaii)	Tubercidin	Pyrrolopyrimidine	not reported
O. acutissima Strain B-1 (U.S., Hawaii)	Acutiphycin	Macrolide	not reported

[a] *See* text for explanation of terminology.

cyanobacteria to be chemically defined. It is the secondary amine, 2-acetyl-9-azabicyclo[4.2.1]non-2-ene (*16,17*) molecular weight 165 daltons (Figure 1). It has been synthesized through a ring expansion of cocaine (*18,19*), from iminium salts (*20–22*), from nitrone (*23,24*), from 4-cycloheptenone or tetrabromotricyclooctane (*25*) by construction of the azabicyclo ring from 9-methyl-9-azabicyclo[3.3.1]nonan-1-ol (*26*), and by starting with 9-methyl-9-aza[4.2.1]nonan-2-one (*27*).

ANTX-A is a potent, postsynaptic, depolarizing, neuromuscular blocking agent that affects both nicotinic and muscarinic acetylcholine (ACH) receptors (*28–31*). Signs of poisoning in field reports for wild and domestic animals include staggering, muscle fasciculations, gasping, convulsions, and opisthotonos (birds). Death by respiratory arrest occurs within minutes to a few hours depending on species, dosage, and prior food consumption. The LD_{50} intraperitoneal (ip) mouse for purified toxin is about 200 μg/kg body weight, with survival time of 4–7 min. Thus, animals need to ingest only a few milliliters to a few liters of the toxic surface bloom to receive a lethal bolus (*32–34*). Detection of ANTX-A, while still primarily by mouse bioassay, is being supplemented by three analytical detection methods. These methods are based on high performance liquid chromatography (HPLC) (*35,36*), gas chromatography-mass spectrometry (GC-MS) (37, Himberg personal communication), and gas chromatography-electron capture detection (GC-ECD) (*38*).

All known occurrences of ANTX-A production have been from Canada or the United States. More recently, ANTX-A has been detected in *A. flos-aquae* blooms from Japan (Watanabe, personal communication), Norway (Skulberg, personal communication), and Finland (*99*).

A neurotoxin more recently isolated and under current study is referred to as anatoxin-a(s) [ANTX-A(S)]. The primary source of this toxin is *A. flos-aquae* strain NRC-525-17 isolated in 1965 from Buffalo Pound Lake in Saskatchewan, Canada. ANTX-A(S) is physiologically and chemically different from ANTX-A. It produces opisthotonos in chicks, as does ANTX-A, but also causes viscous salivation [which gives the terminology its (S) label] and lachrymation in mice, chromodacryorrhea (bloody tears) in rats, urinary incontinence, muscular weakness, fasciculation, convulsion and defecation prior to death by respiratory arrest. Also observed is a dose-dependent fasciculation of limbs for 1–2 min after death. ANTX-A(S) was purified, from *A. flos-aquae* cells, by column chromatography and HPLC (*39,40*), and its structure is given in Figure 1 (*100*). ANTX-A(S) is acid stable, unstable in basic conditions, has very low UV absorbance, gives a positive alkaloid test, and has a molecular weight of 253 $(m+H^+)$ daltons.

The LD_{50} ip mouse for ANTX-A(S) is about 20 μg/kg, ten times more lethally toxic than ANTX-A. At the LD_{50} the survival time for mice is 10–30 min. Mahmood and Carmichael (*41*) concluded from the signs of poisoning, i.e., salivation and muscle twitch potentiation, that the cholinergic system was the primary target of ANTX-A(S). No serum cholinesterase activity was observed in rats dosed with 350 and 600 μg/kg of ANTX-A(S), suggesting an anticholinesterase mechanism. ANTX-A(S) also sensitized the frog rectus abdominus muscle and chick biventer cervicis muscle to exogenous acetylcholine, but had no effect on the action of ANTX-A (a depolaring agent) on frog rectus abdominus. Further work with ANTX-A(S) (*42*) has confirmed and extended the conclusion that this toxin is an anticholinesterase agent. ANTX-A(S) was shown to inhibit in vitro electric eel acetylcholinesterase (AChE, EC 3.1.1.7) and horse serum butyrylcholinesterase (BUChE, EC 3.1.1.8) in a time and concentration dependent fashion by a mechanism similar to the organophosphate anticholinesterases as illustrated in Scheme I.

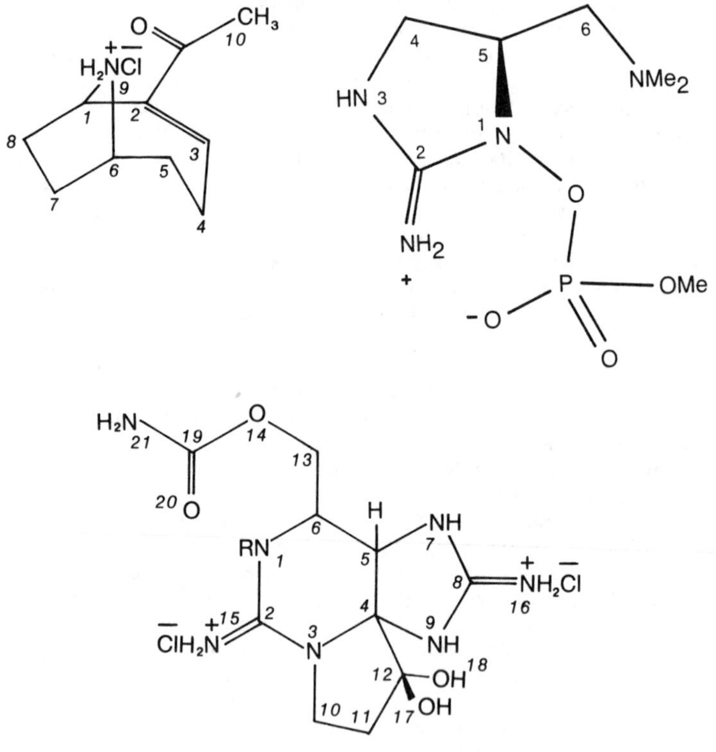

R = H; saxitoxin dihydrochloride
R = OH; neosaxitoxin dihydrochloride

Figure 1. Left: Anatoxin-a (ANTX-A) hydrochloride. Produced by the freshwater filamentous cyanobacterium *Anabaena flos-aquae* NRC-44-1. Right: Anatoxin-a(s). Produced by the freshwater filamentous cyanobacterium *Anabaena flos-aquae* NRC-525-17. Bottom: Aphantoxin-I (neosaxitoxin) and Aphantoxin-II (saxitoxin) produced by certain strains of the filamentous cyanobacterium *Aphanizomenon flos-aquae*.

$$EOH^+IX \underset{k_{-1}}{\overset{k_1}{\rightleftharpoons}} EOH(IX) \overset{k_2}{\underset{HX}{\rightarrow}} EOI \overset{k_3}{\underset{H_2O}{\rightarrow}} EOH^+H^+IO^- \qquad (I)$$

In this scheme, EOH is the enzyme, IX is the inhibitor (either a carbamate or an organophosphate). EOH(IX) is analogous to the Michaelis Menton complex seen with the substrate reaction. EOI is the acyl-enzyme intermediate for carbamates or a phosphoro-enzyme intermediate for the organophosphates. The equilibrium constant for this reaction (K_d) is defined as k_1/k_{-1} and the phosphorylation or carbamylation constant is defined as k_2. In this study (42), ANTX-A(S) was found to be more specific for AChE than BUChE. The double reciprocal and Dixon plot of the inhibition of electric eel AChE indicated that the toxin is a non-competitive inhibitor (V_{max} decreases, k_m remains unchanged) (Figure 2). Since non-competitive inhibitors can be reversible or irreversible, a plot of V_{max} vs. the total amount of enzyme was used to distinguish which of the two ANTX-A(S) resembled. In this type of plot the lines for control and non-competitive inhibitor will pass through the origin, while that of the irreversible inhibitor parallels the control. Figure 3 shows that ANTX-A(S) is an irreversible inhibitor of electric eel AChE. Further comparison of the kinetic result of ANTX-A(S) inhibition of AChE with diisopropyl-fluorophosphate (DFP; an irreversible cholinesterase inhibitor) indicated that inhibition follows the generalized scheme for irreversible inhibitors and that the phosphorylation constant (2 per min) was five times less than DFP (10 per min). The affinity constant of DFP (55.2 μg/mL) is also much weaker (~ 110 fold) than ANTX-A(S) (0.5 μg/mL). The bimolecular inhibition constant ($K_i = k_2/K_d$) indicates that ANTX-A(S) is 22 times more powerful than DFP.

More recent unpublished work is continuing on the molecular mechanism of AChE/ANTX-A(S) binding. This work focuses on two areas: the first is comparison of the kinetic inhibition for different sources of AChE by ANTX-A(S) and then comparing the kinetic constants against that found with DFP.

The second set of investigations is looking at the nature of the interaction of ANTX-A(S) and AChE. Binding of substrate and the irreversible inhibitors of AChE involves attachment at two sites on the AChE molecule (43,44). These two sites are 5 Å apart and comprise the active site of the enzyme. The first is known as the anionic site which contains one or more negatively charged residues surrounded by regions of hydrophobicity (45). This area binds the cationic portion of the molecule, and for uncharged molecules the hydrophobic regions act as an anchoring point (46). The second site is the esteratic site where the ester linkage of the substrate or inhibitor is cleaved by a serine residue made highly reactive by a "charge-relay" system made up of a histidine and an aspartate or tyrosine residue.

There are two ways of producing inhibition by anticholinesterase agents. The first, termed irreversible inhibition, occurs when a carbamate or organophosphate reacts with AChE. The serine residue can become carbamylated with the subsequent hydrolysis of the carbamyl group. This process requires only minutes for reactivation of the enzyme. In the second process, the serine becomes phosphorylated and it requires hours to days for hydrolysis back to the active enzyme. The second type of inhibition, termed reversible, includes true reversible anticholinesterases (i.e., tetramethylammonium, TMA) which block only the anionic site and are displaced by increasing substrate concentration. AChE can also be inhibited by high substrate concentrations where the cationic head of one molecule and the ester linkage of another molecule bind concurrently at the active site and substrate hydrolysis cannot occur resulting in inhibition (44,47). In our studies to date, and reported here, we have used ACh, 3,3'-dimethylbutylacetate (an uncharged

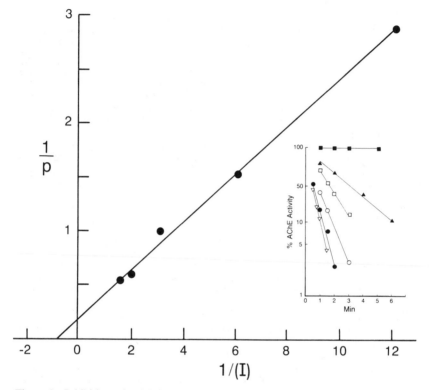

Figure 2. Inhibition of eel AChE by ANTX-A(S) - the 'secondary plot'. P, the first-order rate constant which was the rate of inhibition at that ANTX-A(S) concentration obtained from the 'primary plot' (insert). The intercept on the 1/P axis is $1/k_2$ and the intercept on the 1/[I] axis is $-1/K_d$. Figure insert: Progressive irreversible inhibition of eel AChE by ANTX-A(S). The inactivation followed first-order kinetics. ANTX-A(S) concentrations, $\mu g/mL$: (▲) 0.083; (□) 0.166; (o) 0.331; (•) 0.497; (∇) 0.599; (■) control. Each point represents the mean of 3 or 4 determinations.

carbon isostere of acetylcholine), and TMA to investigate the inhibition mechanism of ANTX-A(S).

In the first group of studies, involving kinetic inhibition studies, comparisons of the equilibrium (K_d), phosphorylation (K_2), and inhibition constant (K_i) for the inhibition of electric eel and human erythrocyte AChE by ANTX-A(S) and DFP were done (Table II). From Table II it is seen that ANTX- A(S) has a higher affinity for human erythrocyte AChE $(K_d=0.253 \ \mu M)$ than electric eel AChE $(K_d=3.67 \ \mu M)$. ANTX-A(S) also shows greater affinity for AChE than DFP $(K_d=300 \ \mu M)$. And finally the bimolecular rate constant, K_i, which indicates the overall rate of reaction, shows AChE is more sensitive toward inhibition by ANTX-A(S) $(K_i=1.36 \ \mu M^{-1}min^{-1})$ than DFP $(K_i = 0.033 \ \mu M^{-1}min^{-1})$. These studies add information to the comparative activity of ANTX-A(S) and other irreversible AChE inhibitors but do not show the site of inhibition.

The second group of studies was designed to investigate the site of inhibition. The results of these protection studies are shown in Figures 4 – 7. Figure 4 is the dose-activity curve to various concentrations of ANTX-A(S) after a 2-min preincubation. Figures 5, 6, and 7 show the same dose-activity curves in the presence of 0.33 mM ACh, 2.6 mM 3,3'-dimethyl butylacetate and 0.11 mM TMA. Calculation of a protection index (PI) (slope of % activity/min in presence of protectant divided by the slope of % activity/min in absence of protectant; PI<1, protection; $\geq$1 no protection) shows that all compounds protect AChE from ANTX-A(S) inhibition. Referenced to the 0.1 μg ANTX-A(S) curve, the protective indices are: ACh, 0.5501; DMBA, 0.6460; and TMA, 0.8598. From this one can see that TMA protects only slightly, indicating that ANTX-A(S) site of attachment is not primarily the anionic site. ACh and DMBA both protect the enzyme although ACh is slightly better than DMBA. Protection by DMBA indicates that ANTX-A(S) does not inhibit by the substrate inhibition mechanism. In addition to these laboratory studies on ANTX-A(S), field poisonings by ANTX-A(S) have now been confirmed in at least one case (*48*). The poisonings involved the death of five dogs, eight pups, and two calves that ingested quantities of *A. flos-aquae*, containing ANTX-A(S), in Richmond Lake, South Dakota, in late summer 1985.

In summary, ANTX-A(S) uses the two-site attachment mechanism analogous to substrate and does not inhibit AChE in the manner of the reversible anticholinesterases.

Aphantoxins. Occurrence of neurotoxins (aphantoxins) in the freshwater filamentous cyanobacterium *Aphanizomenon flos-aquae* was first demonstrated by Sawyer et al. (*49*). All aphantoxins (APHTXS) studied to date have come from waterblooms and laboratory strains of nonfasciculate (non-flake forming) *Aph. flos-aquae* that occurred in lakes and ponds of New Hampshire from 1966 through 1980. Toxic cells and extracts of *Aph. flos-aquae* were shown to be toxic to mice, fish, and waterfleas (*Daphnia catawba*) by Jakim and Gentile (*50*). Chromatographic and pharmacological evidence established that APHTXS consist mainly of two neurotoxic alkaloids that strongly resemble saxitoxin (STX) and neosaxitoxin (neoSTX), the two primary toxins of red tide paralytic shellfish poisoning (PSP) (*51*). The bloom material and toxic strain used in studies before 1980 came from collections made between 1960 and 1970. The more recent work on APHTXS has used two strains (NH-1 and NH-5) isolated by Carmichael in 1980 from a small pond near Durham, New Hampshire (*52,53*). These APHTXS, like neoSTX and STX, are fast-acting neurotoxins that inhibit nerve conduction by blocking sodium channels without affecting permeability to potassium, the transmembrane resting potential, or membrane resistance (*54*). Mahmood and Carmichael (*55*), using the NH-5 strain, showed that batch-cultured cells have a mouse ip LD_{50} of about 5 mg/kg. Each

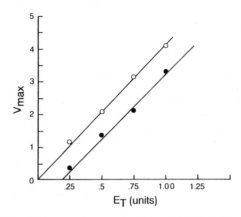

Figure 3. Plot of V_{max} against total enzyme [ET] showing the irreversible inhibition of electric eel acetylcholinesterase (AChE) by ANTX-A(S). The enzymes were incubated with 0.32 μg/mL ANTX-A(S) for 1.0 min and acetylthiocholine (final concentrations 2.5, 4.7, 6.3, and 7.8 x 10^{-4} M) was added. V_{max} was determined from the double reciprocal plots (not shown). Key: (o) control; (•) ANTX-A(S). (Reproduced with permission from Ref. 42. Copyright 1987 Pergamon Press)

Table II. Affinity Equilibrium (K_d), Phopshorylation Rate (k_2) and Biomolecular Rate (k_i) Constants for the Inhibition of AChE by ANTX-A(S) and DFP

	AChE					
	Electric Eel			Human Erythrocytes		
Toxin	K_d*+ (μM)	k_2* (min^{-1})	k_i (μM^{-1}min^{-1})	K_d*+ (μM)	k_2* (min^{-1})	k_i (μM^{-1}min^{-1})
ANTX-A(S)	3.67	2.0	1.36	0.25	21.40	84
DFP	300	10	0.03	-	-	-

* K_d and k_2 were determined from the secondary plots (Figure 2). K_d is the reciprocal of the abscissa intercept and k_2 the reciprocal of the ordinate intercept.

+ An estimated MW of 267 for ANTX-A(S) was used to calculate K_d.

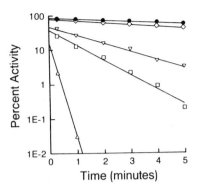

Figure 4. Concentration-dependent inhibition of electric eel AChE by ANTX-A(S). AChE and ANTX-A(S) were incubated for 2 min. before inhibition rate was determined. Key: (Δ) 0.079 μg/mL; (□) 0.032 μg/mL; (∇) 0.016 μg/mL; (◇) 0.0032 μg/mL; (•) .0016 μg/mL. NOTE: Total inhibition occurs when 0.158 μg/mL ANTX-A(S) is preincubated with AChE for two minutes.

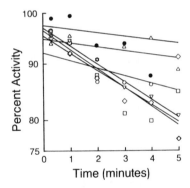

Figure 5. The effects of 0.11 mM acetylcholine on the concentration-dependent inhibition of AChE by ANTX-A(S). Acetylcholine was coincubated with ANTX-A(S) and AChE two minutes before the inhibition rate was determined. Key: (o) 0.158 μg/mL; (Δ) 0.079 μg/mL; (□) 0.032 μg/mL; (∇) 0.016 μg/mL; (◇) 0.0032 μg/mL; (•) .0016 μg/mL. NOTE: Total inhibition occurs when 0.158 μg/mL ANTX-A(S) is preincubated with AChE for two minutes.

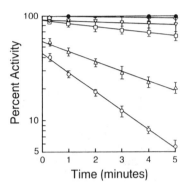

Figure 6.   The effects of 2.6 mM 3,3'-dimethylbutyl acetate on the concentration-dependent inhibition of AChE by ANTX-A(S).   Conditions and symbols are the same as Figure 5.

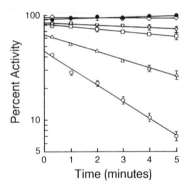

Figure 7.   The effect of 0.11 mM tetramethyl ammonium iodide on the concentration dependent inhibition of AChE by ANTX-A(S). Conditions and symbols are the same as Figure 5.

gram of lyophilized cells yields about 1.3 mg aphantoxin I (neosaxitoxin) and 0.1 mg aphantoxin II (Saxitoxin) (Figure 1). Also detected are three labile neurotoxins that are not similar to any of the known paralytic shellfish poisons.

Shimizu et al. (56) studied the biosynthesis of the STX analog neoSTX using *Aph. flos-aquae* NH-1. They were able to confirm its presence in strain NH-1 and to explain the biosynthetic pathway for this important group of secondary chemicals.

Hepatotoxins. Low-molecular-weight peptide toxins that affect the liver have been the predominant toxins involved in cases of animal poisonings due to cyanobacterial toxins (2,5,57). After almost 25 years of structure analysis on toxic peptides of the colonial bloom-forming cyanobacterium *Microcystis aeruginosa*, Botes et al. (58–60) and Santikarn et al. (61) provided structure details on one of four toxins (designated toxin BE-4) produced by the South African *M. aeruginosa* strain WR70 (= UV-010). They concluded that it was monocyclic and contained three D-amino acids (alanine, erythro-β-methylaspartic acid, and glutamic acid), two L-amino acids (leucine and alanine), plus two unusual amino acids. These were *N*-methyldehydroalanine (Medha) and a nonpolar side chain of 20 carbon atoms that turned out to be a novel β-amino acid; 3-amino-*O*-methoxy-2,6,8-trimethyl-10-phenyldeca-4,6-dienoic acid (ADDA). Based on fast atom bombardment mass spectrometry (FABMS) and nuclear magnetic resonance (NMR) studies, BE-4 toxin is now known to be a cyclic heptapeptide having a molecular weight of 909 daltons (Figure 8).

Instead of calling the BE-4 toxin microcystin, as previous *Microcystis* toxins were called (62-64) and using alphabetical or numerical suffixes to indicate chromatographic elution order or structural differences, Botes (60) proposed the generically derived designation cyanoginosin (CYGSN). This prefix, which indicates that cyanobacterial origin, is followed by a two-letter suffix that indicates the identity and sequence of the two L-amino acids relative to the *N*-Me-dehydroalanyl-D-alanine bond. Thus toxin BE-4 was renamed cyanoginosin-LA since leucine and alanine are the L-amino acids. These L-amino acids were shown by Eloff et al. (65) and Botes et al. (66) to vary between strains and to account for structural differences for different toxic fractions within a single strain. Botes et al. (66) showed that the other three toxins of strain WR-70 all had the same three D-amino acids and two novel amino acids (Medha and ADDA). The L-amino acids were leucine- arginine (CYGSN-LR), tyrosine-arginine (CYGSN-YR), and tyrosine-alanine (CYGSN-YA). They were also able to show that the hepatotoxin isolated by Elleman et al. (67) from waterbloom material collected in Malpas Dam, New South Wales, Australia, contained the five characteristic amino acids plus the L-amino acid variants tyrosine-methionine (therefore it is termed CYGSN-YM).

Microcystin (MCYST) is the term given to the fast death factor (FDF) produced by *M. aeruginosa* strain NRC-1 and its daughter strain NRC-1 (SS-17) (62,68). An absolute structure for the toxin of strain NRC-1 (SS-17) is not yet available but is known to be a peptide (MW 994) containing the variant amino acids leucine and arginine (Carmichael, unpublished). Krishnamurthy et al. (69,70) have shown that the toxin isolated from a waterbloom of *M. aeruginosa* collected in Lake Akersvatn, Norway (71), has a structure similar to that of MYCST from NRC-1 (SS-17) and CYGSN-LR. This toxin has also been found to be the main toxin produced by the Scottish strain of *M. aeruginosa* PCC-7820 and a Canadian *A. flos-aquae* strain S-23-g-1 (69,70) (Figure 8). The identification of a peptide toxin from *A. flos-aquae* S-23-g-1 provides the first evidence that these hepatotoxins are produced by filamentous as well as coccoid cyanobacteria. *A. flos-aquae* S-23-g-1 and toxic *M. aeruginosa* from a waterbloom in Wisconsin also produced a second cyclic

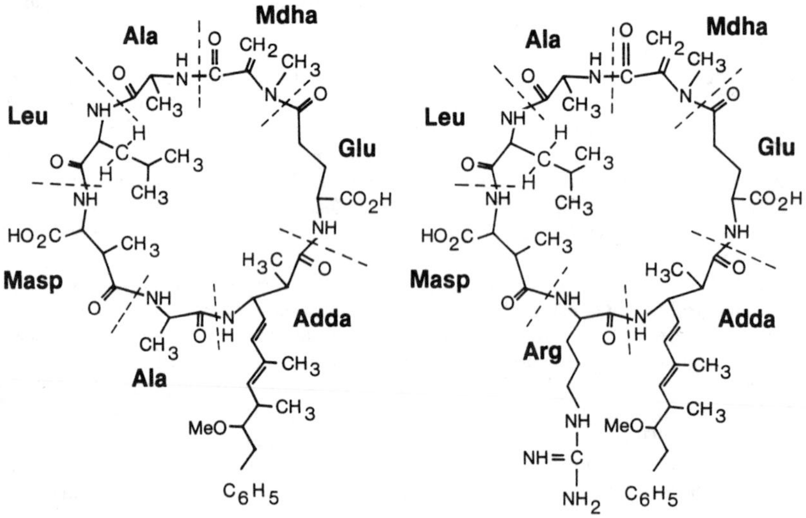

Figure 8.   Left: The cyclic heptapeptide hepatotoxin microcystin-LA (cyanoginosin-LA) produced by the colonial cyanobacterium *Microcystis aeruginosa* strain WR-70 (UV-010).   MW = 909.   Right: The cyclic heptapeptide hepatotoxin microcystin-LR (cyanoginosin-LR) produced by a waterbloom of the colonial cyanobacterium *Microcystis aeruginosa* collected in Lake Akersvatn, Norway, 1984-85; MW=994, (*69,71*).

heptapeptide hepatotoxin, which has been found to have six of the same amino acids, that is, leucine-arginine, but has aspartic acid instead of β-methylaspartic acid (*69*).

The filamentous genus *Oscillatoria* has also been shown to produce a hepato-toxin (*72,73*). From water blooms of *O. agardhii* var. and *O. agardhii* var. *isothrix*, two similar cyclic heptapeptides have been isolated. Both toxins have the variant amino acids arginine-arginine and aspartic acid instead of β-methylaspartic acid. *O. agardhii* var. isothrix also has dehydroalanine instead of methyldehydroalanine (*70*). More recently *M. viridis* (*74*) and *M. aeruginosa* (*75*) have been shown to produce the cyclic heptapeptide with an arginine-arginine "L" amino acid variant.

Nodularia spumigena has also been shown to produce a peptide with hepato-toxic activity. The more recent reports come from Australia (*76*), the German Democratic Republic (*77*), Denmark (*78*), Sweden (*79*), and Finland (*80,81*). Recently structure information on *Nodularia* toxin has been presented by Rinehart (*97*) for waterbloom material collected in Lake Forsythe, New Zealand, in 1984; by Eriksson et al. (*81*) from waterbloom material collected in the Baltic Sea in 1986, and Runnegar et al. (*82*) for a field isolate from the Peel Inlet, Perth, Australia. Structure work by Rinehart, Eriksson, and Runnegar all indicate that the peptide is smaller than the heptapeptide toxins. Rinehart's work (*97*) indicates the toxin is a pentapeptide with a similar structure to the heptapeptides and containing β-methylaspartic acid, glutamic acid, arginine, dehydrobutyrine, and ADDA (MW 824).

The hepatotoxins have been called Fast-Death Factor (*68*), microcystin (*62*), cyanoginosin (*60*), cyanoviridin (*74*), and cyanogenosin (apparently a misspelling of cyanoginosin) (*75*). Because of the different terms used to refer to similar com-pounds it has been proposed (*83*) that the term microcystin (MCYST) plus the suf-fix "XY" (designating the variant L-amino acids) be used as the basis for naming all the existing and future monocyclic heptapeptide toxins of cyanobacteria. Cyclic peptides with fewer or greater than seven peptides or peptide linked components (i.e., the pentapeptide from *Nodularia* is termed nodularin [NODLN]) should be named according to the genus name from which they are originally isolated or by their chemical composition relative to the existing microcystins. The hepatotoxic heptapeptide general structure thus becomes:

```
    1   2         3          4  5   6         7
Cyclo(-D-Ala-L-"X"-D-erythro-β-methylAsp-L-"Y"-ADDA-D-Glu-N-MethyldehydroAla).
```

X = Leucine (L), Arginine (R), Tyrosine (Y)
Y = Arginine (R), Alanine (A), Methionine (M)

"XY" combinations for heptapeptide toxins currently defined: LR; LA; YA; YM; YR; RR

ADDA = 3-amino-9-methoxy-2,6,8-trimethyl-10-phenyldeca-4,6-dienoic acid

Mode of Action for Microcystins. The liver has always been reported as the organ that showed the greatest degree of histopathological change when animals are poisoned by these cyclic peptides. The molecular basis of action for these cyclic peptides is not yet understood but the cause of death from toxin and toxic cells administered to laboratory mice and rats is at least partially known and is con-cluded to be hypovolemic shock caused by interstitial hemorrhage into the liver (*84*). This work with small animal models is currently being extended to larger animals in order to study the uptake, distribution, and metabolism of the toxins (85). There is evidence to show from studies using ^{125}I-labeled MCYST-YM (CYGSN-YM) that the liver is the organ for both accumulation and excretion (86,87). Brooks and Codd (88), using ^{14}C labeled MCYST-LR, showed that 70% of the labeled toxin was localized in the mouse liver after 1 min following ip injection of the toxin.

Studies at both the light and electron microscopic (EM) level of time-course histopathological changes in mouse liver show rapid and extensive centrilobular necrosis of the liver with loss of characteristic architecture of the hepatic cords. Sinusoid endothelial cells and then hepatocytes show extensive fragmentation and vesiculation of cell membranes (89,90). Using microcystin-LR from M. aeruginosa strain PCC-7820, Dabholkar and Carmichael (91) found that at both lethal and sublethal toxin levels hepatocytes show progressive intracellular changes beginning at about 10 min postinjection. The most common response to lethal and sublethal injections is vesiculation of rough endoplasmic reticulum (RER), swollen mitochondria, and degranulation (partial or total loss of ribosomes from vesicles). The vesicles appear to form from dilated parts of RER by fragmentation or separation. Affected hepatocytes remain intact and do not lyse. Use of the isolated perfused rat liver to study the pathology of these toxins shows similar results to the in vivo work. Berg et al. (92) used three structurally different cyclic heptapeptide hepatotoxins (MCYST-LR, desmethyl 3-MCYST-RR, and didesmethyl 3,7-MCYST-RR). All three toxins had a similar effect on the perfused liver system although both "RR" toxins required higher concentrations (5-7 ×) to produce their effect. This finding was consistent with the lower toxicity of the "RR" toxins, which was about 500 and 1000 μg/kg ip mouse compared to 50 μg/kg for MCYST- LR.

In vitro studies on isolated cells including hepatocytes, erythrocytes, fibroblasts, and alveolar cells continue to demonstrate the specificity of action that these toxins have for liver cells (83,86,93). This specificity has led Aune and Berg (94) to use isolated rat hepatocytes as a screen for detecting hepatotoxic waterblooms of cyanobacteria.

The cellular/molecular mechanism of action for these cyclic peptide toxins is now an area of active research in several laboratories. These peptides cause striking ultrastructural changes in isolated hepatocytes (95) including a decrease in the polymerization of actin. This effect on the cells cytoskeletal system continues to be investigated and recent work indirectly supports the idea that these toxins interact with the cells cytoskeletal system (86,96). Why there is a specificity of these toxins for liver cells is not clear although it has been suggested that the bile uptake system may be at least partly responsible for penetration of the toxin into the cell (92).

Summary

Acute poisoning of humans by freshwater cyanobacteria as occurs with paralytic shellfish poisoning, while reported, has never been confirmed. Humans are probably just as susceptible as pets, livestock, or wildlife but people naturally avoid contact with heavy waterblooms of cyanobacteria. In addition, there are no known vectors, like shellfish, to concentrate toxins from cyanobacteria into the human food chain. Susceptibility of humans to cyanobacteria toxins is supported mostly by indirect evidence. In many of these cases, however, if a more thorough epidemiological study had been possible these cases probably would have shown direct evidence for toxicity.

In addition to acute lethal poisonings, episodes of dermatitis and/or irritation from contact with freshwater cyanobacteria are occurring with increasing frequency. This is in part because more eutrophic waters are being used for recreational purposes. Investigations in the United States (*12*), Canada (*14*), Scotland (*4*), and Norway (*13*) have shown that dermatotoxic blooms may be dominated by *Anabaena*, *Aphanizomenon*, *Gloeotrichia*, and *Oscillatoria*, respectively.

In summary, it can be reported that toxic cyanobacteria can produce neurotoxic, hepatotoxic, and dermatotoxic compounds that are a direct threat to animal and human water supplies. This threat increases as water bodies become more eutrophic, thus supporting higher production of toxic and nontoxic cyanobacteria. Presence of these potent natural product toxins poses an increasing threat to the maintenance of quality water supplies for agriculture, municipal, and recreational use.

Acknowledgments

Work with European toxic cyanobacteria was partially supported by a NATO collaborative research grant between W.W. Carmichael and G.A. Codd, University of Dundee, Scotland, and O.M. Skulberg, Norwegian Water Research Institute, Oslo, Norway. Toxin structure work on European and North American peptide toxins is supported in part by U.S. AMRDC contract DAMD17-87-C-7019 to W.W. Carmichael. Portions of the work represent part of the Ph.D. dissertation research of N.A. Mahmood and E.G. Hyde. Their work was supported in part by fellowship support from the Biomedical Ph.D. Program, Wright State University.

Literature Cited

1. Carmichael, W. W. In *The Water Environment—Algal Toxins and Health*; Carmichael, W. W., Ed.; Plenum: New York, 1981; p 1.
2. Carmichael, W. W. In *Advances in Botanical Research*; Callow, E. A., Ed.; Academic: London, 1986; Vol. 12, p 47.
3. Carmichael, W. W. In *Handbook of Natural Toxins, Marine Toxins and Venoms*; Tu, A. T., Ed.; Marcel Dekkar: New York, 1988; Vol. 3, p 121.
4. Codd, G. A.; Bell, S. G. *Water Pollution Control* **1985**, *84*, 225–32.
5. Gorham, P. R.; Carmichael, W. W. In *Algae and Human Affairs*; Lembi, C. A.; Waaland, J. R., Eds.; Cambridge Univ. Press: Oxford, 1988; p 403.
6. Moore, R. E.; Patterson, G. M. L; Mynderse, J. L; Barchi, J., Jr. *Pure Appl. Chem.* **1986**, *58*, 263–71.
7. Mason, C. P.; Edwards, K. R.; Carlson, R. E., Pignatello, J.; Glenson, F. R.; Wood, J. M. *Science* **1982**, *215*, 400–02.
8. Gleason, F. K.; Paulson, J. L. *Arch. Microbiol.* **1984**, *138*, 273–77.
9. Moore, R. E.; Cheuk, C.; Patterson, G. M. L. *J. Am. Chem. Soc.* **1984**, *106*, 6456–57.
10. Barchi, J. J., Jr.; Moore, R. E.; Patterson, G. M. L. *J. Am. Chem. Soc.* **1984**, *106*, 8193–97.
11. Raziuddin, S.; Siegelman, H. W.; Tornabene, T. G. *Eur. J. Biochem.* **1983**, *137*, 333–36.
12. Billings, W. H. In *The Water Environment: Algal Toxins and Health*; Carmichael, W. W., Ed.; Plenum: New York, 1981; p 243.
13. Skulberg, O. M.; Codd, G. A.; Carmichael, W. W. *Ambio* **1984**, *13*, 244–47.
14. Carmichael, W.W.; Jones, C. L. A.; Mahmood, N. A.; Theiss, W. W. In *Critical Reviews in Environmental Control*; Straub, C. P., Ed.; CRC Press: Florida, 1985; p 275.
15. Carmichael, W. W.; Gorham, P. R. *Mitt. Int. Verein. Limnol.* **1978**, *21*, 285–95.
16. Huber, C. S. *Acta Crystallograph.* **1972**, *B28*, 2577–82.

17. Devlin, J. P.; Edwards O. E.; Gorham, P. R.; Hunter, N. R.; Pike, P. K.; Stavric, B. *Can. J. Chem.* **1977**, *55*, 1367–71.
18. Campbell, H. F.; Edwards, O. E., Kolt, R. J. *Can. J. Chem.* **1977**, *55*, 1372–79.
19. Campbell, H. F.; Edwards, O. E.; Edler, J. W., Kolt, R. J. *Pol. J. Chem.* **1979**, *53*, 27–37.
20. Bates, H. A.; Rapoport, H. *J. Am. Chem. Soc.* **1979**, *101*, 1259–65.
21. Peterson, J. S.; Toteberg-Kaulen, S.; Rapoport, H. *J. Org. Chem.* **1984**, *49*, 2948–53.
22. Koskinen, M. P.; Rapoport, H. *J. Med. Chem.* **1985**, *28*, 1301–09.
23. Tufariello, J. J.; Mechler, H.; Senaratne, K. P. A. *J. Am. Chem. Soc.* **1984**, *106*, 7979–80.
24. Tufariello, J. J.; Meckler, H.; Senaratne, K. P. A. *Tetrahedron* **1985**, *41*, 3447–53.
25. Danheiser, R. L.; Morin, J. M., Jr.; Salaski, E. J. *J. Am. Chem. Soc.* **1985**, *107*, 8066–73.
26. Wiseman, J. R.; Lee, S. Y. *J. Org. Chem.* **1986**, *51*, 2485–87.
27. Lindgren, B.; Stjernlof, P.; Trozen, L. *Acta Chemica Scand.* **1987**, *B41*, 180–83.
28. Carmichael, W. W.; Biggs, D. F.; Peterson, M. A. *Toxicon* **1979**, *17*, 229–36.
29. Spivak, C. E.; Witkop, B.; Albuquerque, E. X. *Mol. Pharm.* **1980**, *18*, 384–94.
30. Spivak, C. E.; Waters, J.; Witkop, B.; Albuquerque, E. X. *Mol. Pharm.* **193**, *23*, 337–43.
31. Aronstam, R. S.; Witkop, B. *Proc. Natl. Acad. Sci. USA* **1981**, *78*, 4639–43.
32. Carmichael, W. W.; Gorham, P. R. *J. Phycol.* **1977**, *13*, 97–101.
33. Carmichael, W. W.; Gorham, P. R.; Biggs, D. F. *Can. Vet. J.* **1977**, *18*, 71–5.
34. Carmichael, W. W.; Biggs, D. F. *Can. J. Zool.* **1978**, *56*, 510–12.
35. Astrachan, N. B.; Archer, B. G. In *The Water Environment: Algal Toxins and Health*; Carmichael, W. W., Ed.; Plenum: New York, 1981; p 437.
36. Wong, S. H.; Hindin, E. *Am. Water Works Assoc. J.* **1982**, *74*, 528–29.
37. Smith, R. A.; Lewis, D. *Vet. Hum. Toxicol.* **1987**, *29*, 153–4.
38. Stevens, D. K.; Krieger, R. I. *J. Analytical Tox.* **1988**, *12*, 126–31.
39. Carmichael, W. W.; Mahmood, N. A. In *Seafood Toxins*; Ragelis, E. P., Ed.; American Chemical Soc. Symposium Series 262, Washington D.C., 1984; p 377.
40. Mahmood, N. A. Ph.D. Thesis, Wright State University, Dayton, Ohio, 1985.
41. Mahmood, N. A.; Carmichael, W. W. *Toxicon* **1986**, *24*, 425–34.
42. Mahmood, N. A.; Carmichael, W. W. *Toxicon* **1987**, *25*, 1221–27.
43. Rosenberry, T. In *Advances in Enzymology*; Meister, A., Ed.; John Wiley and Sons: New York, 1975; p 103.
44. Main, A. R. In *Introduction to Biochemical Toxicology*; Hodgson, E.; Guthrie, F., Eds.; Elsevier: New York, 1980; p 193.
45. Wilson, I. B.; Quan, C. *Arch. Biochem. Biophys.* **1958**, *73*, 131–43.
46. Husan, F. B.; Cohen, J. B. *J. Biol. Chem.* **1980**, *255*, 3896–3904.
47. Tomlinson, G.; Mutusi, B.; McLennan, I. *Mol. Pharm..* **1980**, *18*, 33–9.
48. Mahmood, N. A.; Carmichael, W. W.; Pfahler, D. *Am. J. Vet. Res.* **1988**, *49*, 500–03.
49. Sawyer, P. J.; Gentile, J. H.; Sasner, J. J., Jr. *Can. J. Microbiol.* **1988**, *14*, 1199–204.
50. Jakim, E.; Gentile, J. H. *Science* **1968**, *162*, 915–16.
51. Sasner, J. J., Jr.; Ikawa, M.; Foxall, T. L. In *Seafood Toxins*; Ragelis, E. P., Ed.; American Chem. Soc. Symp. Series: Washington D.C., 1984; p 391.
52. Carmichael, W. W. *S. Afr. J. Sci.* **1982**, *78*, 367–72.
53. Ikawa, M.; Wegner, K.; Foxall, T. L.; Sasner, J. J., Jr. *Toxicon* **1982**, *20*, 747–52.

54. Adelman, W. J., Jr.; Fohlmeister, J. F.; Sasner, J. J., Jr.; Ikawa, M. *Toxicon* **1982**, *20*, 513–16.
55. Mahmood, N. A.; Carmichael, W. W. *Toxicon* **1986**, *24*, 175–86.
56. Shimizu, Y.; Norte, M.; Hori, A.; Genenah, A.; Kobayashi, M. *J. Am. Chem. Soc.* **1984**, *106*, 6433–34.
57. Schwimmer, M.; Schwimmer, D. In *Algae, Man and the Environment*; Jackson, D. F., Ed.; Syracuse Univ.: Syracuse, 1968; p 279.
58. Botes, D. P.; Druger, H.; Viljoen, C. C. *Toxicon* **1982**, *20*, 945–54.
59. Botes, D. P.; Viljoen, C. C., Kruger, H.; Wessels, P. L.; Williams, D. H. *J. Chem. Soc. Perkin. Trans.* **1982**, *1*, 2742–48.
60. Botes, D. P. In *Mycotoxins and Phycotoxins*; Steyn, P. S.; Vleggaar, R., Eds.; Elsevier: Amsterdam, 1986; Vol. 1, p 16.
61. Santikarn, S.; Williams, D. H.; Smith, R. J.; Hammond, S. J.; Botes, D. P. *J. Chem. Soc. Chem. Commun.* **1983**, *12*, 652–54.
62. Konst, H.; McKercher, P. D.; Gorham, P. R.; Robertson, A.; Howell, J. *Can. J. Comp. Med. Vet.* **1965**, *29*, 221–28.
63. Murthy, J. R.; Capindale, J. B. *Can. J. Biochem.* **1970**, *48*, 508–10.
64. Rabin, P.; Darbre, A. *Biochem. Soc. Trans.* **1975**, *3*, 428–30.
65. Eloff, J. N. Ph.D. Thesis; Univ. of the Orange Free State, R.S.A., 1987; 8 chapters.
66. Botes, D. P.; Wessels, H.; Kruger, H.; Runnegar, M. T. C.; Santikarn, S.; Smith, R. J.; Barna, J. C. J.; Williams, D. H. *J. Chem. Soc. Perkin, Trans.* **1985**, *1*, 2747–48.
67. Elleman, T. C.; Falconer, I. R.; Jackson, A. R. B.; Runnegar, M. T. *Aust. J. Biol. Sci.* **1978**, *31*, 209.
68. Bishop, C. T.; Anet, E. F. L. J.; Gorham, P. R. *Can. J. Biochem. Physiol.* **1959**, *37*, 453–71.
69. Krishnamurthy, T.; Carmichael, W. W.; Sarver, E. W. *Toxicon* **1986**, *24*, 865–73.
70. Krishnamurthy, T.; Szafraniec, L.; Sarver, E. W.; Hunt, D. F.; Shabanowitz, J.; Carmichael, W. W.; Missler, S.; Skulberg, O.; Codd, G. *Proc. Am. Soc. Mass Spec.* (ASMS) - Cincinnati, OH, 1986; p 93.
71. Berg, K.; Carmichael, W. W.; Skulberg, O. M.; Benestad, C.; Underrdal, B. *Hydrobiologia* **1987**, *144*, 97–103.
72. Ostensvik, O.; Skulberg, O. M.; Soli, N. E. In *The Water Environment: Algal Toxins and Health*; Carmichael, W. W., Ed.; Plenum: New York, 1981; p 315.
73. Eriksson, J.; Meriluoto, J. A. O.; Kujari, H. P.; Skulberg, O. M. *Comp. Biochem. Physiol.* **1988**, *89*, 207–10.
74. Kusumi, T.; Oui, T.; Watanabe, M. M.; Takahogh, H.; Kakisawa, H. *Tetrahed. Letters* **1987**, *28*, 4695–98.
75. Painuly, P.; Perez, R.; Fukai, T.; Shimizu, Y. *Tetrahed. Letters* **1988**, *29*, 11–14.
76. Main, D. C.; Berry, P. H.; Peet, R. L.; Robertson, J. P. *Aust. Vet. J.* **1977**, *53*, 578–81.
77. Kalbe, L.; Tiess, D. *Arch. Exp. Vet. Med.* **1964**, *18*, 535–39.
78. Lindstrom, E. *Dansk. Vet. Tidsskr.* **1976**, *59*, 637–4
79. Edler, L.; Ferno, S.; Lind, M. G.; Lundberg, R.; Nilsson, P. O. *Ophelia* **1985**, *24*, 103–09.
80. Persson, P. E.; Sivonen, K.; Keto, J.; Kononen, K.; Niemi, M.; Viljamaa, H. *Aqua Fenn.* **1984**, *14*, 147–54.
81. Eriksson, J. E.; Meriluoto, J. A. O.; Kujari, H. P.; Osterlund, K.; Fagerlund, K.; Hallbom, L. *Toxicon* **1988**, *26*, 161–66.
82. Runnegar, M. T. C.; Jackson, A. R. B.; Falconer, I. R. *Toxicon* **1988**, *26*, 143–5.

83. Carmichael, W. W.; Beasley, V. A.; Bunner, D. L.; Eloff, J. N.; Falconer, I.;
     Gorham, P. R.; Harada, K-I.; Yu, M-J.; Krishnamurthy, T.; Moore, R. E.;
     Rinehart, K.; Runnegar, M.; Skulberg, O. M.; Watanabe, M. *Toxicon* **1988**, *26*,
     971-3.
84. Theiss, W. C.; Carmichael, W. W.; Wyman, J.; Brunner, R. *Toxicon* **1988**, *26*,
     603-13.
85. Beasley, V. R.; Lovell, R.; Cook, W.; Lunden, G.; Holmes, K.; Hooser, S.;
     Haschek-Hock, W.; Carmichael, W. W. *Proc. Am. Chem. Soc.*, 8th Rocky Mt.
     Regional Meeting, 1986 (Abstract).
86. Falconer, I. R.; Runnegar, M. T. C. *Chem. Biol. Interactions* **1987**, *63*, 215–25.
87. Runnegar, M. T. C.; Falconer, I. R. *Toxicon* **1987**, *24*, 105–15.
88. Brooks, W. P.; Codd, G. A. Pharm. Tox. **1987**, *60*, 187–91.
89. Runnegar, M. T. C.; Falconer, I. R. In *The Water Environment: Algal Toxins
     and Health*; Carmichael, W. W., Ed.; Plenum: New York, 1981; p 325.
90. Foxall, T. L.; Sasner, J. J., Jr. In *The Water Environment: Algal Toxins and
     Health*; Carmichael, W. W., Ed.; Plenum: New York, 1981; p 365.
91. Dabholkar, A. S.; Carmichael, W. W. *Toxicon* **1987**, *25*, 285–92.
92. Berg, K.; Wyman, J.; Carmichael, W. W.; Dabholkar, A. S. *Toxicon* **1988**, **26**,
     827-37.
93. Runnegar, M. T. C.; Andrews, J.; Gerdes, R. G.; Falconer, I. R.   *Toxicon*
     1987, *25*, 1235–39.
94. Aune, T.; Berg, K. *J. Toxicol. Environ. Health* **1986**, *19*, 325–36.
95. Runnegar, M. T. C.; Falconer, I. R. *Toxicon* **1986**, *24*, 105–15.
96. Eriksson, J.; Hagerster, H.; Isomaa, B. *Biochem. Biophysc. Acta* **1987**, *930*,
     304–10.
97. Rinehart, K.; Harada, K.-I; Namikoshi, M.; Chen, C.; Harvis, C.A.; Munroe,
     M.H.G.; Blunt, J.W.; Mulligan, P.E.; Beasley, U.R.; Dahmen, A.M.; Carmi-
     chael, W.W. *J. Am. Chem. Soc.* **1988**, *110*, 8557-8.
98. Francis, G. *Nature* (London) **1878**, *18*, 11-2.
99. Sivonen, K.; Himberg, K.; Luukkamen, R.; Niemelä, S.I.; Poon, G.K.; Codd,
     G.A. *Tox. Assess.*, **1989**, *4*, 339-52.
100. Matsunaga, S.; Moore, R. E.; Niemczurd, W. P.; Carmichael, W. W. *J. Am.
     Chem. Soc.* **1989**, *111*, 8021–23.

RECEIVED August 9, 1989

Chapter 7

Nicotinic Acetylcholine Receptor Function Studied with Synthetic (+)-Anatoxin-a and Derivatives

K. L. Swanson[1], H. Rapoport[2], R. S. Aronstam[3], and E. X. Albuquerque[1]

[1]Department of Pharmacology and Experimental Therapeutics, University of Maryland School of Medicine, Baltimore, MD 21201
[2]Department of Chemistry, University of California, Berkeley, CA 94720
[3]Department of Pharmacology and Toxicology, Medical College of Georgia, Augusta, GA 30912

(+)-Anatoxin-a is the most potent and most stereospecific nicotinic acetylcholine receptor agonist thus far identified. It is also highly selective for nicotinic receptors over muscarinic receptors. The molecular parameters which influence the binding affinity, channel activation, channel blockade, and receptor desensitization are being studied. Modifications of the carbonyl and amine moieties can reduce or nearly eliminate the receptor agonist potency of the compounds and also determine the channel blocking characteristics.

The physiology and pharmacology of the nicotinic acetylcholine receptor (AChR) have historically been beset with complications. The natural neurotransmitter acetylcholine (ACh) binds not only to the nicotinic receptor but also to many types of muscarinic receptors. This has made it difficult to study the functional effects of nicotinic receptors in neuronal systems. One means to avoid the conflict of multiple receptor types was to study the neuromuscular junction because the only postjunctional receptors present are nicotinic. Alternatively, ACh was used in the presence of a selective antimuscarinic agent such as atropine; unfortunately, atropine also has noncompetitive antagonist effects at the nicotinic receptor (1). Furthermore, because ACh is rapidly hydrolyzed by acetylcholinesterase (AChE), anti-AChE agents such as neostigmine were commonly used, but these agents also have noncompetitive effects at the AChR (2). The AChR antagonist α-bungarotoxin (αBGT) was an excellent pharmacological tool for the localization of receptors on the muscle membrane because of irreversible binding to a site overlapping the nicotinic agonist site. However, αBGT does not bind to the same site in the central nervous system (CNS) as does (-)-nicotine (3).

The debut of the selective AChR agonist (+)-anatoxin-a has provided a new tool for AChR physiology and pharmacology. (+)-Anatoxin not only has high affinity for the nicotinic AChR but it also has high selectivity for nicotinic over muscarinic receptors in the mammalian CNS. Recently, the use of (+)-anatoxin-a was essential to the identification of nicotinic receptors on cultured neurons (4). We are studying the features which allow it to bind with high affinity to the peripheral and central nicotinic receptors and the kinetic effects on receptor conformational

0097–6156/90/0418–0107$06.00/0

transitions. These studies are complicated by the need to assess noncompetitive AChR antagonistic and anti-AChE effects of each analog. The single channel recording technique is particularly useful because the AChE has been eliminated during isolation of muscle fibers and because the kinetics can detect specific agonistic and antagonistic effects.

Discovery of Poisoning by Fresh Water Algal Blooms and Signs of Toxicity

Algal blooms in fresh water ponds occasionally poison livestock and waterfowl. Axenic cultures of *Anabaena flos-aquae* NRC 44-1 were shown to produce the toxic principle (5) which can be present in the algae and in the water of mature cultures (6). The discovery of the toxin was fortuitous in the sense that AChR agonists do not have a (known) constructive function in the algae; evolution of the synthetic pathway was likely a by-product of metabolic pathways in the algae. The compound became evident only through its toxic effects on other organisms.

The toxin causes death rapidly in mammals, birds and fish. The LD_{50} (i.p., mouse) is only 0.2 mg/kg (6, 7). In each case paralysis by depolarizing blockade is the predominant sign: early muscle fasciculations in mammals, prolonged opisthotonos in birds, and muscular rigidity in fish (5). Death is due to respiratory paralysis. Poisoned calves were maintained by artificial ventilation for up to 30 hrs without sufficient recovery for independent survival. In vitro the toxin caused depolarization of frog muscles and increased the spontaneous, vesicular release of transmitter (8, 9).

The active principle of the algal blooms was first extracted serially with ethanol, chloroform, and acidic water (5). In another method, a good recovery (73%) was achieved by freeze drying followed by acidic methanol and benzene–chloroform extractions; purification was by preparative thin-layer chromatography (6). The presence of an enone in the structure of the toxin with absorption at 1670 cm^{-1} was used in final stages of isolation (6). The chemical structure, 2-acetyl-9-azabicyclo[4.2.1]nonene, and its absolute configuration (Figure 1) were determined by X-ray crystallography of *N*-acetyl anatoxin-a (10). Subsequent synthesis from l-cocaine confirmed the structure and stereochemistry (11). Other complex methods tackling the problem of bridged alkaloids have synthesized racemic (7, 12, 13) or optically active anatoxin-a (14).

Chemical Model of Nicotinic Agonists

Structure–activity relationships form the basis of receptor pharmacology and drug development research. A preliminary assay for nicotinic activation is to apply a drug to a muscle in a bathing medium and measure the generation of contractile forces. In our experiments we used the frog rectus abdominis muscle (Figure 2). Such an assay cannot be conclusive, i.e. it can also be influenced by desensitization and noncompetitive antagonism. This is due to multiple drug binding sites on the AChR macromolecule (Table 1). Mixed agonistic–antagonistic effects are characteristic of many AChR ligands.

The agonist site on the nicotinic receptor has in general demonstrated only mild stereospecificity—in fact the agonist could be described by a small number of characteristics. A positively charged group, usually an amine, is present which may be alkylated to varying degrees (15). A polar moiety, often a carbonyl group, forms a hydrogen bond at 5.9 or 6.0 Å from the amine (16, 17). The polar group is also contained in a planar region of the agonist molecule (18).

Using synthetic enantiomers, we found that anatoxin-a is highly stereospecific with the (+) isomer having 150-fold greater potency than the (-) isomer (Figure 2) (19). The semi-rigid nature of anatoxin-a undoubtedly facilitates its stereospecificity.

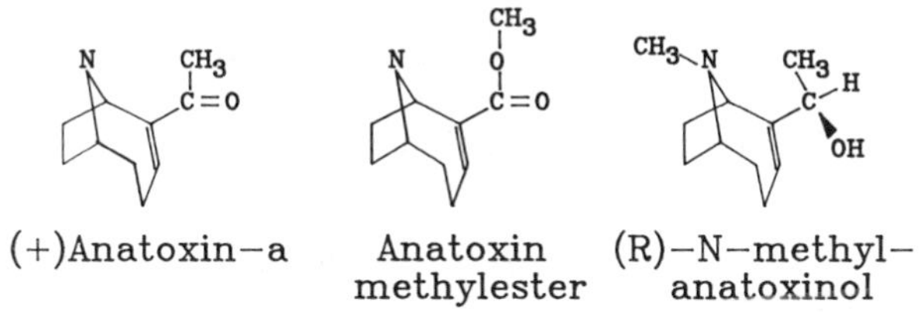

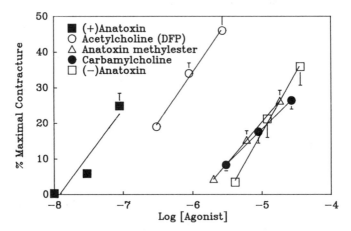

(+)Anatoxin—a   Anatoxin   (R)—N—methyl—
methylester   anatoxinol

Figure 1. Structure of (+)-anatoxin-a and two analogs with agonist and noncompetitive antagonist activity, respectively.

Figure 2. Potency of anatoxin-a analogs to induce contracture in frog *rectus abdominis* muscle. The data from two experiments are combined in this figure. In one, anatoxinmethylester was found to be equipotent with carbamylcholine. In the other, the anatoxin isomers were assayed against ACh [after cholinesterse inhibition by diisopropylfluorophosphate (DFP) followed by washing of the preparation] (*19*). Maximal contracture was measured by depolarization with KCl at the end of each experiment.

Table I. Sites of Chemical Interaction on the Peripheral Nicotinic Receptor

Binding Site and Specific Ligands	Site Localization on Receptor	Pharmacological Effect
1. agonist site ACh, (+)-anatoxin-a & αBGT	1 per alpha subunit, in the extracellular domain	channel activation and depolarization
2. phosphorylation site	delta subunit	desensitization
3. high affinity site HTX, PCP	1 per AChR ion channel formed by all subunits	open channel blockade desensitization[a]
4. low affinity site chlorpromazine	many sites at protein-lipid interface	desensitization[a]

[a]Desensitization is accompanied by increased agonist-receptor affinity and stimulation of channel activity at low concentrations.

Stereospecificity suggests that there are three geometrically distinct determinants of the structure which are important to agonist efficacy and potency, whereas previous models only utilized two features. A hydrophobic pocket may also be important (20); this could be supplied by the [4.2.1]nonene structure. Our intentions are to explore the domains of the agonist molecule to determine the specific determinants of agonist activity.

Two regions of the toxin are targets for this initial study: the carbonyl and the amine regions. The few compounds reported on here do not provide a comprehensive evaluation of structural parameters, but they do serve to introduce the variety of questions which we anticipate will be addressed by contrasts between a larger number of compounds. In the first analog, anatoxin methylester, modification of the carbonyl moiety altered the hydrogen bonding potential and the molecular dimensions, but may not have altered the amine–carbonyl separation or the planarity of the carbonyl region. Thus, by creating an ester, this relatively minor modification could make the compound more similar to the natural transmitter than anatoxin. The other type of alteration, reduction of the carbonyl to an alcohol generated two analogs of S and R configuration. The alcohol moiety would have significantly lower hydrogen bonding potential. Furthermore, the toxin loses planarity in that region. At the other target of investigation, methylation of the amine could also change the potency of the toxin. In the related molecule, norferruginine, methylation increased the potency (21). Preliminary data indicate the N, N-dimethyl anatoxin may only have very weak agonistic properties along with antagonistic properties. Also, it is important that multiple alkylation to form quaternary compounds could change the ability of the analogs to penetrate CNS tissue.

Structure and Function of the Nicotinic AChR

The AChR of *Torpedo* or immature muscle is a complex formed with 5 polypeptide chains, 2 alpha, and 1 each of beta, gamma, and delta chains (22). Each alpha chain contains a binding site for ACh (and αBGT). Although the primary structures of the alpha peptide chains are identical, they can be distinguished with antibodies because they have different 3-dimensional configurations in the protein complex (23). Each of the other chains are similar, yet different enough to possess other binding sites (Table I). Several transmembrane alpha-helices of each peptide line the ion channel.

It is necessary for two molecules of agonist (A) to bind to the receptor (R) in order to initiate a conformational shift from the closed ion channel configuration (A_2R) to the open channel configuration (A_2R^*). In mature muscle the open ionic channel has a conductance of 30–32 pS; this is a constant property of the receptor. [A lower conductance state occurs with receptors found in immature or denervated muscles (24–26).] The properties which depend upon the agonist are the rates of binding and dissociation and the rates at which conformational shifts occur.

Agonist model: $$2A + R \rightleftarrows A + AR \rightleftarrows A_2R \rightleftarrows A_2R^*$$

Using the cell-attached patch clamp technique on frog muscle fibers (19), one can observe only two conditions: the open, conducting state of the receptor and a nonconducting state of unknown identity. The transitions behave according to stochastic principles; the lifetimes of any particular condition are distributed exponentially. The open state has a mean duration that is the inverse of the rate of channel closing. Because channel open time depends only upon a conformational shift, agonist concentration does not influence the parameter. It is, however, influenced by the chemistry of the two agonists which are bound to the receptor and the transmembrane potential. The biophysical properties which determine the open duration are as yet to be determined.

Agonistic Effects of (+)-Anatoxin-a

(+)-Anatoxin-a has high potency by virtue of high affinity for the agonist binding site of the receptor. Assay of contracture potency demonstrated the overall relative potency of a variety of agonists (Figure 2). This data can be compared with the concentrations of the toxins which inhibit the binding of the antagonist αBGT to *Torpedo* electroplaque receptors (Table II, αBGT IC_{50}) and their ability to activate the ion channel, reducing the barriers to the binding of histrionicotoxin (HTX) and thus increasing specific binding (Table II, HTX ED_{50}; *see* discussion of HTX binding under **Allosteric Antagonism of AChR Function by** (+)-**Anatoxin-a Analogs**).

With agonists bound to the receptor, the channel may open and then close; after closure (while the agonist remains bound) the rates for channel reopening and agonist dissociation compete. When ACh is the agonist, the dissociation rate is higher and therefore typical channel activity consists of single, well-separated openings (Figure 3). When (+)-anatoxin-a is the agonist, the dissociation rate is slower and the channel is likely to reopen; this condition of the receptor can be recognized experimentally by the presence of short duration closures, "flickers". Thus with (+)-anatoxin-a, the flickers are significantly more frequent and channel activity consists of several openings separated by only a fraction of a millisecond. It is "pharmaco-gnomonic" of agonists that the form of the burst is independent of agonist concentration. Thus, the average burst in response to 20 or 200 nM (+)-anatoxin-a is the same; the individual open times (Figure 3), the total burst duration, the flicker durations, and the number of openings per burst are all the same at these two concentrations (*19*). The rate of ion channel closure, which determines the mean open time, is more rapid for (+)-anatoxin-a than for ACh. Perhaps such comparison can facilitate understanding of the biophysical properties which determine the open duration.

Desensitizing Properties of (+)-Anatoxin-a

Another characteristic of agonists is the ability to desensitize the receptor, i.e., to make a population of receptors become non-responsive to further agonist. When a high concentration of agonist produces this effect it is known as depolarizing blockade. There are, in addition to the agonist model drawn above, one or more desensitized states of the receptor. Agonists, in general, have a higher affinity for the alpha-subunit sites of these desensitized states than of the normal receptor. This effect occurs by the allosteric binding of agonist or desensitizing agent to allosteric sites or by phosphorylation of the delta-subunit protein (*27*). Noncompetitive antagonists such as HTX and phencyclidine also promote desensitization by allosteric mechanisms (*28, 29*). Because of these allosteric mechanisms, desensitizing potency is not directly correlated with agonist potency. Although (+)-anatoxin-a is much more potent than ACh as an agonist, desensitization occurs more slowly with (+)-anatoxin-a than with an equipotent (as agonist) concentration of ACh (*19*).

Allosteric Antagonism of AChR Function by (+)-Anatoxin-a Analogs

Several drugs, the most well-known being local anesthetics and histrionicotoxin (HTX) (*28*), bind to an allosteric site on the AChR (relative to the agonist binding site). Biochemically, this site is identified by high affinity binding of $[^3H]H_{12}$-HTX (Table I). It may be located at the ion channel and coordinates between several of

Table II. Binding Constants for (+)-Anatoxin-a and Related Molecules at Nicotinic and Muscarinic Receptor Sites

Ligands	Torpedo Electric Organ			Rat Brain
	αBGT IC_{50}	HTX ED_{50}	HTX K_i	Scopolamine K_i
(+)-Anatoxin-a	0.085	0.032	--	9.3
Acetylcholine	0.30	0.15	--	--
Anatoxin methylester	2.4	0.22	320	--
(-)-Anatoxin-a	4.4	1.6	--	--
(S)-N-Methylanatoxinol	>100	--	117	0.33
(R)-N-Methylanatoxinol	>100	--	8.5	4.4

NOTE: All values are in μM and are defined as follows: αBGT IC_{50} = concentration producing 50% inhibition at $[^{125}I]\alpha$BGT binding; HTX ED_{50} = concentration of toxin producing 50% stimulation of $[^3H]$HTX binding; HTX K_i = inhibition constant in units of toxin concentration which represent inhibition of the binding of $[^3H]$HTX which was enhanced by 1 μM carbamylcholine; scopolamine K_i = inhibition constant in units of toxin concentration which represent inhibition of the binding of $[^3H]$scopolamine.

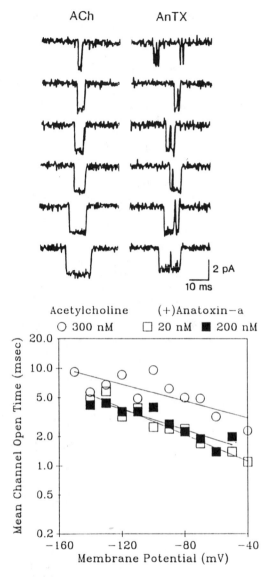

Figure 3. (+)-Anatoxin-a (AnTx) and ACh induced single ion channel currents in isolated frog muscle fibers. Open channels with 32 pS conductance are downward deflections (inward current at hyperpolarized potentials). The currents shown on the left are all at one potential. The duration of channel open events had a similar voltage-dependence for both ACh and (+)-anatoxin-a. With ACh, the events were most often singular, while with (+)-anatoxin-a the events were shorter and were more frequently paired so that the mean duration of the exponentially distributed open times and selected membrane holding potentials was approximately one-half, independent of the concentration of the agonist applied.

the polypeptide chains (30, 31). This site was not thought to have stereospecificity (32, 33).

The binding of antagonists is dependent upon the presence of agonist. In the presence of an agonist such as carbamylcholine, allosteric antagonists inhibit the binding of HTX to the ion channel site (Table II; HTX K_j). A sequential model of ion channel blockade is commonly used as a standard against which to characterize the effects of blockers. Once in the open configuration (A_2R^*), the drug (D) is able to bind and block the channel. The resulting state A_2R^*D has no conductance; it can be distinguished from A_2R only by statistical analysis.

Sequential model: $$A_2R \leftrightarrows A_2R^* + D \leftrightarrows A_2R^*D$$

The primary characteristic of a sequential blocker, as observed with the patch clamp technique, is that the reciprocal of the mean duration of the lifetime equals the normal channel closing rate plus the rate constant of channel blockade times the drug concentration. Therefore, increasing the drug concentration shortens the mean channel open time.

The (+)-anatoxin-a analogs (R)- or (S)-N-methylanatoxinol, both possess the ability to block the ion channel [the voltage-dependence of their characteristics differ in a way which is the topic of another report (34)]. As is often the case, the blocked state induced by (R)-N-methylanatoxinol and the closed state differed significantly in duration, such that the blocked state was associated with short closed periods within groups of openings called bursts (Figure 4). The number of short closed periods per burst increased with the concentration of the drug because the likelihood of channel blockade increased relative to the likelihood of channel closure.

The duration for which the receptor remains in the blocked state is a property of the drug, because it is a simple dissociation reaction. It often depends upon voltage, although the strength of the voltage relationship is a characteristic of the drug. In the case of (R)-N-methylanatoxinol, the voltage dependence is insignificant; the mean short closed duration was a few milliseconds at all potentials. We believe this may be due to a large degree of hydrophobic binding utilizing van der Waals forces. In contrast, the more polar S isomer binds rapidly but also dissociates rapidly due to reliance on coulombic interactions. The slower dissociation of the (R)-N-methylanatoxinol isomer is the factor which determines the greater potency to inhibit HTX binding (Table II).

Neuronal Nicotinic AChR

In the neuronal tissues, describing the function of nicotinic AChR has been considerably more complicated than at the neuromuscular junction (35). The lack of a suitable agonist which was selective for the AChR was confounded by the difficulty that αBGT was also an unsuitable antagonist (3). Apparently, αBGT prefers to bind to a different receptor site in the CNS than does nicotine. Only recently has optically active [3H]nicotine become available to make selective agonist binding studies feasible. The K_i of (+)-anatoxin-a for the inhibition of (-)-[3H]nicotine binding to rat brain membranes was 0.34 nM (36) and the IC_{50} for inhibition of [3H]ACh binding was 4.5 nM (37). (+)-Anatoxin-a was a 2- to 20-fold better ligand than (-)-nicotine, in these respective studies. It also has high selectivity for nicotinic over muscarinic receptors; the K_i for inhibition of a muscarinic antagonist, [3H]scopolamine, binding to rat brain synaptosomes was 9.3 μM (Table II).

(+)-Anatoxin-a acts presynaptically as an agonist in the CNS, as evidenced by

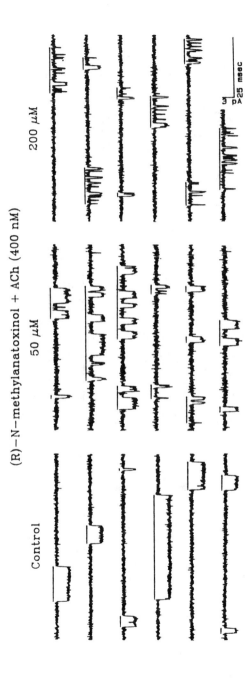

Figure 4. Concentration-dependent ion channel blockade by (R)-*N*-methylanatoxinol. The patterns identified as bursts and separated by long (>8 msec) closed intervals are indicated with a bar, the figure was designed to show approximately 2 bursts per trace. The dose-related decrease in mean channel open time resulted from the blockade of the open channel by the (R)-*N*-methylanatoxinol. The channel amplitude is related to membrane voltage (as was given in Figure 3) by the slope conductance such that 1 pA is equivalent to 30 mV. *Continued on next page.*

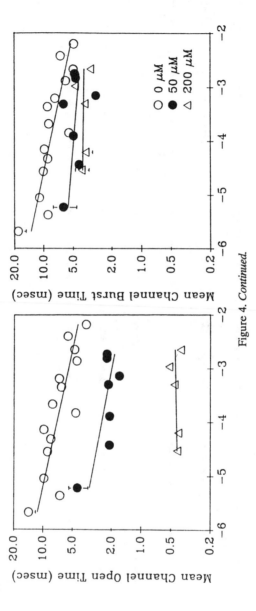

Figure 4. *Continued.*

the stimulation of neurotransmitter release (*38*). In this case, 1 μM anatoxin was as effective as 10 μM nicotine. Aracava et al. (*4*) disclosed the presence of nicotinic AChR on cultured, neonatal brain stem and hippocampal neurons. Single ion channel currents were recorded from the base of the apical dendrites of pyramidal cells using the patch clamp technique with ACh or (+)-anatoxin-a as the agonists. These nicotinic AChR had low conductance and the openings were brief, with characteristics which are similar to those observed for embryonic muscle receptors. Also found in retinal ganglion cells, the neuronal nicotinic receptors responded with the kinetics of ion channel blockade, including enhancement upon hyperpolarization as seen with peripheral receptors; the single channel currents were sensitive to phencyclidine and also showed a possible blockade by micromolar concentrations of (+)-anatoxin-a (*39*). While many details of the function of the central nicotinic receptors remain to be elucidated, the initial results suggest a large degree of functional homology between peripheral and central receptors (*40*).

Conclusion

(+)-Anatoxin-a has already proven its usefulness as a research tool in our laboratories. It is facilitating the understanding of the biophysical properties of the AChR and of the localization of the AChR in the CNS. The toxin or derivatives of it could be useful therapeutically in diseases of nicotinic receptor pathology (myasthenia gravis or Alzheimer's disease), because as a secondary amine (+)-anatoxin-a can penetrate into the CNS.

Acknowledgments

This work was supported by NIH Grant NS 25296 and U.S. Army Research and Development Command Contract DAMD17-88-C8119.

Literature Cited

1. Adler, M.; Albuquerque, E. X.; Lebeda, F. J. *Mol. Pharmacol.* **1978,** *14*, 514.
2. Aracava, Y.; Deshpande, S. S.; Rickett, D. L.; Brossi, A.; Shonenberger, B.; Albuquerque, E. X. *New York Acad. Sci.* **1987,** *505*, 226.
3. Clarke, P. B. S *Trends in Pharmacological Sciences*, **1987,** *8*, 32.
4. Aracava, Y.; Deshphande, S. S.; Swanson, K. L.; Rapoport, H.; Wonnacott, S.; Lunt, G.; Albuquerque, E. X. *FEBS Letters* **1987,** *222*, 63.
5. Carmicael, W. W.; Biggs, D. F.; and Gorham, P. R. *Science* (Washington, D.C.) **1975,** *187*, 542-544.
6. Devlin, J. P.; Edwards, O. E.; Gorham, P. R.; Hunter, N. R.; Pike, R. K.; Stavric, B. *Can. J. Chem.* **1977,** *55*, 1367.
7. Bates, H. A.; Rapoport, H. *J. Amer. Chem. Soc.* **1979,** *101,* 1259.
8. Biggs, D. F.; Dryden, W. F. *Proc. West. Pharmacol. Soc.* **1977,** *20*, 461.
9. Spivak, C. S.; Witkop, B.; Albuquerque, E. X. *Mol. Pharmacol.* **1980,** *18*, 384.
10. Huber, C. S. *Acta. Cryst.* **1972,** *B28*, 2577.
11. Campbell, H. F., Edwards, O. E.; and Kolt, R. *Can. J. Chem.* **1977,** *55*, 1372.
12. Campbell, H. F.; Edwards, O. E.; Elder, J. W.; Kolt, R. J. *Polish J. Chem.* **1979,** *53*, 27.
13. Vernon, P.; Gallagher, T. *J. Chem. Soc., Chem. Commun.* **1987,** 245.
14. Petersen, J. S.; Fels, G.; Rapoport, H. *J. Am. Chem. Soc.* **1984,** *106*, 4539.
15. Spivak, C. C.; Albuquerque, E. X. In *Progess in Cholinergic Biology: Model Cholinergic Synapses*; Hanin, I.; Goldberg, M., Eds.; Raven Press: New York, **1982,** 323.
16. Beers, W. H.; Reich, E. *Nature (Lond.)* **1970,** *228,* 917.

118   MARINE TOXINS: ORIGIN, STRUCTURE, AND MOLECULAR PHARMACOLOGY

17. Koskinen, A. M. P.; Rapoport, H. *J. Med. Chem.* **1985**, *28*, 1301.
18. Chothia, C.; Pauling, P. *Proc. Natl. Acad. Sci. USA* **1970**, *65*, 477.
19. Swanson, K. L.; Allen, C. N.; Aronstam, R. S.; Rapoport, H.; Albuquerque, E. X. *Mol. Pharmacol.* **1986**, *29*, 250.
20. Waksman, G.; Changeux, J. -P.; Roques, B. P. *Mol. Pharmacol.* **1980**, *18*, 20.
21. Albuquerque, E. X.; Spivak, C. E. In *Natural Products and Drug Development*; Krogsgaard-Larsen, P.; Brøgger Christensen, S.; Kofod, H., Eds.; Alfred Benson Symposium 20; Munksgaard, Copenhagen, **1984**; pp 301-321.
22. Stroud, R. M. *Neuroscience Commentaries* **1983**, *1*, 124.
23. Maelicke, A.; Fels, G.; Plümer-Wilk, R.; Wolff, E. K.; Covarrubias, M; Methfessel, C. In *Ion Channels in Neural Membranes*, Allan R. Liss, **1986**, 275.
24. Allen, C. N.; Albuquerque, E. X. *Exp. Neurol.* **1986**, *91*, 532.
25. Akaike, A.; Ikeda, S. R.; Brookes, N.; Pascuzzo, G. J.; Rickett, D. L.; Albuquerque, E. X. *Mol. Pharmacol.* **1984**, *25*, 102.
26. Aracava, Y.; Ikeda, S. R.; Daly, J. W.; Brookes, N.; Albuquerque, E. X. *Mol. Pharmacol.* **1984**, *26*, 304.
27. Steinbach, J. H.; Zempel, J. *Trends in Neuroscience* **1987**, *10*, 61.
28. Changeux, J. -P.; Nevillers-Thiéry, A.; Chemouilli, P. *Science* **1984**, *225*, 1335.
29. Oswald, R. E.; Heidmann, T.; Changeux, J. -P. *Biochem.* **1983**, *22*, 3128.
30. Heidmann, T.; Oswald, R. E.; Changeux, J. -P. *Biochem.* **1983**, *22*, 2112.
31. Changeux, J. -P.; Revah, F. *Trends in Neuroscience* **1987**, *10*, 245.
32. Spivak, C. E.; Maleque, M. A.; Takahashi, K.; Brossi, A.; Albuquerque, E. X. *FEBS Lett.* **1983**, *163*, 189.
33. Henderson, F.; Prior, C.; Dempster, J.; Marshall, I. G. *Mol. Pharmacol.* **1986**, *29*, 52.
34. Swanson, K. L; Aracava, Y.; Sardina, F. J.; Rapoport, H.; Aronstam, R. S.; Albuquerque, E. X. *Mol. Pharmacol.* **1989**, *35*, 223.
35. Rovira, C.; Ben-Ari, J.; Cherubini, E.; Krnjevic, K.; Roper, N. *Neurosci.* **1983**, *8*, 97.
36. Macallan, D. R. E.; Lunt, G. G.; Wonnacott, S.; Swanson, K. L.; Rapoport, H.; Albuquerque, E. X. *FEBS Lett* **1987**, *226*, 357.
37. Zhang, X.; Stjernlof, P.; Adem, A.; Nordberg, A. *Eur. J. Pharmacol.* **1987**, *135*, 457.
38. Lunt, G.; Wonnacott, S.; Thorne, B. *Neurosc. Abst.* **1987**, *13*, 940.
39. Albuquerque, E. X.; Alkondon, M.; Lima-Landman, M. T.; Deshpande, S. S.; Ramoa, A. S. In *Neuromuscular Junction*; Fernström Foundation Series Vol. 13; Sellin, L. C.; Libelius, R.; Thesleff, S.; Eds.; Elsevier Science Publ.: Cambridge, U.K., **1988**, pp. 273-300.
40. Aracava, Y.; Swanson, K. L.; Rozental, R.; Albuquerque, E. X. In *Neurotox* Elsevier Science Publ.: Cambridge, U.K., **1988**, pp. 157-84.

RECEIVED May 17, 1989

POLYETHER TOXINS

Chapter 8

Polyether Toxins Involved in Seafood Poisoning

Takeshi Yasumoto and Michio Murata

Department of Food Chemistry, Faculty of Agriculture, Tohoku University, Tsutsumidori Amamiya, Sendai 980, Japan

Recent studies on marine toxins point to the involvement of an increasing number of novel polyether compounds in seafood poisonings. Causative toxins for ciguatera, ciguatoxin, scaritoxin, and maitotoxin seem to have polyether skeletons, although their structures remain unknown. Palytoxin was found to be responsible for the fatal poisonings caused by ingestion of xanthid crabs and trigger fish. Three classes of polyethers, okadaic acid derivatives, pectenotoxins, and yessotoxin were isolated from bivalves in connection with diarrhetic shellfish poisoning. The etiology of the toxins, toxicological properties, and determination methods are described.

Ciguatera

Ciguatera is a term given to intoxication caused by eating a variety of fish inhabiting or feeding on coral reefs. Its occurrence is most prevalent in the Caribbean and the tropical Pacific, affecting probably over 10,000 people annually. Although ciguatera has been known to the scientific world since 17th century, chemical studies of causative toxin(s) have been handicapped by the extreme difficulty in obtaining toxic materials. Thus, the disease had been defined symptomatologically rather than chemically, until Scheuer's group characterized some properties of the principle toxin named ciguatoxin. The variability of ciguatera symptoms, which include digestive, neurological, and cardiovascular disorders, and the diversity of fish species involved, however, led scientists to explore additional toxins, mostly in herbivorous fish. Consequently, scaritoxin and maitotoxin were found in parrotfish and surgeonfish, respectively.

Ciguatoxin. The toxin was isolated from moray eels and purified to crystals by Scheuer's group (1). Structural determination of the toxin by x-ray or NMR analyses was unsuccessful due to the unsuitability of the crystals and due to the extremely small amount of the sample. The toxin was presumed to have a molecular formula of $C_{60}H_{86}NO_{19}$ from HRFAB-MS data (MH$^+$, 1111.5570) and to have six hydroxyls, five methyls, and five double bonds in the molecule (2). The number of unsaturations (18 including the five double bonds) and the abundance of oxygen atoms in the molecule point to a polyether nature of the toxin. The toxin, or a closely related toxin if not identical, is believed to be the principal toxin in ciguatera. Ciguatoxin was separable on an alumina column into two interconvertible entities presumably differing only in polarity (3).

Scaritoxin. During a survey of ciguatera intoxication in the Gambier Islands, Bagnis et al. observed that patients poisoned by parrotfish (Scaridae) suffered a longer period than conventional ciguatera symptoms, and postulated the presence of a

0097–6156/90/0418–0120$06.00/0

new toxin, scaritoxin (4). Later, two toxins code-named SG1 and SG2 were isolated from the flesh of the parrotfish *Scarus gibbus* and were suggested to correspond to scaritoxin and ciguatoxin, respectively (5). Scaritoxin was distinguished from ciguatoxin on a DEAE cellulose column, which did not adsorb scaritoxin from chloroform solution but did absorb ciguatoxin. The two toxins were also separable by thin layer chromatography, in which scaritoxin showed a higher Rf value than ciguatoxin. The two toxins had the same pharmacological properties (6). Scaritoxin was suggested to be the less polar entity of the two interconvertible forms of ciguatoxin (7). The hypothesis, however, remains to be verified. The toxin has been detected in some snappers, but the amount was never as significant as in parrotfish (8).

Assays of ciguatoxin. Determination of ciguatoxin levels in fish was carried out in many laboratories by mouse assays. Enzyme immunoassay to screen inedible fish has been proposed by Hokama (9). No specific chemical assay has been developed, as information on functional groups suitable for fluorescence labeling is not available. Analyses conducted in the authors' laboratory on remnant fish retrieved from patients' meals indicated that ciguatoxin content as low level as 1 ppb could cause intoxication in adults. An extremely high sensitivity and a sophisticated pretreatment method will be required for designing a fluorometric determination method for the toxin.

Maitotoxin (MTX). The toxin was first detected in the surgeonfish *Ctenochaetus striatus* and thus bears the Tahitian name of the fish, maito (10). The toxin probably explains the epidemiological observation of the different symptomatology in patients intoxicated by surgeonfish (4). Subsequently the origin of the toxin was identified as the dinoflagellate *Gambierdiscus toxicus*, and the toxin was isolated from cultures of the organism as shown in Figure 1.

Maitotoxin judged as pure by TLC and HPLC was obtained as a colorless solid (11); $[\alpha]_D^{21}$ +16.8 (c 0.36, MeOH-H$_2$O 1:1); UV$_{max}$ (MeOH-H$_2$O, 1:1) 230 nm (ϵ 9600); mouse lethality, 0.13 μg/kg (i.p.); soluble in H$_2$O, MeOH, and dimethylsulfoxide, practically insoluble in CHCl$_3$, acetone and MeCN. The toxin reacted positively to Dragendorff's reagent, but not to ninhydrin reagent. In FAB mass spectra (Figure 2) a fragment ion shows up at 3299 (M-SO$_3$Na$_2$+H))$^-$, suggesting that the toxin is disulfated compound. Determination of SO$_4^{2-}$ liberated by solvolysis also supported the presence of two sulfate ester groups in the molecule. The molecular weight of the toxin as disodium salt was thus estimated to be 3,424.5±0.5. ^{13}C NMR spectra of the toxin (Figure 3) indicated the presence of 160±5 carbon signals. Twenty-one methyls, about 36 methylenes, and five methines were observed in the aliphatic region. No quaternary carbon appeared in this range. In the region for oxygenated carbons, one methylene, approximately 74 methines, and fifteen quaternary carbons were observed. Eight olefinic carbons were observed. Absence of signals assignable to acetal/ketal or to carbonyl suggested that MTX has no repeating units, such as amino acids or sugars. Maitotoxin presumably has no side chains other than methyls or an exomethylene, nor does it have any carbocyles, since all 22 trialkylated carbons are accounted for by 5 methyl doublets, 15 methyl singlets on oxygenated carbons, one singlet methyl on an aliphatic carbon, and one exomethylene. The presence of many hydroxyls and ether rings is reminiscent of palytoxin (Figure 4). However, MTX exceeds palytoxin in the number of carbons, ether rings, and quaternary methyls.

MTX exceeds any other marine toxins known in the mouse lethality (0.13 μg/kg) and is 80 times more potent than commercial saponin (Merck) in hemolytic activity. According to Terao (12), MTX induced severe pathomorphological change in the stomach, heart and lymphoid tissues in mice and rats by i.p. injection of 200

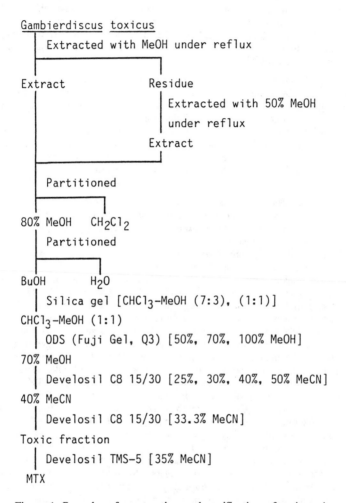

Figure 1. Procedure for extraction and purification of maitotoxin.

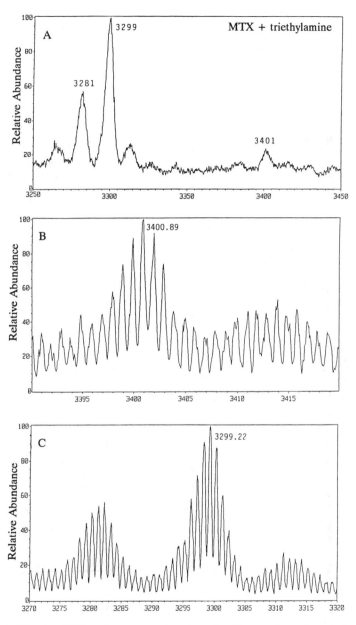

Figure 2. Negative FAB mass spectra of maitotoxin. The numbers denote the mass number at the centroid of each peak. A. A survey scan at a low resolution (R=300). B. Resolution enhanced spectrum (R=3000) for ion clusters at around *m/z* 3300. C. Resolution enhanced spectrum (R=3000) for ion clusters at around *m/z* 3400.

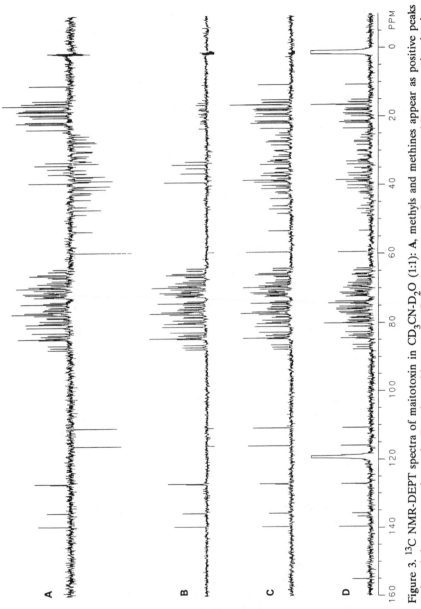

Figure 3. ^{13}C NMR-DEPT spectra of maitotoxin in CD_3CN-D_2O (1:1): A, methyls and methines appear as positive peaks and methylenes as negative peaks; B, only methines appear; C, no quaternary carbons appear; and D, a conventional noise-decoupled spectrum.

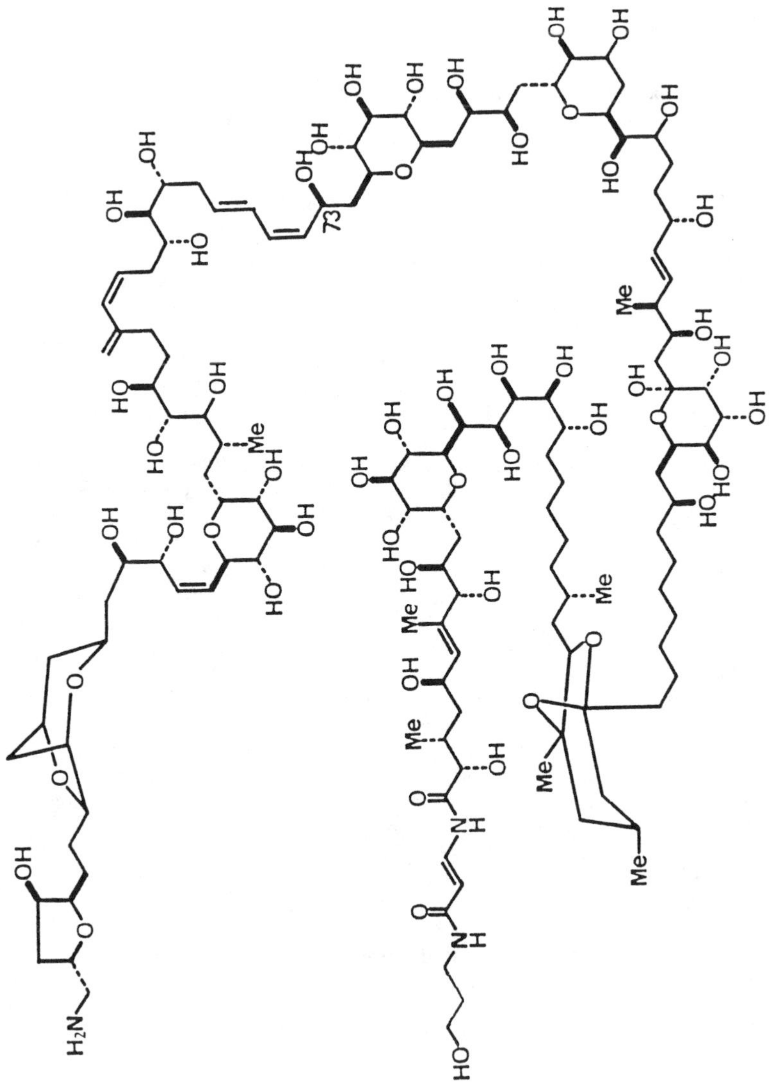

Figure 4. Structure of palytoxin.

or 400 ng/kg body weight. The toxic effects, especially those on immune systems, point to the necessity of paying more attention to a potential hazard due to continuous intakes of the toxin, even at subintoxication levels. The specific action of MTX to increase Ca^{2+} influx through cell membranes made the toxin a unique tool for investigating biological phenomena induced by Ca^{2+} influx, as described by Ohizumi in this book.

Palytoxin Poisoning

Palytoxin isolated from zoanthids *Palythoa* spp. is known to be extremely lethal (0.5 μg/kg, mouse, i.p.) and to have the most complicated natural product structure (Figure 4) ever elucidated *(13,14)*. The toxin was later revealed to be a potent tumor promoter *(15)*. Yet, health risks due to the toxin have remained unclear, as the zoathids are the most unlikely organisms to be regarded as foodstuff. Recently, however, data are being compiled to indicate the wide distribution of this toxin among marine biota.

Distribution. The first indication of the occurrence of palytoxin in fish was presented in 1969 *(16)*. The filefish *Altera scripta* belonging to family Monacathidae was traditionally known in Okinawa, Japan, to contain a toxic substance in the gut and, thus, to kill pigs when fed to them. The presence of fragments of *Palythoa* sp. in the guts and the resemblance in solubility between the fish toxin and palytoxin led the authors to a conclusion that the toxic principle in the filefish viscera was palytoxin. Incidence of human intoxication due to eating the filefish was not confirmed.

Trigger fishes of the family Balistidae are common in tropical shores and eaten in many areas. Their large and oily livers are often esteemed as specialities among fishermen. Yet, potential dangers of eating the fishes frequently appear in the literature and their sale is prohibited in several places *(17–20)*. Although the poisoning caused by trigger fishes has been regarded as a form of ciguatera *(18)*, the acuteness and severity of the poisoning *(19,21)* suggested the toxin to be different from ciguatoxin. In a subsequent study, the trigger fish *Melichthys vidua* was proven to contain significant amounts of palytoxin, or one of its congeners, in the flesh and the viscera *(22)*. The viscera was much higher than the flesh in toxin content. The parrotfish *Ypsiscarus ovifrons* inhabiting the western part of Japan occasionally causes severe intoxication when the livers are eaten. The causative toxin was indistinguishable from palytoxin in chromatographic and pharmacological properties *(23)*.

Occasional outbreaks of fatal crab poisoning have been known to occur in the Philippines *(24)* and Singapore *(25)*. Examination of three xanthid crabs, *Lophozozymus pictor*, *Demania alcalai*, and *D. reynaudii*, resulted in identification of palytoxin as the causative toxin *(26,27)*. The toxin contents in these crabs were extraordinarily high, amounting to 13 mg in one specimen. All tissues and organs of the crabs were toxic, but high toxicity was observed in gills, livers, gonads, and carapaces.

Two palytoxin analogs were isolated form a red alga, *Chondria armata* *(28)*, which also produces domoic acid and thus was used in the past as an folk medicine against helminths *(29)*. The yield of the two analogs was about one milligram each from 400 g of dried alga. No incidence connecting human intoxication with the alga is known. These finding suggest diverse sources and multiple links in the food chain of palytoxin.

Determination of palytoxin. A mouse bioassay method proposed by Teh and Gardiner *(25)* for determination of *L. pictor* toxin (=palytoxin) was useful in our studies on the palytoxin-containing animals. The toxin amount in mouse units (MU, equivalent to 10 ng of the toxin) was calculated from the equation $Y = 225.19X^{-0.99}$, where Y and X denote MU and death time in minutes, respectively. Purification of the toxin followed the procedure used by Uemura et al. *(30)*. Identification of small amount of the toxin was facilitated by TLC *(22)* and by HPLC on reversed phase and gel permeation columns *(26)*.

Diarrhetic Shellfish Poisoning

Diarrhetic shellfish poisoning is an illness caused by polyether toxins produced by dinoflagellates and accumulated in shellfish *(31)*. Patients suffer from diarrhea, nausea, and stomach pain but recover within three days without serious aftereffects. Despite the relatively mild symptoms, careful attention should be paid to the occurrence of okadaic acid analogs because of their potent tumor promoting activity *(32)*. Economic damage to the shellfish industry is also grave, as harvesting shellfish is banned during the infestation period, which may last for a half year. The disease has a worldwide distribution but is most prevalent in Europe and Japan, where shellfish culture is extensively carried out.

Toxic constituents. Three groups of toxins comprising 11 polyether compounds have been isolated from digestive glands of the scallop *Patinopecten yessoensis* collected in Japan during infestation period. Structures of eight components (Figure 5) have been determined. The first group includes okadaic acid and its derivatives. Okadaic acid (OA) is a polyether fatty acid previously isolated from two sponges, *Halichondria okadaii* and *H. melanodocia* *(33)*. Dinophysistoxin-1 (DTX1 = 35-methylokadaic acid) was named after the causative dinoflagellate, *Dinophysis fortii* *(34,35)*. Pectenotoxin-1, -2, -3 and -6 (PTX1-3, PTX6) share the same macrolide skeleton. The relative stereochemistry of PTX1 was determined by x-ray analysis *(36)* and the structure of other analogs, which differ from each other only in the functionality at C-43, were determined by comparison of NMR data *(37,38)*. Pectenotoxin-4 and -5 were isolated in pure forms, but attempts for structural determination were unsuccessful due to the small amount of the samples. Yessotoxin (YTX) consists of contiguously transfused ether rings *(39)*, which partly resemble the brevetoxins *(40,41)*, isolated as potent ichthyotoxins from a dinoflagellate *Ptychodiscus brevis* (=*Gymnodinium breve*). Yet, the toxin is distinct in having a terminal side chain of nine carbons, and two sulfate esters and in lacking carbonyl groups.

Not all these polyether occur together in the same shellfish samples. OA was the major toxin in the mussel specimens from most of the European countries *(42)*, while DTX1 was the major toxin in mussel in Japan and in Sogndal, Norway *(43)*. Scallops in Japan show the most complicated toxin profile. Furthermore, the relative ratio of the toxins varied regionally, seasonally, and annually. Pectenotoxins were detected, however, only in Japanese shellfish. Distribution of toxins is summarized in Table I.

Causative organism. Previous study on shellfish toxicity and dinoflagellate species occurring in the environment indicated a good correlation between the population density of *Dinophysis fortii* and shellfish toxicity. Toxicity tests on plankton samples obtained by a size-selecting method also gave a good correlation between the amount of *D. fortii* cells and the toxicity *(35)*. These results strongly suggested that *D. fortii*, was the source of toxins. Because all the *Dinophysis* species have been uncultivable under laboratory conditions, further confirmation of the toxigenicity of suspected species was not possible until a fluorometric HPLC determination method, which will be described later, was developed. The high sensitivity of the method enabled us to determine toxins in small numbers of *Dinophysis* cells purified

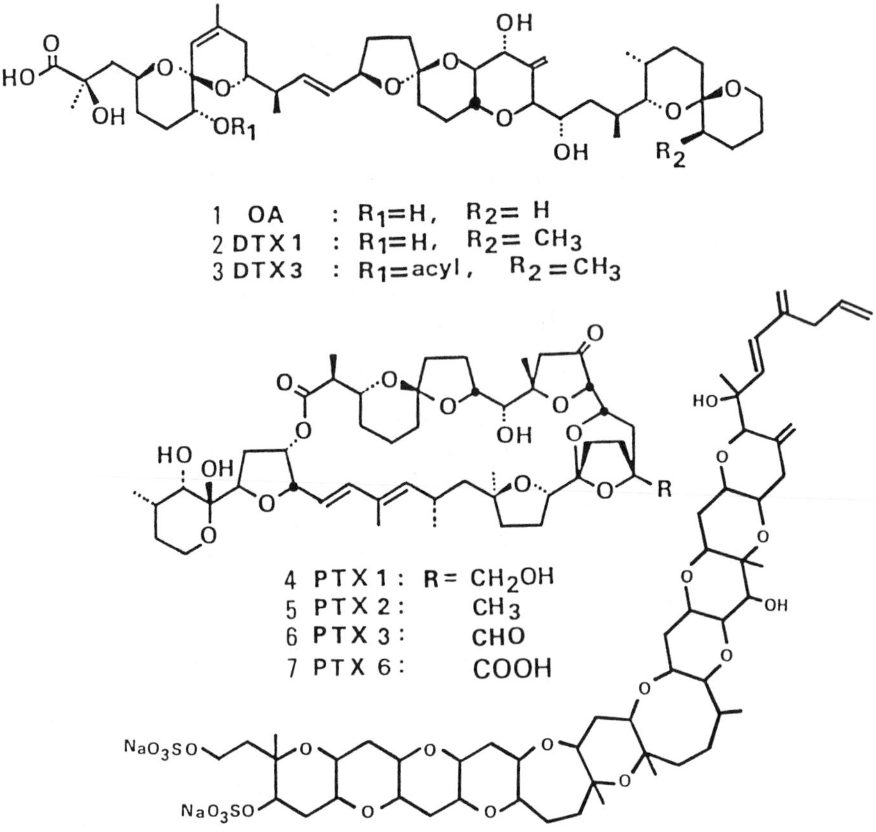

1 OA   : R₁=H,   R₂= H
2 DTX1 : R₁=H,   R₂= CH₃
3 DTX3 : R₁=acyl, R₂ = CH₃

4 PTX 1 : R = CH₂OH
5 PTX 2 :   CH₃
6 PTX 3 :   CHO
7 PTX 6 :   COOH

8 YTX

Figure 5. Diarrhetic shellfish toxins.

Table I. Distribution of Diarrhetic Shellfish Toxins

Toxin	Shellfish	Area	Toxicity (μg/kg) (LD_{99}, mouse, i.p.)
okadaic acid	mussel	Netherlands	200
	mussel	France	
	mussel	Sweden	
	mussel	Spain	
	mussel	Norway (south)	
dinophysistoxin-l	mussel	Japan	160
	mussel	Norway (Sogndal)	
	scallop	Japan	
dinophysistoxin-3	scallop	Japan	ca. 500
pectenotoxin-1-6	scallop	Japan	160-770
yessotoxin	mussel	Norway (Sogndal)	100
	scallop	Japan	

under a microscope by capillary manipulation. The following species were confirmed to produce OA and/or DTX1: *D. fortii. D. acuminata, D. acuta, D. norvegica,* and *D. mitra* (Lee and Yasumoto, unpublished data). Worthy of note is that *D. fortii* produced DTX1 in northern areas of Japan but that in southern areas produced no detectable amount of DTX1. Similar intraspecies variations in toxin profiles may occur in other species. *D. fortii* produced PTX2, but not other PTXs suggesting that a series of oxidations at C-43 of PTX2 takes place in shellfish. The origin of YTX is unknown at present.

Prorocentrum lima also produces OA and its derivatives (*44,45*) and is distributed in mussel farming areas (Fraga, Lee and Yasumoto, unpublished data). Contribution of *P. lima* to mussel toxicity awaits further studies.

Toxicology. Diarrheagenicity of OA and its derivatives was first demonstrated by Hamano et al. (*46*). Terao et al. (*47*) showed that DTX1 caused severe injuries on the intestinal mucosa, while PTX1 caused acute liver injuries. Similar observation was also made by Ishige et al. (*48*). Another important aspect of OA and DTX1 is their potent tumor promoting activity (*32*). Attention should be paid to continuous uptake of subacute levels of OA and DTX1 through seafoods. Toxicological effects of YTX, the latest member of the diarrhetic shellfish toxins, has not been elucidated.

Determination methods. The official regulation in Japan requires the toxin level in shellfish to be below 5 MU per 100 g meat (*49*). The method defines the toxicity by lethality and thus cannot distinguish toxins of different toxicological effects. Recently a fluorometric HPLC method for determination of OA and DTX1 was established (*50*) by labeling the carboxylic group of the toxin with 9-anthryldiazomethane (ADAM, Funakoshi Yakuhin Co.). Another fluorescent labeling reagent, l-anthroylnitrile (Wako Pure Chemicals Co.) seems to be a promising reagent to derivatized the primary hydroxyl group of PTXl (Lee et al., unpublished data). Typical examples of chromatograms are shown in Figure 6.

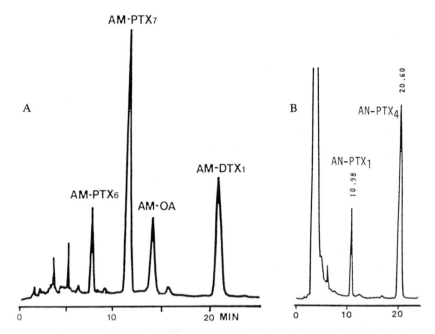

Figure 6. Fluorometric HPLC of diarrhetic shellfish toxins: A, AM (9-anthrylmethyl) esters of constituents which have a carboxylic acid; column, Develosil ODS (Nomura Chem.); solvents, $CH_3CN-CH_3OH-H_2O$ (8:1:1); monitor ex. 365, em. 412 nm; and B, AN (anthroyl) esters of PTX1 and PTX4; column and solvents were the same as those in A; monitor ex. 365, em. 465.

Discussion

A number of polyether toxins are accumulated by fish and shellfish through the food chain. The diversity of the structures and the toxicological effects of the toxins are apparent. In investigating these toxins, the priority should be given to the structural elucidation. The knowledge of structures will help design highly sensitive and specific determination methods such as fluorometric HPLC. Development of immunoassay methods also relies on the information of functional groups suitable for preparing toxin-protein conjugates to be used as antigens. Once sensitive analytical methods are established, they will be useful not only for monitoring toxin levels in seafoods but for surveying the distribution of the toxins in various seafoods and for identifying their origins. Collaboration among research groups of different disciplines and among those in different regions seems indispensable for further progress in seafood toxin research.

Literature Cited

1. Tachibana, K. Ph. D. Thesis, University of Hawaii, Honolulu, 1980.
2. (a) Legrand, A.M.; Litaudon, M.; Genthon, J.N.; Bagnis, R. *J. Appl. Phycol.* **1989**, in press. (b) Murata, M.; Legrand, A.M.; Tasumoto, T. *Tetrahedron Lett.*, in press.
3. Nukina, M.; Kuroyanagi, L. M.; Scheuer, J. P. *Toxicon,* **1984**, *22*, 169-76.
4. Bagnis, R.; Loussan, E.; Thevenin, S. *Med. Trop. ,* **1974**, *34*, 523- 527.

5. Chungue, E.; Bagnis, R.; Fusetani, N.; Hashimoto, Y. *Nippon Suisann Gakkaishi,* **1977**, *15,* 89-93.
6. Tatsumi, M.; Kajiwara, A.; Yasumoto, 7.; Ohizumi, Y. *J. Pharmacol. Exp. Ther.,* **1985**, *235,* 783-787.
7. Joh, Y-G.; Scheuer, P. J. *Marine Fisheries Review,* **1986**, *48,* 19-22.
8. Hashimoto, Y.; Kamiya, H.; Konjo, K.; Yoshida, C. *Nippon Suisan Gakkaishi.* **1975**, *41,* 903-905.
9. Hokama, Y.; Abad, M. A.; Kimura, L. H. *Toxicon,* **1983**, *21,* 817-824.
10. Yasumoto, T.; Bagnis, R.; Vernoux, J. P. *Nippon Suisan Gakkaishi* **1976**, *42,* 359-365.
11. Yokoyama, A.; Murata, M.; Oshima, Y.; Iwashita, T.; Yasumoto, T. *J. Biochem.,* **1988**, *104,* 184-187.
12. Terao, K.; Ito, E.; Sakamaki, Y.; Igarashi, K.; Yokoyama, A.; Yasumoto, T. *Toxicon,* **1988**, *26,* 395-402.
13. Uemura, D.; Ueda, K.; Hirata, Y.; Naoki, H.; Iwashita, T. *Tetrahedron let.* **1981**, *22,* 2781-8⁴.
14. Moore, R. E.; Bartolini, G. J. Am. Chem. Soc., **1981**, *103,* 2419- 94.
15. Fujiki, H.; Suganuma, M.; Nakayasu, M.; Hakii, H.; Horiuchi, T.; Takayama, S.; Sugimura, T. *Carcinogenesis,* **1986**, *7,* 707-710.
16. Hashimoto, Y.; Fusetani, N.; Kimura, S. *Nippon Suisan Gakkaishi,* **1969**,35, 1086-93.
17. Bagnis, R.; Mazllier, P.; Bennett, J.; Christian, E. In *Fishes of Polynesia;* les Education du Pacifique, Papeete, Tahiti, **1982**, p 210.
18. Halstead, B. W. In *Poisonous and Venomous Marine Animals;* United States Government Office, Washington D. C., 1967, p 124 and p 215.
19. Herre, A. W. C. T. *Philippine J. Sci.* 1924, *25,* 422-439.
20. Tinker, S. W. *Fishes of Hawaii;* Hawaiian Service Inc., Honolulu, **1978**, p 470.
21. Bagnis, R. *Clin. Toxic.* **1970**, *3,* 579.
22. Fukui, M.; Murata, M.; Inoue, A.; Gawel, M.; Yasumoto, T. *Toxicon,* **1987**, 25, 1121-1124.
23. Noguchi, T.; Hwang, D-H.; Arakawa, O.; Daigo, K.; Sato, S.; Ozaki, H.; Kawai, N.; Ito, M.; Hashimoto, K. In *Progress in Venom and Toxin Research;* Gopalakrishnakone, P.; Tan, C. K. Ed.; Faculty of Medicine, National University of Singapore: Singapore, 1987, 325-335.
24. Gonzales, R. B.; Alcala, A. C. *Toxicon,* **1977**, *12,* 169-170.
25. Teh, Y. F.; Gardiner, J. E. *Toxicon,* **1974**, *12,* 603-601.
26. Yasumoto, T.; Yasumura, D.; Ohizumi, Y.; Takahashi, M. *Agric. Biol. Chem.* **1986**, *50,* 163-167.
27. Alcala, A. C.; Alcala, L. C.; Garth, J. S.; Yasumura, D.; Yasumoto, T. *Toxicon,* **1988**, *26,* 105-107.
28. Maeda, M.; Kodama, T.; Tanaka, T.; Yoshizumi, H.; Nomoto, K.; Takemoto, T.; Fujiki, T. Symposium papers, 27th Symp. Chem. Nat. Prod. Oct., 1985, p 616-623. *Chem. Abstr. 104* 183260t.
29. Daigo, K. *Yakugaku Zasshi,* **1959**, *79,* 350-360.
30. Uemura, D.; Hirata, Y.; Iwashita, T.; Naoki, T. *Terahedron.* **1985**, *41,* 1007-1111.
31. Yasumoto, T.; Murata, M.; Oshima, Y.; Matsumoto, G.K.; Clardy, J. In *Seafood Poisoning,* ACS Symposium Series 262; Ragelis, E.P. Ed.; American Chemical Society: Washington, DC, 1984, 208-214.
32. Suganuma, M.; Fujiki, H.; Suguri, H.; Yoshizawa, S.; Hirota, M.; Nakayatsu, M.; Ojika, M.; Wakamatsu, K.; Yamada, K.; Sugimura, T. *Proc. Natl. Acad. Sci. USA,* **1988**, *85,* 1768-1771.
33. Tachibana, K.; Scheuer, P.J.; Tsukitani, Y.; Kikuchi, H.; Engen, D.V.; Clardy, J.; Gopichand, Y.; Schmitz, F.J. *J. Am. Chem. Soc.,* **1981**, *103,* 2469-71.

34. Murata, M.; Shimatani, M.; Sugitani, H.; Oshima, Y.; Yasumoto, T. *Nippon Suisan Gakkaishi,* **1982**, *48,* 549-522.
35. Yasumoto, T.; Oshima, Y.; Sugawara, W.; Fukuyo, Y.; Oguri, H.; Igarashi, T.; Fujita, N. *Nippon Suisan Gakkaishi,* **1980**, *46,* 14051411.
36. Yasumoto, T.; Murata, M.; Oshima, Y.; Sano, M.; Matsumoto, G.K.; Clardy, J. *Tetrahedron,* **1985**, *41,* 1019-25.
37. Murata, M.; Sano, M.; Iwashita, T.; Naoki, H.; Yasumoto, T. *Agric. Biol. Chem.* **1986**, *50,* 2693-95.
38. Lee, J.S.; Murata, M.; Yasumoto, T. Abstracts of Papers, Annual Meeting of Jpn. Soc, Sci. Fish. **1988**, p 81.
39. Murata, M., Kumagai, M.; Lee, J.S.; Yasumoto, T. *Tetrahedron Lett.* **1987**, *28,* 5869-72.
40. Lin, Y.Y.; Risk, M.; Ray, S.M.; Engen, D.V.; Clardy, J.; Golik, J.; James, J.C.; Nakanishi, K. *J. Am. Chem. Soc.,* **1981**, *103,* 6773-75.
41. Shimizu, Y.; Chou, H.N.; Bando, H.; Duyne, V.; Clardy, J. *J. Am. Chem. Soc.,* **1986**, *108,* 514-515.
42. Kumagai, M.; Yanagi, T.; Murata, M.; Yasumoto, T.; Lassus, R.; Rodoriquez-Vazquez, J.A. *Agric. Biol. Chem.,* **1986**, *50,* 2853-57.
43. Lee, J.S.; Tangen, K.; Yasumoto, T. *Nippon Suisan Gakkaishi,* **1988**, *54,* 1953-57.
44. Murakami, Y.; Oshima, Y.; Yasumoto, T. *Nippon Suisan Gakkaishi,* **1982**, 69-72.
45. Yasumoto, T.; Seino, N.; Murakami, Y.; Murata, M. *Biol. Bull.,* **1987**, *172,* 128-131.
46. Hamano, Y.; Kinoshita, Y.; Yasumoto, T. *J. Food Hyg. Soc. Jpn.,* **1986**, *27,* 375-379.
47. Terao, K.; Ito, E.; Yanagi, T.; Yasumoto, T. *Toxicon,* **1986**, *24,* 1141-51.
48. Ishige, M.; Satoh, N.; Yasumoto, T. *Hokkaido Eiseikenkyushoho,* **1988**, *38,* 15-19.
49. Yasumoto, T. *Shokuhin Eisei Kenkyu,* **1981**, *31,* 515-522.
50. Lee, J.S.; Yanagi, T.; Kenma, R.; Yasumoto, T. *Agric. Biol. Chem.,* **1987**, *51,* 877-881.

RECEIVED May 30, 1989

Chapter 9

Ca-Dependent Excitatory Effects of Maitotoxin on Smooth and Cardiac Muscle

Yasushi Ohizumi and Masaki Kobayashi

Mitsubishi Kasei Institute of Life Sciences, 11 Minamiooya,
Machida, Tokyo 194, Japan

Maitotoxin (MTX), a principal toxin of ciguatera seafood poisoning
and known as the most potent marine toxin, caused concentration-
dependent vasoconstrictive, cardiotonic and cardiotoxic effects on rabbit aorta or guinea pig and rat cardiac tissues. These effects were
markedly suppressed by verapamil or divalent cations, but were little
affected by tetrodotoxin, various receptor blockers, or reserpine. In
the isolated or cultured rat myocytes, MTX produced various
arrhythmic contractions and this action of MTX was abolished by
verapamil. MTX increased the tissue Ca content, ^{45}Ca uptake by cultured cardiac myocytes, and intracellular free Ca^{2+} concentration of
isolated myocardial cells. The enzyme activities of cardiac Na^+, K^+-
ATPase, cyclic AMP phosphodiesterase, and sarcoplasmic reticulum
Ca^{2+}-ATPase were uninfluenced by MTX. The electrophysiological
experiments performed by whole-cell patch-clamp method showed that
MTX did not increase the usual Ca channel current of a guinea pig
cardiac myocyte. Instead, MTX produced a steadily flowing current,
which was abolished by Cd^{2+}. The voltage dependence curve for the
MTX-induced current was almost linear. These results suggest that
MTX directly increases a current flowing through the cardiac muscle
membrane which is carried by Ca^{2+}. This could elevate the intracellular free Ca^{2+} concentration and thus cause excitatory effects on
smooth and cardiac muscles.

A number of marine toxins such as tetrodotoxin (1), saxitoxin (2), ciguatoxin (3,4),
and sea anemone toxins (5-7) have been studied extensively because of their pharmacological action on specific channel sites on the cell membrane. In tropical or
subtropical regions, ciguatera is well known to be a seafood poisoning caused by
ingesting a variety of poisonous reef fishes. Ciguatera poisoning produces various
symptoms such as cardiovascular, gastrointestinal, sensory, and motor disturbances.
Ciguatoxin (8), a lipophilic substance, and maitotoxin (MTX) (9), a more polar
compound, have been isolated from ciguateric poisonous fish as principal toxins.
MTX is the most potent marine toxin known with the minimum lethal dose of
0.17 $\mu g/kg$ i.p. in mice, and is believed to be produced by toxic dinoflagellate *Gambierdiscus toxicus* and transmitted to reef fishes through the food chain. The chemical structure of MTX has been only partially determined, although it is considered
to be a non-peptidic substance having a large molecular weight (9). We have
shown that MTX produces a Ca^{2+}-dependent release of norepinephrine from a rat
pheochromocytoma cell line PC12 (10,11) and Ca^{2+}-dependent contraction of

0097–6156/90/0418–0133$06.00/0
© 1990 American Chemical Society

smooth muscle (12,13) and cardiac muscle (14,15). It has also been reported that MTX may exhibit excitatory effects on voltage-sensitive Ca channels of insect skeletal muscle (16), cardiac muscle (17) and cultured neuronal cells (18). Here we report the mechanism of excitatory action of MTX on smooth and cardiac muscles.

Methods

Mechanical Response. Rabbit aorta, guinea pig left and right atria and rat ventricle strips were excised and mounted vertically in an organ bath containing a Krebs-Ringer bicarbonate solution (aorta 37°C, others 30°C). Left atria and ventricle strips were electrically stimulated at a frequency of 2 Hz by rectangular pulses of 5 msec at supramaximal intensities. Isometric contractions were measured with a force-displacement transducer. Isolated rat myocardial cells were prepared by collagenase digestion as described previously (19) and were driven by a stimulator through a pair of platinum electrode. Rat cultured cardiac myocytes were prepared according to the method reported (20). The beating activity of isolated or cultured myocytes was observed under a phase contrast microscope equipped with a thermostatically controlled stage (37°C) and was recorded either with a video recording system or with a high-speed movie camera.

Ca^{2+} Movements. Rabbit aorta were cut into helical strips and suspended in the Krebs-Ringer solution. Guinea pig left atria were cut in half, suspended in the same solution and stimulated electrically. After a 30-min incubation, the tissues were blotted, weighed and then ashed with perchloric acid in a quartz tube at 500°C for 7 hr. The amount of Ca in the solution of ashed samples was determined with an atomic absorption spectrophotometer. The ^{45}Ca uptake of cultured cells was measured by the method previously described (10). The reaction was initiated by adding the assay medium containing ^{45}Ca and stopped by removing the medium after the desired incubation period at 37°C. The cells were washed quickly 3 times with the same medium, solubilized with Triton X-100 and used for scintillation counting. The intracellular free Ca^{2+} concentration in isolated cells was measured as described previously (21). The cells were suspended in the assay medium containing Quin 2 acetoxymethyl ester and incubated at 30°C for 1 hr. The suspension of Quin 2-loaded cells in a quartz cuvette was placed in a spectrofluorometer equipped with a magnetic stirrer and a temperature controller (37°C). Monochromater settings were 339 nm for excitation and 500 nm for emission. The fluorescence signal was converted to the intracellular free Ca^{2+} concentration using a Quin 2-Ca^{2+} dissociation constant of 115 nM.

Enzyme Assay. Na^+, K^+-ATPase, and sarcoplasmic reticulum Ca^{2+}-ATPase were prepared from rat hearts (22) and dog hearts (23), respectively. Bovine heart cyclic AMP phosphodiesterase was purchased from Sigma. The enzyme reaction was carried out after 5-min pretreatment with the drug, and the amount of inorganic phosphate liberated during the reaction period was determined.

Electrophysiological Experiments. Guinea pig myocardial cells prepared as described previously (24) were superfused at 37°C with a Tyrode solution. Electrical properties of the myocytes were examined by the patch-clamp methods (25) using fire-polished pipettes. The current was measured by means of a patch-clamp amplifier, stored on the tape through a digital PCM data recording system, and analyzed with a computer.

Results

Mechanical Response. At concentrations above 10^{-10} g/mL, MTX caused a sustained contraction in the rabbit aorta. The MTX-induced contraction lasted for at least 3 hr. As shown in Figure 1, the contractile response was increased with MTX concentrations in the range between 10^{-10} to 3×10^{-8} g/mL in a dose-dependent manner. The concentration–contractile response curve for MTX was shifted to the right in a parallel manner by a Ca channel blocker, verapamil (10^{-6} M), (Figure 1) and was slightly shifted to the right by an α-adrenorecptor blocker, phentolamine (10^{-6} M), whereas the curve was not affected by a Na channel blocker, tetrodotoxin (10^{-6} M), a histamine receptor blocker, chlorpheniramine (10^{-6} M), a serotonin receptor blocker, methysergide (10^{-6} M), or a prostaglandin synthesis inhibitor, indomethacin (3×10^{-6} M). As shown in Figure 2, the MTX (10^{-8} g/mL)-induced contraction of the aorta increased in a linear fashion with increasing concentration of external Ca^{2+} from 0.03 to 1.2 mM. Following incubation in a Ca^{2+}-free solution, the contractile responses of the aorta induced by re-introduction of Ca^{2+} were potentiated by the presence of MTX (10^{-8} g/mL). After treatment with verapamil (10^{-6} M) or $MgCl_2$ (10 mM), the concentration–contractile response curve for Ca^{2+} in the presence of MTX was shifted to the right in a parallel manner, indicating competitive antagonism. On the other hand, the concentration–contractile curve for Ca^{2+} in the presence of a Ca^{2+} ionophore A23187 (3×10^{-5} M) was totally unaffected by verapamil (10^{-6} M). At lower concentrations of 10^{-10} to 4×10^{-9} g/mL, MTX caused a sustained cardiotonic effect on guinea pig left atria in a concentration-dependent manner (Figure 3). In this concentration range, the ceffect of MTX continued for over 1 hr without an increase in diastolic tension of the atria. In the rat ventricle strips, MTX also caused an increase in the developed tension at concentrations of 10^{-9} to 10^{-8} g/mL. The 50% effective concentration of MTX were 10^{-9} g/mL for the left atria and 2×10^{-9} g/mL for the ventricles, indicating the approximate equipotent effects of the toxin on both tissues. After treatment of spontaneously beating right atria of the guinea pig with MTX (3×10^{-10} to 4×10^{-9} g/mL), dose-dependent increases in the contractile force and heart rate were observed. The contractile response to MTX (2×10^{-9} g/mL) was markedly suppressed by the administration of Co^{2+} (2 mM), or verapamil (3×10^{-7} M) in the low-Ca^{2+} (0.3 mM) or normal (1.2 mM) solution. However, the elevation of the external Ca^{2+} concentration to 3 mM abolished the inhibitory effect of Co^{2+} or verapamil. Following treatment with tetrodotoxin (5×10^{-7} M), a β-adrenorecptor blocker, propranolol (10^{-6} M), or a catecholamine depleting drug, reserpine (2 mg/kg i.p.), MTX still profoundly increased the twitch tension of the atria.

The application of high concentrations (above 5×10^{-9} g/mL) of MTX caused a biphasic inotropic change and gradual rise in diastolic tension of the atria. Furthermore, MTX produced various arrhythmic movements and cardiac arrest in the guinea pig left and right atria (Figure 4).

In the electrically stimulated rat myocytes, MTX (10^{-8} g/mL) produced a time-dependent increase in the degree and the rate of longitudinal contractions. Then, the cellular motion was changed into an irregular beating, followed by gradual shortening. In spontaneously beating cultured rat cells, MTX at concentrations above 10^{-11} g/mL caused various abnormal cell movements such as arbitrary beating, after-contraction, and faint fibrillatory movements with a high frequency. As shown in Figure 5, the percentage of arrhythmically beating cells among total moving cells increased gradually after the application of MTX. The potency of this effect increased in a concentration-dependent manner. After the administration of MTX (3×10^{-10} g/mL), the beating rate of the cells increased to reach an almost constant level of about twice the control, as shown in Figure 6. When the cultured cells were pretreated with verapamil (10^{-7} M), the tachyarrhythmia induced by MTX was prevented.

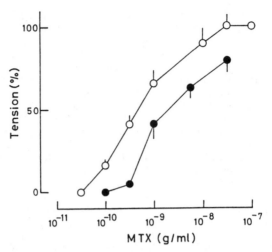

Figure 1. The log concentration–contractile response curves for maitotoxin (MTX) in the rabbit aorta in the presence (●) or absence (○) of verapamil (10^{-6} M). Verapamil was administered 15 min before the application of MTX. The maximum response to MTX (3 × 10^{-8} g/mL) is expressed as 100%. Vertical lines indicate the standard error of mean (n=7). (Reproduced with permission from Ref. 13. Copyright 1983 Press Syndicate of the University of Cambridge)

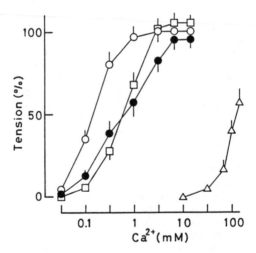

Figure 2. Effect of maitotoxin (MTX) on the log concentration–contractile response curve for Ca^{2+} in the presence or absence of verapamil or $MgCl_2$ in the rabbit aorta. Control (△), 10^{-8} g/mL MTX (○), 10^{-6} M verapamil plus 10^{-8} g/mL MTX (●), and 10 mM $MgCl_2$ plus 10^{-8} g/mL MTX (□). The aorta was incubated in a Ca^{2+}-free solution for 1 hr before the cumulative application of $CaCl_2$. Verapamil or $MgCl_2$ and MTX were added 45 and 30 min before the application of $CaCl_2$, respectively. The maximum response to norepinephrine (3 × 10^{-6} M) is expressed as 100%. Vertical lines indicate the standard error of mean (n=7). (Reproduced with permission from Ref. 13. Copyright 1983 Press Syndicate of the University of Cambridge)

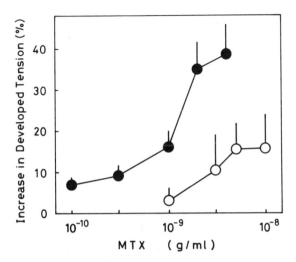

Figure 3. The log concentration–contractile response curves for maitotoxin (MTX) on the developed tension of guinea pig left atria (●) and rat ventricle strips (○). Vertical lines indicate the standard error of mean (n=6). (Reproduced with permission from Ref. 14. Copyright 1985 Macmillan)

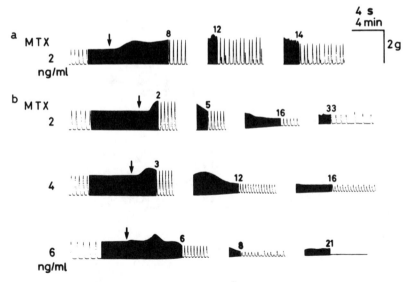

Figure 4. Effect of maitotoxin (MTX, 2-6 × 10^{-9} g/mL) on the contractile response of guinea pig left (a) and right (b) atria. MTX was administered at the arrow. Numbers above the tracing indicate the times after the application of MTX. The horizontal calibration indicates 4 min for the tracing recorded at a slower sweep speed and 4 sec for that recorded at faster speed. (Reproduced with permission from Ref. 20. Copyright 1987 Elsevier)

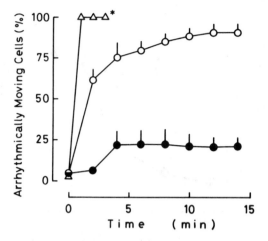

Figure 5. Effect of maitotoxin (MTX) on the contractile activity of cultured rat myocardial cells. The value (mean of three sets of experiments) is the percentage of arrhythmically beating myocytes among total moving cells (43-69 myocytes). MTX concentration used were 10^{-11} g/mL (o), 10^{-10} g/mL (●), and 10^{-9} g/mL (Δ). Vertical lines indicate the standard error of mean. The asterisk indicates that all the cells were arrested within 1 min. (Reproduced with permission from Ref. 20. Copyright 1987 Elsevier)

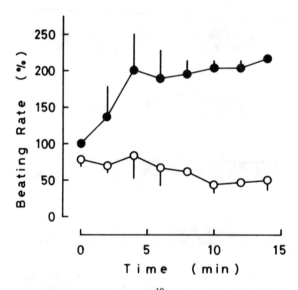

Figure 6. Effect of maitotoxin (MTX, 3 × 10^{-10} g/mL) on the beating rate of cultured rat myocardial cells in the absence (●) or presence (o) of verapamil (10^{-7} M). Verapamil was administered 5 min before the application of MTX. Vertical lines indicate the standard error of mean (n=5). (Reproduced with permission from Ref. 20. Copyright 1987 Elsevier)

Ca^{2+} **Movements.** In rabbit aorta, the tissue Ca content was increased by MTX (10^{-8} g/mL) from 1.45 ± 0.10 (control) to 1.90 ± 0.10 μmol/g wet weight, a 31% increase, under low-Ca^{2+} (0.3 mM) conditions, and this effect of MTX was inhibited 58% by verapamil (2×10^{-6} M). As shown in Figure 7, MTX caused a concentration-dependent increase in the tissue Ca content of guinea pig left atria at concentrations of 2×10^{-9} to 3×10^{-8} g/mL. The tissue Ca content continued to increase linearly during 120 min after the application of MTX (4×10^{-9} g/mL). In the normal solution MTX (4×10^{-9} g/mL) increased the tissue Ca content by about 49%. The MTX (4×10^{-9} g/mL)-induced increase in the Ca content was inhibited approximately 78% and 25% by treatment with Co^{2+} (2 mM) and verapamil (10^{-5} M), respectively. When the concentration of external Ca^{2+} was decreased to 0.3 mM, MTX (4×10^{-9} g/mL) elevated the Ca content 32% and this effect of MTX was suppressed about 86% and 93% by the administration of Co^{2+} (2 mM) and verapamil (10^{-5} M), respectively. In addition, even in Na^+-free medium the Ca content was clearly increased by MTX (4×10^{-9} g/mL).

To certify the important role of Ca^{2+} in the excitatory action of MTX, the effect of MTX on the Ca movements in cardiac muscle was examined at the cellular level. Figure 8 shows the time course of the ^{45}Ca influx in the presence or absence of MTX (10^{-9} g/mL). The ^{45}Ca uptake in the control experiment increased with time, to reach a saturation level about 5 min after administration of ^{45}Ca. When MTX (10^{-9} g/mL) and ^{45}Ca were applied simultaneously, the increase in ^{45}Ca uptake at 5 min was 31% larger than that of the control. Furthermore, when the intracellular Ca^{2+} concentration of isolated cardiac myocytes was determined from the Quin 2 fluorescence, MTX (10^{-9} g/mL) caused a marked increase in the free Ca^{2+} concentration from 122 ± 9 nM (control) to 380 ± 23 nM, as shown in Figure 9.

Effects on Enzyme Activities. MTX at concentrations up to 10^{-8} g/mL had no effect on the enzymes related to the possible mechanism of cardioexcitatory action, such as cardiac Na^+,K^+- ATPase, cyclic AMP phosphodiesterase, and sarcoplasmic reticulum Ca^{2+}-ATPase.

Electrophysiological Experiments. In order to clarify the mechanism of excitatory action of MTX, electrophysiological effects of MTX on isolated guinea pig cardiac myocytes were examined by means of whole-cell patch-clamp techniques. MTX did not increase the voltage-dependent Ca channel current. Instead, MTX produced a steadily flowing current in the cells superfused with Na^+-free, K^+-free Tyrode solution. The exposure to MTX (3×10^{-10} g/mL) caused a gradual inward shift of the holding current level at -80 mV. This inward current was abolished by treatment with Cd^{2+} (1 mM). The voltage dependence of the steady current was examined with a series of clamp steps applied to vary the holding potential from -80 to +40 mV. As shown in Figure 10, the current–voltage relation for the MTX-induced current was almost linear. In the K^+-free (replaced with Cs^+) Tyrode solution, the reversal potential of MTX (10^{-9} g/mL)-activated current was about -30 mV. When the Na^+ concentration was reduced 10 times, the reversal potential shifted by about 20 mV in the negative direction. Furthermore, when the Ca^{2+} concentration was raised 12 times, the reversal potential shifted by about 28 mV in the positive direction. By means of the constant field equation, assuming that Ca^{2+}, Na^+, and Cs^+ are the only charge carriers of MTX-activated current under these conditions, the permeability for this current of Ca^{2+} was estimated to be 50 and 14 times higher than those of Na^+ and Cs^+, respectively. These results suggest that the dominant charge carrier of this current is Ca^{2+}.

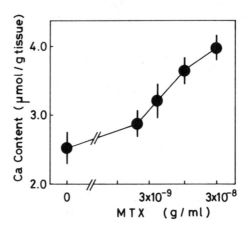

Figure 7. The log concentration–response curve for maitotoxin (MTX) on the tissue Ca content of guinea pig left atria. Vertical lines indicate the standard error of mean (n=6). (Reproduced with permission from Ref. 14. Copyright 1985 Macmillan)

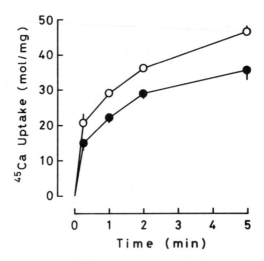

Figure 8. Effect of maitotoxin (MTX) on the time course of an increase in ^{45}Ca uptake of cultured rat cardiac myocytes. Control ($\bullet$), and 10^{-9} g/mL MTX ($\circ$). Vertical lines indicate the standard error of mean (n=3). (Reproduced with permission from Ref. 20. Copyright 1987 Elsevier)

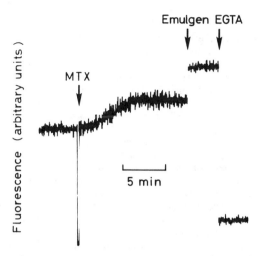

Figure 9. Typical fluorescence signals obtained from a suspension of isolated rat cardiac myocytes after the application of maitotoxin (MTX). The arrow indicates the addition of MTX (10^{-9} g/mL), a detergent Emulgen 810 (1%), which frees all vesicular Ca^{2+}, or EGTA (3.5 mM), a chelator that removes all free Ca^{2+} in the cuvette. The intensity of Quin 2 fluorescence is expressed in arbitrary units. (Reproduced with permission from Ref. 20. Copyright 1987 Elsevier)

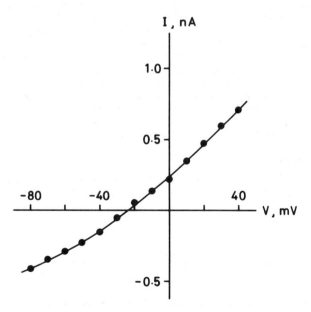

Figure 10. The current–voltage relationship of the MTX-induced steadily flowing current. Each value is the difference between the steady-state current before and 4 min after the administration of MTX (10^{-9} g/mL). (Reproduced with permission from Ref. 24. Copyright 1987 Macmillan)

Discussion

During recent decades there has been a dramatically growing awareness that the Ca^{2+} movements in the cell can play an essential role in many cellular events. From this point of view, substances that influence cellular Ca^{2+} movements provide valuable probes for the elucidation of various cellular functions.

MTX caused a contraction of vascular smooth muscle and positive inotropic, positive chronotropic and arrhythmogenic effects on cardiac muscle. The effect of MTX was little affected by various receptor blockers, a Na channel blocker or a catecholamine depleting agent. Further, MTX had no effect on the enzymes which were related to Ca^{2+} movements, such as Na^+, K^+-ATPase, cyclic AMP phosphodiesterase, and sarcoplasmic reticulum Ca^{2+}-ATPase. These results would eliminate the possible involvement of an indirect action elicited by the release of chemical mediators and direct modifications of their receptors, Na channels, or various enzymes as a major mechanism of action of MTX.

In the present experiments, MTX markedly increased the tissue Ca content, ^{45}Ca uptake, and intracellular free Ca^{2+} concentration of smooth or cardiac muscles. These Ca^{2+}-mobilizing effects of MTX as well as its vasoconstrictive, cardiotonic, and cardiotoxic effects were profoundly suppressed or abolished by Ca^{2+} entry blockers, polyvalent cations, or Ca^{2+}-free medium. It has been reported that MTX produces Ca^{2+}-dependent excitatory effects on neuronal (10,11,18) or pituitary (26) cells and smooth (12,13), cardiac (14,15,17), or skeletal (16) muscles, and that all these actions of MTX were antagonized by Ca antagonists or polyvalent cations. These observations suggest that the enhanced Ca^{2+} influx and the subsequent increase in cytoplasmic free Ca^{2+} concentration play a dominant role in the excitatory effects of MTX.

The electrophysiological experiments reported here and done with patch-clamp techniques support this idea. The external application of MTX to isolated cardiac myocytes caused a sustained inward current which was carried by Ca^{2+}. MTX did not increase the voltage-dependent Ca channel current, and both the time dependence and voltage dependence of the MTX-induced current were clearly different from those of the usual Ca channel current. These results suggest that the MTX-induced steady current is different from the usual voltage-dependent Ca channel current, and that this is possibly a current which flows through a new type of Ca^{2+}-permeable channel. The steady current described here may be responsible for the highly enhanced Ca influx induced by MTX and could account for the excitatory action of MTX on smooth and cardiac muscles.

Acknowledgments

We are grateful to Prof. T. Yasumoto of Tohoku University for generously supplying maitotoxin. We are indebted to Ms. Y. Murakami of this Institute for secretarial assistance.

Literature Cited

1. Narahashi, T. *Physiol. Rev.* **1974**, *54*, 813.
2. Catterall, W.A. *Annu. Rev. Pharmacol. Toxicol.* **1980**, *20*, 15.
3. Ohizumi, Y.; Shibata, S.; Tachibana, K. *J. Pharmacol. Exp. Ther.* **1981**, *217*, 475.
4. Ohizumi, Y.; Ishida, Y.; Shibata, S. *J. Pharmacol. Exp. Ther.* **1982**, *221*, 748.
5. Romey, G.; Abita, J.P.; Schweitz, H.; Wunderer, G.; Lazdunski, M. *Proc. Natl. Acad. Sci. U.S.A.* **1976**, *73*, 4055.
6. Ohizumi, Y.; Shibata, S. *Br. J. Pharmacol.* **1981**, *72*, 239.

7. Ohizumi, Y.; Shibata, S. *Am. J. Physiol.* **1982**, *243*, C237.
8. Scheuer, P.J.; Takahashi, W.; Tsutsumi, J.; Yoshida, T. *Science* **1967**, *21*, 1267.
9. Yasumoto, T.; Nakajima, I.; Oshima, Y.; Bagnis, R. In *Toxic Dinoflagellate Blooms*; Taylor, D.L.; Seliger, H., Eds.; Elsevier: New York, 1979; p 65.
10. Takahashi, M.; Ohizumi, Y.; Yasumoto, T. *J. Biol. Chem.* **1982**, *257*, 7287.
11. Takahashi, M.; Tatsumi, M.; Ohizumi, Y.; Yasumoto, T. *J. Biol. Chem.* **1983**, *258*, 10944.
12. Ohizumi, Y.; Kajiwara, A.; Yasumoto, T. *J. Pharmacol. Exp. Ther.* **1983**, *227*, 199.
13. Ohizumi, Y.; Yasumoto, T. *J. Physiol.* **1983**, *337*, 711.
14. Kobayashi, M.; Ohizumi, Y.; Yasumoto, T. *Br. J. Pharmacol.* **1985**, *86*, 385.
15. Kobayashi, M.; Miyakoda, G.; Nakamura, T.; Ohizumi, Y. *Eur. J. Pharmacol.* **1985**, *111*, 121.
16. Miyamoto, T.; Ohizumi, Y.; Washio, H.; Yasumoto, T. *Pflügers Arch.* **1984**, *400*, 439.
17. Legrand, A.M.; Bagnis, R. *J. Mol. Cell. Cardiol.* **1984**, *16*, 663.
18. Freedman, S.B.; Miller, R.J.; Miller, D.M.; Tindall, D.R. *Proc. Natl. Acad. Sci. U.S.A.* **1984**, *81*, 4582.
19. Kobayashi, M.; Kondo, S.; Yasumoto, T.; Ohizumi, Y. *J. Pharmacol. Exp. Ther.* **1986**, *238*, 1077.
20. Kobayashi, M.; Goshima, K.; Ochi, R.; Ohizumi, Y. *Eur. J. Pharmacol.* **1987**, *142*, 1.
21. Powell, T.; Tatham, P.E.R.; Twist, V.W. *Biochem. Biophys. Res. Commun.* **1984**, *122*, 1012.
22. Pitts, B.J.R.; Schwartz, A. *Biochim. Biophys. Acta* **1975**, *401*, 184.
23. Harigaya, S.; Schwartz, A. *Circ. Res.* **1969**, *25*, 781.
24. Kobayashi, M.; Ochi, R.; Ohizumi, Y. *Br. J. Pharmacol.* **1987**, *92*, 665.
25. Hamill, O.P.; Marty, A.; Neher, E.; Sakmann, B.; Sigworth, F.J. *Pflügers Arch.* **1981**, *391*, 85.
26. Schettini, G.; Koike, K.; Login, I.S.; Judd, A.M.; Cronin, M.J.; Yasumoto, T.; MacLeod, R.M. *Am. J. Physiol.* **1984**, *247*, E520.

RECEIVED June 8, 1989

Chapter 10

X-ray Crystallographic Studies of Marine Toxins

Gregory D. Van Duyne

Department of Chemistry, Baker Laboratory,
Cornell University, Ithaca, NY 14853

This chapter deals with single crystal x-ray diffraction as a tool to study marine natural product structures. A brief introduction to the technique is given, and the structure determination of PbTX-1 (brevetoxin A), the most potent of the neurotoxic shellfish poisons produced by *Ptychodiscus brevis* in the Gulf of Mexico, is presented as an example. The absolute configuration of the brevetoxins is established via the single crystal x-ray diffraction analysis of a chiral 1,2-dioxolane derivative of PbTX-2 (brevetoxin B).

Single crystal x-ray diffraction analysis has assumed an increasingly important role in the study of marine natural products in recent years. The various stages involved in determining a molecular structure from the information gathered in an x-ray diffraction experiment can be conveniently described with an analogy to light microscopy. With an ordinary microscope, visible light with wavelengths of a few microns is scattered by matter in a sample. In any given direction, the net scattering of light is the sum of the scattering in that direction from all the different parts of the sample. The resulting pattern of scattered radiation is called the diffraction pattern. In some directions, the scattered waves may be in phase, resulting in diffraction maxima, and in other directions the scattered radiation may be out of phase, resulting in diffraction minima. The diffracted light waves are collected by a lens and recombined to form a magnified image of the sample. The highest resolution that can be observed corresponds roughly to the wavelength of light used—a few microns or a few thousand Angstroms. In order to observe this resolution limit, the light waves of highest diffraction angle must be gathered by the lens.

In an x-ray diffraction analysis, the sample is a very special form of matter, a crystal, and the radiation used has a wavelength of about 1 Å. The x-rays are scattered by the electrons in the crystal and the intensities of the individual diffracted beams are measured. The scattering from all parts of the crystal contributes to a given diffraction direction. This is the stage where the analogy with light microscopy breaks down. There are no x-ray lenses to recombine the diffracted waves to form a magnified image of a molecule. Instead, the crystallographer uses a computer to try to recombine the waves and form a recognizable image of the electron density. The highest resolution that can be observed in the image is roughly 1 Å, which corresponds to the distances between atoms in a molecule. In order to observe 1 Å resolution, the x-ray waves of highest diffraction angle must be measured.

The data measurement in an x-ray analysis is typically performed using a computer controlled diffractometer. Except for the crystal mounting and some preliminary procedures, the measurement of intensities is fully automated. It is the next

0097–6156/90/0418–0144$06.50/0

stage, a series of numerical procedures ultimately leading to the electron density computation, that can often be difficult. The electron density and the diffraction pattern of a crystal are linked mathematically. If one is known completely, the other can be computed in straightforward manner. In an x-ray diffraction study, the diffraction pattern is the observed data and the electron density is the desired result. Unfortunately, the intensities of the diffracted x-ray beams represent only half of the information needed to calculate the electron density. The relative phases of the scattered waves are also required, but in general cannot be measured experimentally.

The most favorable case for solving this so-called "phase problem" includes compounds that crystallize in such a way that the crystal contains inversion centers relating one molecule to another. Since only racemic or achiral compounds can crystallize in this way, most natural products do not fall in this category. Another favorable case includes compounds that contain one or more atoms the size of sulfur or larger. Solving the phase problem is then broken into two steps—first the location of the heavy atoms, and then the lighter ones. Knowledge of the positions of the heavy atoms is of great use in approximating phases which lead to the remainder of the structure. While some naturally occurring compounds do contain heavy atoms, many do not, and it is these compounds that present the greatest difficulty in solving the phase problem. Some excellent reviews are available which explain the direct methods of crystallography, that is, the process of obtaining phases from the observed intensities (*1–3*).

Once the phase problem has been solved, the structure is improved by careful least squares fitting of the molecular model to the observed data. The x-ray diffraction analysis is complete after least squares refinement, producing as final results the coordinates and a description of the thermal motion for each of the atoms in the molecule. With the final coordinates available, bond lengths, bond angles, dihedral angles, tests for planarity, and a number of other geometric results are easily done. Perhaps the most impressive display of the results of an x-ray diffraction analysis is a plot of the molecule showing its three-dimensional structure, usually a much different representation than the common chemical line drawings.

No other technique can provide the large quantity of precise data that is available from an x-ray diffraction analysis. Besides the wealth of information obtained, an advantage of x-ray diffraction over other techniques is the small sample size required. A typical crystal weighs roughly 10 μg, leaving the remainder of a typically scarce sample available for other purposes. An x-ray diffraction analysis is also independent of any other technique. In fact, incorrect structure predictions based on spectral methods usually hinder the structure solution process by x-ray diffraction methods more than knowledge of the correct structure helps the process. The major disadvantage of the x-ray diffraction technique is the requirement of a single crystal. This problem cannot be overstated; some compounds simply won't crystallize. Derivitization often leads to crystalline material, but also increases the work involved and the sample size required

The development of new NMR techniques and the more widespread use of molecular mechanics in recent years has shown that the structure resulting from an x-ray diffraction analysis is often the best starting point for other studies. The most interesting marine natural products are those with biological activity and the relationship between molecular conformation and biological activity is always of interest. Since the conformation of a molecule inside a crystal represents one out of perhaps many local energy minima, the atomic coordinates provide an excellent starting point for studying solution conformation by NMR and for exploring other conformations with molecular mechanics techniques.

Structural Studies of the Brevetoxins

In contrast to the water-soluble paralytic shellfish toxins typified by saxitoxin, the brevetoxins are lipid-soluble compounds produced by the dinoflagellate *Ptychodiscus brevis* in the Gulf of Mexico. During blooms, the brevetoxins cause neurotoxic shellfish poisoning in humans, not nearly as serious a health threat as the PSP toxins. Although their human toxicity is relatively low, the brevetoxins are extremely potent ichthyotoxins, responsible for massive fish kills during *Ptychodiscus brevis* blooms. The sanitation problems and economic hardships on fishing and tourism industries can be substantial during Florida red tides because of the tons of dead fish that result. The brevetoxins also differ from saxitoxin in their mode of action; they enhance, rather than block, ion conduction through sodium channels of excitable membranes (*4*). Thus the brevetoxins provide yet another probe for the study of neuromuscular sodium channels.

The toxicity of the brevetoxins is impressive. The LD_{100} (lethal dose required to kill 100% of a test group) values after 1 h against the common zebra fish *Brachydanio rerio* are 3 ng/mL, 16 ng/mL, and 60 ng/mL for brevetoxins A, B, and C, respectively (*5*). The brevetoxins' neurotoxic activity is achieved by altering the properties of membrane-bound sodium channels so that a significant number are in an active state at resting membrane potential. From studies with other types of neurotoxins, four separate receptor sites have been identified for the sodium channel protein complex in human neuroblastoma cells and rat brain synaptosomes (*6*). Receptor site 1 binds tetrodotoxin and saxitoxin, which block ion transport. Receptor site 2 binds grayanotoxin and the alkaloids veratridine, batrachotoxin, and aconitine, which cause repetitive firing and persistent activation of sodium channels. Receptor site 3 binds the α-scorpion toxins and sea anemone toxins, which slow sodium channel inactivation and enhance activation of the sodium channel by toxins acting at receptor site 2. Receptor site 4 binds β-scorpion toxins, which cause repetitive action potentials and shift the voltage dependence of sodium channel activation. The brevetoxins have recently been shown to bind to a different receptor site, increasing the total number of sites to five (*7, 9*). The different receptor sites interact with one another, making the sodium channel a complex allosteric system.

Numerous attempts have been made to isolate and identify the individual brevetoxins and the results have been reviewed (*5, 8*). The difficulties involved in culturing the organisms to obtain useful amounts of toxins and the microscale purifications have, until recently, severely hampered efforts towards complete characterization. In addition, considerable confusion has arisen over the naming of the brevetoxins and of the toxin-producing organism. The main reason for the confusion over the toxin names has been the varying degrees of purity obtained by different groups while working with what later proved to be similar or identical compounds. Nakanishi has summarized the relationships between names and compounds (*5*). A new naming scheme has been recently proposed (*9*) which uses the prefix PbTx and the numbering scheme suggested by Shimizu (*10*). The most recently proposed name for the organism, *Ptychodiscus brevis*, has replaced the former name, *Gymnodinium breve*, in much of the literature, but the appropriateness of this taxonomical change has been questioned (*11, 12*).

It was not until 1981 that the first brevetoxin structure was determined (*13*). PbTX-2, the most plentiful of the brevetoxins, was purified to crystallinity and the structure determined by x-ray crystallography. This important first step was the result of a three-group collaboration among Clardy, Lin, and Nakanishi. The yields of the major components from 50 L of culture (ca. 5×10^8 cells) were 0.8 mg of PbTX-1, 5.0 mg of PbTX-2, and 0.4 mg of PbTX-8. The structure of PbTX-2, shown below, is a completely unprecedented ladder-like array of eleven trans-fused rings with an oxygen atom in each ring and an α,β-unsaturated aldehyde and lactone. With the remarkable structure of PbTX-2 known, several of the closely related minor toxins were soon identified by spectral comparisons and x-ray crystallography (*14–16*). The list of toxins shown represents those compounds with the PbTX-2 skeleton that are known at this time.

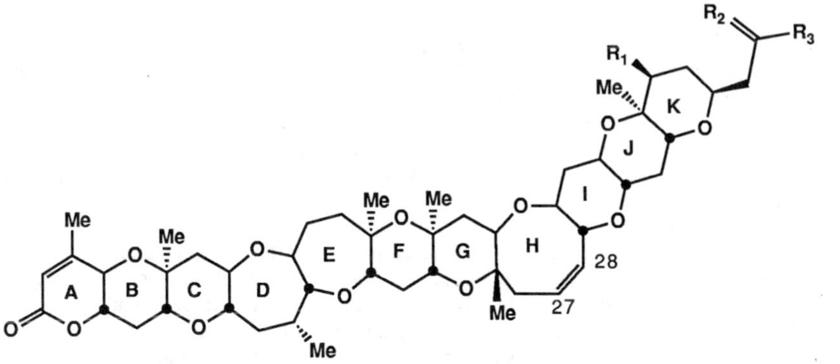

Name	R₁	R₂	R₃
PbTx-2 (BTX-B)	OH	CH₂	CHO
PbTx-8 (BTX-C)	OH	O	CH₂Cl
PbTx-3 (GB-3)	OH	CH₂	CH₂OH
PbTx-4 (GB-5)	OAc	CH₂	CHO
PbTx-5 (GB-6)	OH	CH₂	CHO

(27,28-β-epoxide)

Until 1986, the structure of PbTX-1, the most potent of the brevetoxins, remained unknown. Similarities in the ^{1}H and ^{13}C NMR spectra between PbTX-2 and PbTX-1 indicated that the two shared some structural features, but a complete structural assignment was made difficult by severely broadened spectral lines. Tentative structures for PbTX-1 were proposed (*5, 17*), but were shown to be incorrect upon completion of the work described in this chapter.

PbTX-1 (Brevetoxin-A)

The opportunity to work on the PbTX-1 structures arose through collaboration with Professor Yuzuru Shimizu at the University of Rhode Island. He and his coworkers isolated PbTX-1 from the methylene chloride extracts of cultured cells of *Ptychodiscus brevis*. The extract was partitioned between petroleum ether and 90% methanol, and the methanolic extract was chromatographed on SiO_2 with methylene chloride–benzene–methanol (40:5:1) and then methylene chloride–ethyl acetate–methanol (5:3:1). PbTX-2 was removed from the toxin mixture by crystallization and PbTX-1 was purified by normal phase HPLC using isooctane–99% isopropanol (4:1) to yield 1.2 mg from 10^9 cells. The pure PbTX-1 could be crystallized from acetonitrile, mp 197-199°C/218-220°C (double melting point). High resolution FAB mass spectrometry gave the molecular formula $C_{49}H_{70}O_{13}$ (MH$^+$, m/z 867.49). The IR, ^{1}H NMR, and ^{13}C NMR spectra showed the presence of secondary and tertiary methyl groups, an α-methylene aldehyde, two cis-substituted double bonds, and a lactone carbonyl. Based on spin-spin decoupling, proton-proton coupling correlation, and proton-carbon correlation experiments, small fragments of the structure were reported (18), but the fragments could not be connected with certainty due to the absence of several signals in the COSY spectrum.

We were able to grow crystals of PbTX-1 that measured up to 0.5 mm in the largest dimension from acetonitrile by slow evaporation of the solvent. The crystals were irregularly shaped with no prominent faces or axial directions, but were strongly birefringent when viewed with a polarized light microscope. A suitable crystal was chosen and glued with epoxy to the end of a thin glass fiber for x-ray diffraction analysis. The crystal diffracted strongly and preliminary x-ray photographs revealed a monoclinic unit cell containing four molecules of PbTX-1. A disappointing observation as we began a partial data collection was that the intensities followed a pseudo-centering trend. All reflections with $h+k=2n$ were strong and those with $h+k=2n+1$ were systematically weak. This meant that the real space group was not C2 with one molecule of PbTX-1 in the asymmetric unit, but rather $P2_1$ (systematic absence 0k0: $k=2n+1$) with two molecules comprising the asymmetric unit. Space group $P2_1/m$ was ruled out because it requires that the natural product be optically inactive, an unlikely possibility in view of the complexity and enantiomeric purity of the known brevetoxins. A data set of all reflections with $2\theta < 114°$ was measured at room temperature with Cu Kα x-rays over a period of 9 days. Two crystals were used for the complete data collection and for each, three reflections were checked periodically to monitor crystal decomposition. Each crystal showed a slow, anisotropic decomposition throughout the data collection. A summary of the crystallographic data is given in Table I.

Table I. Crystal Structure Data for PbTX-1

formula	$C_{49}H_{70}O_{13}$	formula weight	867.08
crystal size, mm	$0.5 \times 0.5 \times 0.1$	space group	P21
a, Å	26.407(8)		
b, Å	10.548(3)	β, deg	128.074(28)
c, Å	21.348(6)		
V, Å^3	4681.1(28)	Z	4
calculated density, g/cm^3	1.23	temperature, °C	20
λ(Cu Kα), Å	1.54178	μ(Cu Kα), cm^{-1}	6.8
2θ range, deg	0-114	scan type	ω
scan width, deg	1.0	scan speed, deg/min	1.5-30.0
data measured	6705	unique data	6488
R$_{merge}$	0.073	obs data with $|Fo| \geq 3\sigma(F_o)$	3848
refinement	block diagonal	weights	$1/\sigma^2$
R	0.142	R$_w$	0.190
number of parameters	496	goodness of fit	1.91

Attempts to solve the crystal structure of PbTX-1 using direct methods failed miserably. There were three major problems. First, the asymmetric unit contained 124 non-hydrogen atoms, a number well beyond the "automatic" or "routine" range of existing direct methods approaches. Second, pseudo-translational symmetry is known to make the direct phasing procedure much more difficult. Some groups have specifically addressed pseudo-translational symmetry in direct methods and have tailored direct methods procedures to deal with the problem (19, 20). Third, a rather low number of reflections had intensities significantly above background level and the data were known to suffer from an anisotropic decomposition. In addition, we assumed by analogy with the PbTX-2 skeleton and the partial fragments deduced from NMR experiments that the PbTX-1 structure contained a large number of fused rings. The presence of any kind of repeating structural motif (such as rings) in the asymmetric unit merely adds more non-crystallographic symmetry. This kind of problem is often identified by large peaks in the Patterson map resulting from the overlap of many interatomic vectors with low individual weights. The Patterson map of PbTX-1 showed several such strong peaks corresponding to 1,2 and 1,3 interatomic vectors.

Recognizing the difficulties involved with the structure of PbTX-1, we entertained the idea of chemically modifying the structure in hopes that the crystalline packing arrangement would change. With this in mind, Shimizu succeeded in making a dimethyl acetal of the PbTX-1 aldehyde group by treating the natural product with methanol in the presence of Dowex 50W-X8 (H^+ form) at 45°C for 1 h. The dimethyl acetal also crystallized from acetonitrile by slow evaporation of the solvent, mp 233–235°C. We were able to grow crystals of comparable size to that of the natural product from the same solvent and x-ray photographs showed that the crystals diffracted equally well. Autoindexing of low resolution reflections of the dimethyl acetal revealed a cell with nearly identical edge lengths and angles as the natural product. After confirming the apparent monoclinic symmetry with axial photographs, we measured a subset of the data to determine if the cell was or was not centered. Fortunately, the cell was centered and the space group was assigned as C2 (systematic absences: h+k=2n+1) with four molecules of PbTX-1 dimethyl acetal in the unit cell, one molecule in the asymmetric unit. With only one molecule comprising the asymmetric unit, the size of the problem was halved with respect to the underivitized PbTX-1 structure and the level of difficulty in solving the structure decreased substantially. Space groups Cm and C2/m were ruled out because they both require that the natural product be optically inactive. Data were measured to a maximum 2θ of 114° with Cu Kα x-rays at room temperature with no observable decay in standard intensities. A summary of the crystallographic data is given in Table II.

Table II. Crystal Structure Data for PbTX-1 Dimethyl Acetal

formula	$C_{51}H_{76}O_{14}$	formula weight	913.15
crystal size, mm	0.1 × 0.2 × 0.4	space group	C2
a, Å	27.249(5)		
b, Å	10.572(2)	β, deg	128.854(11)
c, Å	21.739(4)		
V, Å^3	4877.1(16)	Z	4
calculated density, g/cm^3	1.24	temperature, °C	20
λ(Cu Kα), Å	1.54178	m(Cu Kα), cm^{-1}	6.9
2θ range, deg	0-114	scan type	ω
scan width, deg	1.0	scan speed, deg/min	1.5-30
data measured	3623	unique data	3504
R$_{merge}$	0.018	obs data with $|Fo|\pm3\sigma(F_o)$	3203
refinement	block diagonal	weights	$1/\sigma^2(F_o)$
R	0.061	R$_w$	0.075
number of parameters	585	goodness of fit	1.12

The large percentage of observed data, an important consideration with large, non-centrosymmetric structures, was an indication that the structure could be solved with direct methods. The initial focus in solving the structure was to identify an outstanding direct methods solution and hopefully identify most of the structure from initial electron density maps. After a number of trials using both the MULTAN (20) phase permutation and RANTAN (21) random starting set procedures, in each case varying the number of phases determined and the size of the starting sets, it was clear that the best direct methods solutions would provide only parts of the structure. The pattern common to all the best phase sets was a zig-zag chain of ten or more atoms that passed through otherwise uninterpretable electron density. The most remarkable feature of these maps was the absence of rings. This was not expected, given the known tendency of fused ring systems to overexpress themselves in the form of infinite sheets of hexagons on the electron density maps—the so-called "chicken wire disaster."

The starting point for partial structure expansion was a RANTAN phase set with the highest combined figure of merit where phases were determined for the best linked 300 out of 350 large E values. In this RANTAN attempt and those that followed, random phases were assigned initial weights of 0.25 and one phase seminvariant estimates (Σ_1 relationships) were ignored. A 17-atom fragment resembling a linear hydrocarbon chain was used to estimate values and weights for the three phase invariants (22) in a subsequent RANTAN job, again with 300 out of 350 phases determined. The electron density map computed from phase sets with either the best or the second best combined figure of merit contained four fused, six-membered rings, three of which would be consistent with the I-K rings of PbTX-2. Three of the rings plus one strong peak (15 atoms in all) were used to once again estimate values and weights for the triplet phase relationships in a RANTAN job where 300 of 350 phases were determined. The electron density map computed from the phase set with the highest combined figure of merit contained the previously determined 15 atoms plus 10 additional strong peaks. These 25 atoms were used to calculate phases for recycling with the tangent formula (23). Starting phases were refined twice with the tangent formula and then fixed during the determination of all phases with $|E| > 1.3$. The resulting electron density map showed many more chemically sensible atoms. After one cycle of weighted Fourier recycling using 42 atoms from tangent formula recycling, 61 of the 65 non-hydrogen atoms were identified. The remaining four atoms were found without difficulty from difference electron density maps following refinement of the 61-atom model.

The structure was refined by block-diagonal least squares in which carbon and oxygen atoms were modeled with isotropic and then anisotropic thermal parameters. Although many of the hydrogen atom positions were available from difference electron density maps, they were all placed in ideal locations. Final refinement with all hydrogen atoms fixed converged at crystallographic residuals of $R = 0.061$ and $R_w = 0.075$.

The structure of PbTX-1 dimethyl acetal is shown below. Like PbTX-2, the structure is composed of a nearly linear array of rings, all trans-fused, with an -O-C-C- repeating pattern through the backbone. There are 5, 6, 7, 8, and 9-membered rings present, an unprecedented occurrence in any known natural product. Rings H-J are the same as in PbTX-2, except for the absence of an axial methyl group on C38. The crystallographic structure assignment agrees well with all available spectral data. Crystallographic drawings of PbTX-1 dimethyl acetal are shown in Figure 1. Since the x-ray diffraction analysis resulted in an arbitrary choice for the enantiomer, the absolute configuration was chosen to be consistent with that tentatively assigned to PbTX-2.

PbTx-1 dimethyl acetal

A conspicuous feature of the structure evident in the crystallographic drawing is the molecular twist present at ring G. The two parts of the molecule on either side of the G ring appear to be turned roughly perpendicular to one another. The dihedral angle between the best least squares plane through rings A-F and the best plane through rings H-J is 122°. The rms deviations from the planes are 0.69 Å for A-F and 0.25 Å for H-J. The 8-membered G-ring is in the boat-chair conformation, one of two conformational energy minima suggested by inspection of Dreiding models. The other is a crown conformation which would make PbTX-1 dimethyl acetal much more planar, like the PbTX-2 molecule, although it would introduce a turn at ring G. The E and F-rings, containing 8 and 9 atoms respectively, each contain a double bond and as a result appear less flexible than saturated rings B and G. The E and/or the F-ring could exist in a conformation with their double bonds pointing toward the opposite face of the molecule. The E-ring appears unstrained in either conformation, but in the F-ring, the second conformation appears reasonable only if the G-ring is in a crown conformation. Ring B is also a saturated, 8-membered ring, but is in the crown conformation. A boat-chair conformation appears quite reasonable based on steric arguments. Ring D contains 7 atoms and has an alternate conformation corresponding to a change in sign of the C12-C13-C14-C15 torsional angle. This conformational change would bring the A and part of the B rings out of the molecular plane and would introduce unfavorable steric interactions between C47 methyl and C23 methylene groups. Rings C, H, I, and J are 6-membered and are quite rigid in the Dreiding model. The isotropic equivalent thermal parameters for the atoms in the entire ring framework range from 3.4-5.4 Å^2, corresponding to mean square vibrational amplitudes of 0.043-0.068 Å^2. The lack of high thermal parameters that stand out in the polycyclic framework suggests that all the rings are locked into their particular conformations and that little or no positional disorder is present.

With the chemical structure of PbTX-1 finally known and coordinates for the molecule available from the dimethyl acetal structure, we wanted to return to the natural product crystal structure. From the similarities in unit cells, we assumed that the structures were nearly isomorphous. Structures that are isomorphous are crystallographically similar in all respects, except where they differ chemically. The difference between the derivative structure in space group C2 and the natural product structure in P2$_1$ (a subgroup of C2) was that the C-centering translational symmetry was obeyed by most, but not all atoms in the natural product crystal. We proceeded from the beginning with direct methods, using the known orientation of the PbTX-1 dimethyl acetal skeleton (assuming isomorphism) to estimate phase

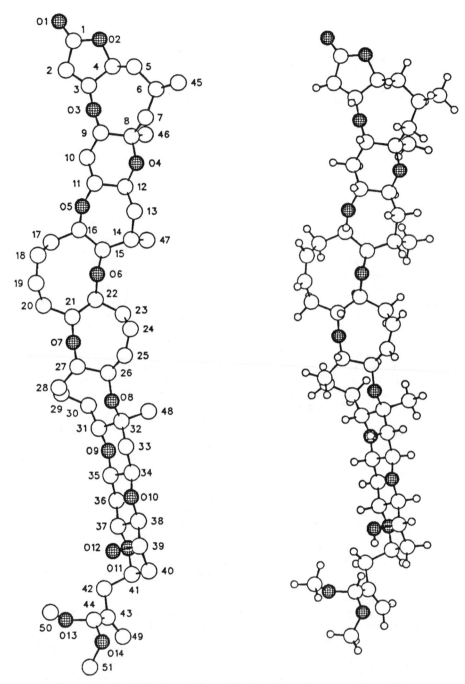

Figure 1. A labeled crystallographic drawing and a crystallographic drawing including hydrogen atoms of PbTX-1 dimethyl acetal. Oxygen atoms are cross-hatched.

relationships, and compute phase sets using RANTAN. The best phase set led to an electron density map which gave most of the backbone atoms for both molecules. Two cycles of weighted Fourier refinement provided the coordinates for nearly all atoms in the two molecular skeletons, but the R value was still suspiciously high (ca. 50%). Repeated refinement of the known positions followed each time by difference maps revealed the α,β-unsaturated aldehyde side chains. An alternative method that achieved the same result was to estimate a subset of phases as in tangent formula recycling but to assign random phases to the rest in the multisolution RANTAN procedure (24).

The structure was refined with block diagonal least squares. In cases of pseudo-symmetry, least squares refinement is usually troublesome due to the high correlations between atoms related by false symmetry operations. Because of the poor quality of the data, only those reflections not suffering from the effects of decomposition were used in the refinement. With all non-hydrogen atoms refined with isotropic thermal parameters and hydrogen atoms included at fixed positions, the final R and R_w values were 0.142 and 0.190, respectively. Refinement with anisotropic thermal parameters resulted in slightly more attractive R values, but the much lower data to parameter ratio did not justify it.

Crystallographic drawings of the two independent molecules in the asymmetric unit of PbTX-1 are shown in Figure 2 along with a chemical drawing containing the numbering scheme. An obvious difference between the two is a rotation about the C42-C43 bond. The values of the C41-C42-C43-C44 torsional angles are -67° and 118°, a difference of 185°! This is not a surprising result given the flexibility of the side chain and it does reveal at least part of the structure that is responsible for violating the pseudo-C-centering condition. The other major difference was at first unnoticed, but turned out to be rather interesting. One of the independent molecules resembles PbTX-1 dimethyl acetal, with the double bonds on rings E and F pointing in the same direction. The other independent molecule, however, has the double bond in ring E pointing opposite to that of ring F. A superposition of the two independent molecules using all non-hydrogen atoms except C43, C44, C49, and O13 gave an average distance between atom pairs of 0.42 Å. The only atoms that disagree strongly between the two are C17 through C20. It is remarkable that the two E ring conformations leave the remainder of the molecule essentially unchanged. A plot of the two independent PbTX-1 molecules superimposed is shown in Figure 3.

Dihydro PbTX-1 (= PbTX-7)

A closely related natural product from *Ptychodiscus brevis* is dihydro PbTX-1, the allylic alcohol resulting from reduction of the C44-O13 α,β-unsaturated aldehyde. We obtained dihydro PbTX-1 from Shimizu and found that it also crystallized nicely from acetonitrile. Given the possibility of different conformations in the larger rings of PbTX-1, we decided to characterize the crystals of the alcohol in hopes of seeing something very different from the unit cells of the first two compounds. A thin, rectangular prism was glued to a glass fiber for analysis and the same preliminary crystallographic procedures were followed as for PbTX-1 and PbTX-1 dimethyl acetal. Much to our surprise, the alcohol crystallized in C2 with cell constants very similar to the dimethyl acetal—quite different from the aldehyde. We had hoped that the packing scheme would be different and that an alternative molecular conformation might exist in the crystal. Although it appeared that this

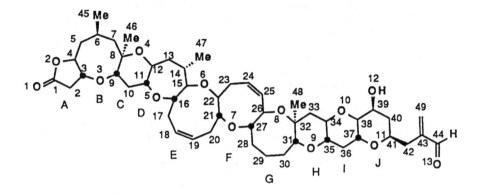

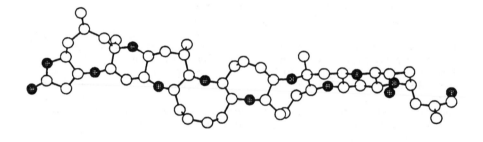

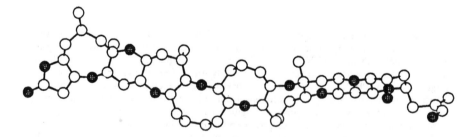

Figure 2. A chemical drawing with the numbering scheme and crystallographic drawings of the two independent molecules of PbTX-1. Oxygen atoms are crosshatched.

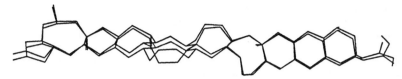

Figure 3. A superposition of the two independent molecules of PbTX-1.

was not the case, another interesting question arose dealing with the relationship among the three crystal structures involved—PbTX-1 dimethyl acetal, PbTX-1, and dihydro PbTX-1. PbTX-1 and dihydro PbTX-1, two molecules differing very little in terms of packing requirements are crystallographically different, yet PbTX-1 dimethyl acetal and dihydro PbTX-1, two molecules with very different sized side chains, are quite similar.

To satisfy our curiosity over the packing schemes and to verify the dihydro PbTX-1 structure assignment, we decided to measure a complete set of data. Data for the alcohol were measured at room temperature with Cu Kα x-rays to a maximum 2θ value of 114°, monitoring for possible decomposition every fifty reflections. No significant decomposition was observed. The crystallographic data are given in Table III.

Solution of the alcohol structure began with the assumption that dihydro PbTX-1 was isomorphous with PbTX-1 dimethyl acetal. Tangent formula recycling using the polycyclic skeleton from the known structure and the largest 500 E values provided all but 4 non-hydrogen atomic positions. After isotropic refinement of the 58 located atoms, the last 4 positions corresponding to the allylic alcohol side chain were located on difference electron density maps. Isotropic refinement followed by anisotropic refinement of 62 non-hydrogen atoms with hydrogen atoms fixed at ideal positions converged at residuals of R=0.087 and R_w=0.099.

The isotropic equivalent thermal parameters are on the whole larger than in the PbTX-1 dimethyl acetal structure or the structure of the natural product. The B values for atoms on the fused ring skeleton range from 4.7 to 12.6 Å^2 (mean square amplitudes of 0.059 and 0.16 Å^2). Curiously, the largest values are associated with C17-C20 of the 9-membered E ring—the ring that adopts two conformations in crystalline PbTX-1. The acyclic atoms do not have appreciably higher thermal parameters, with the exception of hydroxyl O13, which has a B of 22.4 Å^2.

Table III.   Crystal Structure Data for Dihydro PbTX-1

formula	$C_{49}H_{72}O_{13}$	formula weight	869.10
crystal size, mm	0.1 × 0.2 × 0.2	space group	C2
a, Å	26.399(7)		
b, Å	10.520(3)	β, deg	127.663(16)
c, Å	21.335(6)		
V, Å^3	4690.6(22)	Z	4
calculated density, g/cm^3	1.23	temperature, °C	20
λ (Cu Kα), Å	1.54178	μ (Cu Kα), cm^{-1}	6.8
2θ range, deg	0-114	scan type	ω
scan width, deg	1.0	scan speed, deg/min	1.5-30
data measured	3612	unique data	3376
R_{merge}	0.026	obs data with $|Fo|\pm3\sigma(F_o)$	2582
refinement	block diagonal	weights	$1/\sigma^2(F_o)$
R	0.087	R_w	0.099
number of parameters	558	goodness of fit	1.71

The difference electron density map following the last cycle of least squares refinement did not show evidence for a simple disorder model to explain the anomalously high B for the hydroxyl oxygen. Attempts to refine residual peaks with partial oxygen occupancies did not significantly improve the agreement index.

A search for intermolecular bonds resulted in one possible hydrogen bond between hydroxyl O13 and lactone carbonyl O1. The distance between O1 and O13 is 2.85 Å, a value well within the range expected for OH-O hydrogen bonds (25). The hydrogen atom position for hydroxyl O13 was chosen to be along the O13-O1 vector. The hydrogen position was not evident in the difference electron density map, presumably due to problems modeling the O13 position.

Discussion of Structures

The crystal structures of PbTX-1 dimethyl acetal, PbTX-1, and dihydro PbTX-1 provide a total of four independent pictures of the same brevetoxin skeleton. It is rare that this quantity of structural data is available for a natural product of this size. A comparison of torsional angles shows that all four molecules have approximately the same conformations in all rings, except, of course, for the aldehyde side chain and the E-ring in one of the independent molecules of PbTX-1. Least squares superposition fits among the four molecules gave the following average distances:

0.19 Å for PbTX-1 dimethyl acetal/PbTX-1 (I),
0.23 Å for PbTX-1 dimethyl acetal/PbTX-1(II),
0.12 Å for PbTX-1 dimethyl acetal- /dihydro PbTX-1,
0.32 Å for PbTX-1(I)/PbTX-1(II),
0.14 Å for PbTX-1(I)/dihydro PbTX-1, and
0.21 Å for PbTX-1(II)/dihydro PbTX-1.

In each case, C1-C42, C45-C48, and O1-O12 were used in the refinement. Figure 4 contains crystallographic drawings of each of the molecules for comparison.

The broadened signals in the NMR spectra of PbTX-1 have been assigned to the part of the molecule associated with the E and G rings. Given the alternative conformations available from examination of molecular models, a plausible explanation is that slow interconversions between available conformations take place in solution, thus broadening the affected ^{1}H and ^{13}C signals. Such conversions would only be reasonable, however, if the interconverting conformations do not differ by much steric energy.

In order to investigate this possibility for the G ring, we decided to compare the conformational energies of the boat-chair and the crown by molecular mechanics (26). As a starting point for the calculations, we compared the boat-chair and crown forms of cyclooctane. Early calculations had led to the conclusion that the boat-chair conformation with C_s symmetry was slightly lower in energy than any of the symmetry definable crown conformations (D_{4d}, 2.3 kcal; C_{2v}, 1.5 kcal; D_2, 0.2 kcal) (27). When we repeated the cyclooctane calculations, no symmetry constraints were imposed and the boat-chair conformation was lower in energy by a mere 1.0 kcal. When one of the -CH_2- groups was replaced by an oxygen, the boat-chair conformation of the cyclic ether was also favored over the crown, this time by 1.1 kcal. For these unsubstituted 8-membered rings, the boat-chair conformation appears to be favored slightly.

To compare conformations of the G-ring in PbTX-1, we used a sub-structure composed of rings C-H and the C47,C48 methyl groups. Rings C and H were chosen as starting and ending rings because they are the nearest rigid parts of the molecule on either side of the G-ring. The starting conformations for all rings but the G-ring were those found in crystalline PbTX-2 dimethyl acetal. In the first minimization, the G-ring was placed in the boat-chair conformation (as in the crystalline derivatives of PbTX-1) and in a second minimization it was placed in a

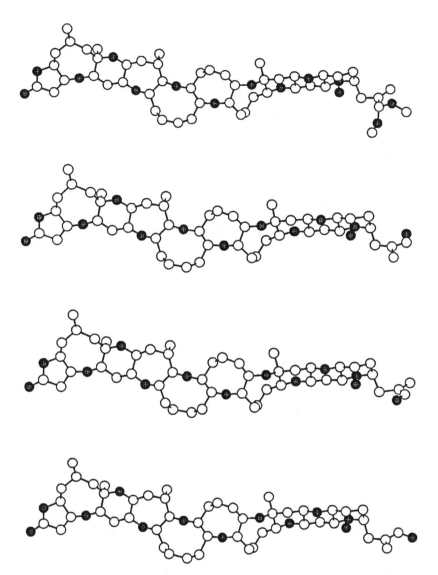

Figure 4. Crystallographic drawings of PbTX-1 dimethyl acetal, PbTX-1 molecules I and II, and dihydro PbTX-1 (top to bottom). Oxygen atoms are cross-hatched.

crown configuration. Plots of the two energy minimized structures are shown in Figure 5. Both conformations are local energy minima that differ by 1.6 kcal with the boat-chair conformation favored. All of the energy difference is attributable to the presence of the C48 axial methyl group. When C48 is replaced by a hydrogen atom and the above calculations repeated, the crown conformation for the G-ring in PbTX-1 is favored by 1.5 kcal. A least squares superposition of the energy-minimized rings C-H and the G-ring in the boat-chair conformation with the same non-hydrogen atoms from the PbTX-1 dimethyl acetal crystal structure gave an average distance between atom pairs of 0.45 Å. Superposition plots clearly show that while the exact distances and angles do not match perfectly, the energy minimized conformations in each ring are essentially the same as those observed in the crystal structure.

The other ring of interest with respect to the broadened spectral lines in PbTX-1 is ring E. In PbTX-1 dimethyl acetal, dihydro PbTX-1, and PbTX-1 molecule I, the E-ring double bond is syn to the F-ring double bond and in PbTX-1 molecule II, the double bonds are anti. When we performed an energy minimization on rings C-H with the E-ring double bond anti and all other rings in the crystalline PbTX-1 dimethyl acetal conformations, we found that the energy differed from the syn configuration by only 0.1 kcal. The two conformations are essentially equal in energy and should be close to equally populated in solution at room temperature. Plots of the energy minimized structures are shown in Figure 6. A least squares fit of rings C-H (and methyls C47 and C48) and ring E in the anti configuration with the corresponding molecule II from the PbTX-1 crystal structure gave an average distance between atom pairs of 0.31 Å.

The 8-membered, saturated B-ring in PbTX-1 is in a crown conformation. After performing molecular mechanics calculations on the G-ring, this observation seemed puzzling. One might think that ring B should be in a boat-chair, as is the case for ring G. To compare the two conformations, we used as a model rings A, B, and C from PbTX-1 and the methyl groups C45 and C46. The crown conformation for ring B ended up 0.4 kcal lower in energy than the boat chair, a difference that might well be viewed as zero. This result is in agreement with the qualitative observation from model building that the two conformations are quite comparable and emphasizes the effects of ring substitution on conformation.

It is dangerous to draw too many conclusions from the numerical results of molecular mechanics calculations. In this case, we wanted only to show that the two likely conformations for the E and G-rings of PbTX-1 are roughly equal in energy and that it is plausible that they could both exist in solution at room temperature. It would be very useful to be able to estimate the energy of activation required for the conversions between conformations. Unfortunately, estimation of the energy barrier is beyond the realm of the empirical force field method for all but the simplest cases.

The crystalline packing arrangement of the PbTX-1 structures has proved to be quite interesting. Packing diagrams of PbTX-1 dimethyl acetal are shown in Figure 7 with views along the a, b, and c* axes. After looking at the cell plots perpendicular to the b axis, it is not at all surprising that the 020 reflection had the largest intensity in all three structures studied. The dihydro PbTX-1 packing is identical to that shown in Figure 7. The PbTX-1 structure is, as expected, slightly different. In order make comparisons between structures easier, we chose to work in a non-standard setting of $P2_1$, with the 2_1 screw axis at x=1/4. Packing diagrams of PbTX-1 are shown in Figure 8. In projection perpendicular to b, the pseudo-two-fold axis at the origin appears very real if the aldehyde side chain is ignored.

The views perpendicular to b, on the other hand, show considerably more asymmetry. The lack of a two-fold axis in the crystal is most evident in the view along c*. In all three of the PbTX-1 structures studied, a short intermolecular distance is calculated between the O12 hydroxyl and the O5 ether oxygens. The two molecules involved in the short contact are related by a C-centering (or pseudo C-centering) operation. In the PbTX-1 dimethyl acetal crystal, the O12-O5 distance is 2.87 Å, in PbTX-1 it is 2.86 Å, and in dihydro PbTX-1 it is 2.84 Å. Dihydro

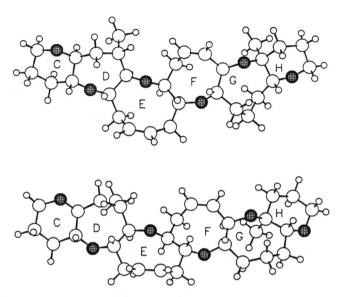

Figure 5. Plots of rings C-H of PbTX-1 resulting from molecular mechanics energy minimization. The two molecules are identical except for ring G. Oxygen atoms are cross-hatched.

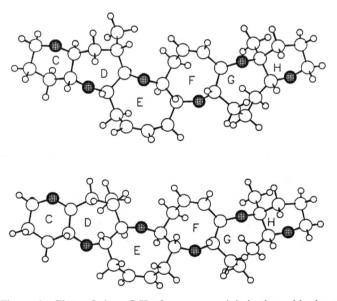

Figure 6. Plots of rings C-H after energy minimization with the ring E double bond syn to the ring F double bond (top) and the ring E double bond anti to the ring F double bond (bottom). Oxygen atoms are cross hatched.

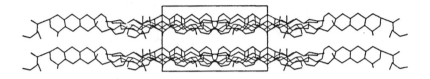

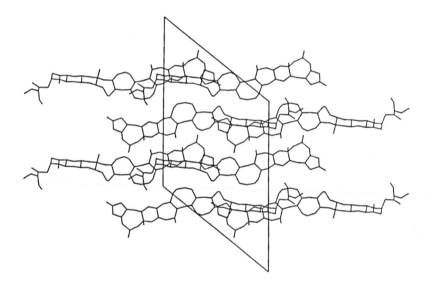

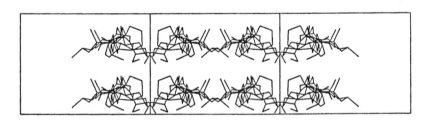

Figure 7. Packing diagrams with views along a (top), b (middle), and c* (bottom) for PbTX-1 dimethyl acetal.

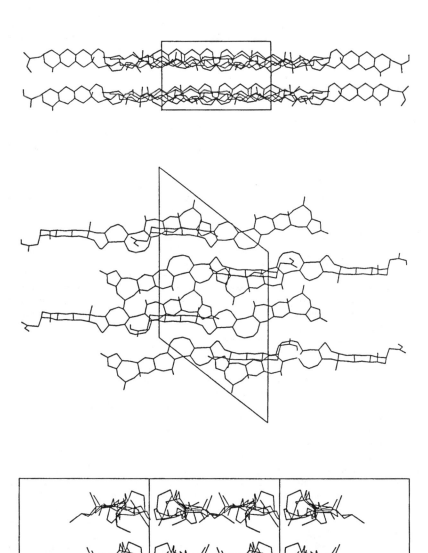

Figure 8. Packing diagrams with views along a (top), b (middle), and c* (bottom) for PbTX-1.

PbTX-1 also contains the intermolecular hydrogen bond between hydroxyl O13 and carbonyl O1 already discussed. Only for dihydro PbTX-1 was it possible to locate an O12 hydroxyl hydrogen atom position from difference electron density maps. In that structure a peak appeared at 1.1 Å from O12 and 2.1 Å from symmetry related O5.

Absolute Configuration of the Brevetoxins

Neither the x-ray crystallographic studies of PbTX-2 nor the work on PbTX-1 just discussed has provided an absolute configuration. In each case, the structures have been drawn to agree with the absolute configuration originally assigned to PbTX-2 (13). Nakanishi's group made the assignment by applying the exciton chirality method (28) to the cis-27,28-dihydroxy derivative of PbTX-2. After reduction of the α,β-unsaturated aldehyde and the α,β-unsaturated lactone and acetylation, the derivative was prepared by osmylation of the C27-C28 double bond. The di-p-bromobenzoate was then prepared and showed the expected split circular dichroism curve. The positively split peak/trough pair established the O-C27-C28-O torsional angle as positive and the absolute configuration shown.

The absolute configuration assignment was based on two critical assumptions: (i) the osmylation of the C27-C28 double bond takes place from the least hindered side of the molecule, and (ii) the 8-membered H-ring is in a crown conformation. Although model building and molecular mechanics calculations suggested that both conditions would be met, the conformational variations observed in the PbTX-1 structures cast some doubt on the issue. With a half a dozen or so groups working towards the synthesis of the brevetoxins and the continuing pharmacological interest in the compounds, an unequivocal absolute configuration assignment seemed essential.

Successful preparation of the acetal derivative of PbTX-1 suggested a similar route for preparing a suitable derivative of PbTX-2 in order to determine the absolute configuration. Shimizu prepared the 1,3-dioxolane derivative of the PbTX-2 aldehyde by treating the natural product with optically active (2R,3R)-(-)-butane-2,3-diol in the presence of Dowex 50W-X8 (H^+ form), in benzene, at 75°C for 1 h. The PbTX-2 derivative could be crystallized from dichloromethane/diethyl ether or acetonitrile as thin leaflets that diffracted x-rays poorly. After several attempts, we finally grew crystals from acetonitrile by very slow evaporation that measured up to 0.05 mm on the thinnest edge. A suitable platelike crystal was carefully placed in a 0.5-mm glass capillary and examined by x-ray photographs. As is often the case with extremely thin crystals, the autoindexing procedure produced axial solutions in only two dimensions. Longer rotation photograph exposures provided the needed weak reflections to define a unit cell. A low resolution data collection of +h,±k,±l confirmed the choice of unit cell, the monoclinic symmetry, and the space group as $P2_1$ with two molecules in the unit cell. A summary of the crystal data is given in Table IV.

The unit cell constants did not resemble those of the natural product, so a structural solution based on an isomorphous natural product was not possible. A RANTAN approach eventually led to the correct structure. Using the best linked 300/350 E values, weights of 0.25, and no Σ_1 relationships, a suitable set of phases could be found if negative quartet relationships were actively employed in the phase refinement. The resulting electron density map showed 36 sensible atomic positions. The 36-atom fragment was used to calculate phases for 198 reflections, which were refined twice by the tangent formula and then fixed during phase extension to the top 500 E values. The electron density map showed 46 correct positions. One cycle of weighted Fourier recycling resulted in the location of all 69

Table IV. Crystal Structure Data for PbTX-2 Dioxolane

formula	$C_{54}H_{78}O_{15}$	formula weight	967.20
crystal size, mm	0.05 × 0.2 × 0.5	space group	$P2_1$
a, Å	11.379(8)		
b, Å	13.320(3)	β, deg	100.50(5)
c, Å	17.876(5)		
V, Å^3	2664.2(22)	Z	2
calculated density, g/cm^3	1.21	temperature, °C	20
λ(Cu Kα), Å	1.54178	μ(Cu Kα), cm^{-1}	6.7
2θ range, deg	0-100	scan type	ω
scan width, deg	1.0	scan speed, deg/min	1.5-30
data measured	3128	unique data	2879
R_{merge}	0.053	obs data with $|Fo|\pm3\sigma(F_o)$	2084
refinement	block diagonal	weights	$1/\sigma^2(F_o)$
R	0.073	R_w	0.072
number of parameters	621	goodness of fit	1.13

non-hydrogen atoms. Block diagonal refinement of the 69 atoms with isotropic then anisotropic thermal parameters and hydrogen atoms fixed at ideal positions converged to crystallographic residuals R=0.073 and R_w=0.072.

Chemical and crystallographic drawings of PbTX-2 dioxolane are shown in Figure 9. The absolute configuration was chosen so as to give an R,R configuration for the dioxolane, thus establishing the absolute configuration of PbTX-2 by the relative stereochemistry. The assignment agrees with that proposed using the exciton chirality method. The close structural resemblance of rings H-J in PbTX-1 and rings I-K in PbTX-2 suggests the absolute configuration of PbTX-1 as well.

Summary

The structural work on the brevetoxins just described raises many new questions. The most potent toxin, PbTX-1, is also the most flexible. At a cost of about 2 kcal, the molecule can adopt a very different conformation from that in the crystal and from PbTX-2 by a change in conformation at ring G. At essentially no cost in energy, the E-ring can flip its double bond in the opposite direction. The consequences of these flexible regions with respect to the binding of toxin to the sodium channel receptor will no doubt be very interesting; the idea of a conformational change in the toxin upon binding to the channel protein is an attractive one. The part or parts of the brevetoxin molecules critical for binding are unknown at this time. The aldehyde ends of PbTX-1 and PbTX-2 are nearly identical, but the lactone ends are quite different. Perhaps the chemically modified structures and analogs that are likely to result from synthetic efforts will be of some use in identifying the essential structural components for neurotoxic activity.

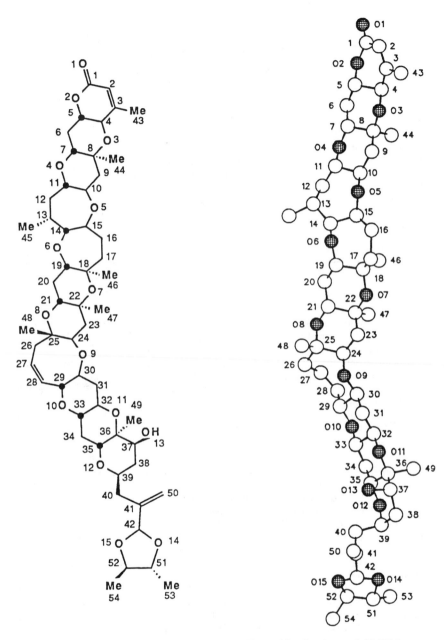

Figure 9. Chemical and labeled crystallographic drawings of PbTX-2 dioxolane. Oxygen atoms are cross-hatched. The configurations at C51 and C52 are R and R.

Acknowledgments

I am grateful for the advice and collaboration of Professors Jon Clardy and Yuzuru Shimizu and for financial support from National Institutes of Health grants GM 28754 (Y.S.) and CA 24487 (J.C.) and the New York State Sea Grant (J.C.)

Literature Cited

1. Woolfson, M. M. *Acta. Cryst.* **1987**, A43, 593-612.
2. Karle, J. *Angew. Chem. In. Ed. Engl.* **1986**, *25*, 614–629.
3. Hauptman, H. *Science* **1986**, *233*, 178–183.
4. Westerfield, M.; Moore, J. W.; Kin, Y. S.; Padilla, G. M. *Am. J. Physiol.* **1977**, *232*, C23.
5. Nakanishi, K. *Toxicon* **1985**, *23*, 473.
6. Catterall, W. A. *Science* **1984**, *223*, 653.
7. Catterall, W. A.; Gainer, M. *Toxicon* **1985**, 23.,497.
8. Baden, D. G. *Int. Rev. Cytol.* **1983**, *82*, 99.
9. Poli, M. A.; Mende, T. J.; Baden, D. G. *Mol. Pharmacol.* **1986**, *30*,
10. Shimizu, Y. *Pure Appl. Chem.* **1982**, *54*, 1973.
11. Steidinger, K. A. In *Toxic Dinoflagellate Blooms*; Taylor, D. L., Seliger, H. H., Eds.; Elsevier/North Holland: New York, 1979; p 435–442.
12. Dodge, J. D. In *Marine Dinoflagellates of the British Isles*; Her Majesty's Stationary Office: London, 182; p 108.
13. Lin, Y.; Risk, M.; Ray, S. M.; Van Engen, D.; Clardy, J.; Golik, J.; James, J. C.; Nakanishi, K. *J. Am. Chem. Soc.* **1981**, *103*, 6773.
14. Chou, H. N.; Shimizu, Y. *Tetrahedron Lett.* **1982**, *23*, 5521.
15. Golik, J.; James, J.C.; Nakanishi, K.; Lin, Y. *Tetrahedron Lett.* **1982**, 23, 2535.
16. Chou, H. N.; Shimizu, Y.; Van Duyne, G.; Clardy, J. *Tetrahedron Lett.* **1985**, *26*, 2865.
17. Tempesta, M.; Golik, J.; James, J. C.; Nakanishi, K.; Pawlak, J.; Iwashita, T.; Gross, M. L.; Tomer, K. B. In *The 1984 International Chemical Congress of Pacific Basin Societies*; Honolulu, HI, Dec 16–21, 1984; Abstr. 10E 45.
18. Chou, H. N.; Shimizu, Y.; Van Duyne, G. D.; Clardy, J. In *Toxic Dinoflagellates*; Anderson, D., White, A., Baden, D., Eds.; Elsevier: New York, 1985; p 305–308.
19. Beurskens, P. T. In *Crystallographic Computing 3: Data Collection, Structure Determination, Proteins and Databases*; Sheldrick, G. M., Kruger, C., Goddard, R., Eds.; Oxford: New York, 1985; p 216–226.
20. Fan, H. F.; Yao, J. X.; Main, P.; Woolfson, M. M. *Acta Cryst.* **1983**, A39, 566.
21. Yao, J. X. *Acta Cryst.* **1981**, A37, 642.
22. Main, P. In *Crystallographic Computing Techniques*; Ahmed, F. R., Ed; Copenhaen: Munksgaard, 1976; p. 97–105.
23. Karle, J. *Acta Cryst* **1968**, B24, 182.
24. Yao, J. X. *Acta Cryst.* **1983**, A39, 35.
25. Pimentel, G. C.; McClellan, A. L. The *Hydrogen Bond*; Freeman: San Francisco, 1960; Ch. 9.
26. Burkert, U.; Allinger. N. L. *Molecular Mechanics*; American Chemical Society: Washington, DC, 1982.
27. Hendrickson, J. B. *J. Am. Chem. Soc.* **1964**, *86*, 4854.
28. Harada, N.; Nakanishi, K. *Acct. Chem. Res.* **1972**, *5*, 257.

RECEIVED August 29, 1989

Chapter 11

Brevetoxins: Unique Activators of Voltage-Sensitive Sodium Channels

Vera L. Trainer[1], Richard A. Edwards[2], Alina M. Szmant[2],
Adam M. Stuart[1], Thomas J. Mende[1], and Daniel G. Baden[1,2]

[1]Department of Biochemistry and Molecular Biology, School of Medicine,
University of Miami, P.O. Box 016129, Miami, FL 33101
[2]Rosenstiel School of Marine and Atmospheric Science, Division of Biology
and Living Resources, University of Miami, 4600 Rickenbacker Causeway,
Miami, FL 33149

Eight polyether toxins isolated from Florida's red tide dinoflagellate
Ptychodiscus brevis are based structurally on two different carbon backbones. All toxins examined exert their toxic effects by specific binding
to Site 5 associated with voltage-sensitive sodium channels. Exposure
to brevetoxins leads to activation of sodium channels at normal resting potential. Specific binding of tritiated brevetoxin PbTx-3 to synaptosomes has been undertaken in rats, fish, and turtles. Dissociation
constants are comparable in the nanomolar concentration range in
each system, and binding maxima are in the pmol/mg protein range in
all cases. Derivative brevetoxins specifically displace tritiated PbTx-3
from its site of action and IC_{50} data for derivative brevetoxins are
comparable in each of the three species. Derived inhibition constants
are in the nanomolar concentration range, with the most toxic
brevetoxins being most most potent at displacing labeled toxin from
its site.

The marine dinoflagellate *Ptychodiscus brevis* (formerly *Gymnodinium breve*) is
responsible for toxic red tides along the Gulf of Mexico coast of Florida and Texas
(*1*), and entrained blooms have been transvected around the Florida peninsula and
up the Eastern coast of the United States as far north as North Carolina. Toxicity
of this red tide organism is due to synthesis and intracellular maintenance of a
multiplicity of potent polyether neurotoxins (*2*), known as the brevetoxins (Figure
1) (*3*). Since this organism is toxic in laboratory culture, most of the recent
brevetoxin work has been undertaken employing laboratory cultures of *P. brevis* and
the toxins derived from these cultures (*2–5*).

The toxicological consequences of *P. brevis* red tides are mass mortality of
fishes exposed to the red tide; toxic shellfish which, if consumed, result in human
neurotoxic shellfish poisoning; and an irritating aerosol which results from contact
with *P. brevis* cell particles entrapped in seaspray. In all cases, the threshhold levels
for intoxication are in the picomolar to nanomolar concentration ranges, implying a
specific locus or loci of action for brevetoxins (reviewed in *6*).

Electrophysiological protocols utilizing crayfish and squid giant axons revealed
that external application of brevetoxin caused a concentration-dependent

0097–6156/90/0418–0166$06.00/0
© 1990 American Chemical Society

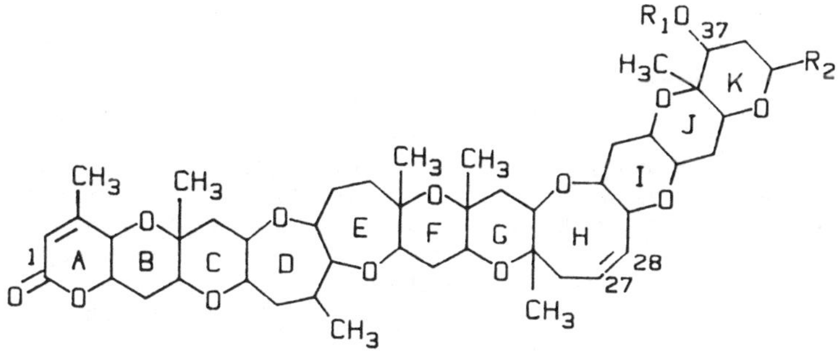

Type-1

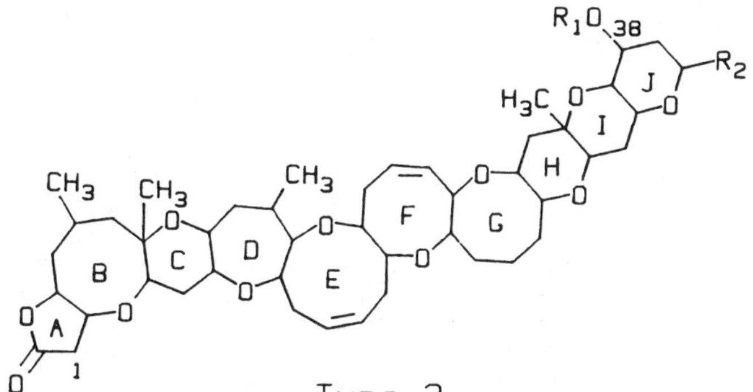

Type-2

Figure 1. The brevetoxins are based on two different backbone structures, as indicated (4). Type 1 toxins (top) include:

PbTx-2 [R_1=H, R_2=CH$_2$C(=CH$_2$)CHO)];
PbTx-3 [R_1=H, R_2=CH$_2$C(=CH$_2$)CH$_2$OH)];
PbTx-5 [R_1=Ac, R_2=CH$_2$C(=CH$_2$)CHO];
PbTx-6 [R_1=H, R_2=CH$_2$C(=CH$_2$)CHO, 27,28 epoxide];
PbTx-8 [R_1=H, R_2=CH$_2$COCH$_2$Cl].

Type 2 toxins (bottom) include:

PbTx-1 [R=CHO];
PbTx-7 [R=CH$_2$OH].

No structual information is available on PbTx-4.

depolarization, repetitive discharges, and a depression of the action potential leading to a block of excitability (Figure 2). Voltage clamp experiments illustrated that only sodium currents were affected (7). Early experiments utilizing neuroblastoma cells illustrated that application of brevetoxin-A (PbTx-1) (8) induced an influx of ^{22}Na$^+$ in a dose-dependent manner. This work was followed by experiments using brevetoxin PbTx-3 and rat brain synaptosomes, again illustrating a dose-dependent uptake of ^{22}Na$^+$ following toxin application (Figure 3) (4).

Catterall and Risk (8) demonstrated that brevetoxins did not interfere with binding of sodium channel-specific neurotoxins which bind at sites 1–3, and Catterall and Gainer illustrated the lack of brevetoxin interaction at site 4 (9). That brevetoxins bind at a unique site associated with voltage-sensitive sodium channels (VSSC) was suggested by this data (8, 9). Specific binding of brevetoxins to synaptosomes was first demonstrated by Poli et al. (Figure 4) (4), by utilizing brevetoxin PbTx-2 synthetically reduced with sodium borotritiide to yield tritiated PbTx-3 with specific activities approaching 20 Ci/mmol (10). Poli demonstrated saturability, competition for specific binding sites by nonradioactive brevetoxin agonists , binding maxima in the pmol/mg protein concentration range, reversibility of radioactive toxin binding, half times for association and dissociation consistent with specific binding, distinct brain regional and subcellular distribution, the presence of a pharmacological response at appropriate concentrations for binding, tissue linearity, and temperature dependence (11). Dissociation constants, binding maxima, and competitive displacement curves for brevetoxin at site 5 parallel those constants derived for saxitoxin binding at site 1 (Figure 5).

Protocols for Binding Assays

Excitable tissue preparations were obtained fresh daily from live animals using the technique described by Dodd et al. (12). Protein was measured on each synaptosome preparation using the Coomassie Brilliant Blue dye technique described by Bradford (13); results were expressed as "toxin bound per mg synaptosome protein".

Binding of tritiated PbTx-3 was measured using a rapid centrifugation technique. All binding experiments were conducted in a binding medium consisting of 50 mM HEPES (pH 7.4), 130 mM choline chloride, 5.5 mM glucose, 0.8 mM magnesium sulfate, 5.4 mM potassium chloride, 1 mg/mL bovine serum albumin, and 0.01% Emulphor EL-620 as an emulsifier (4). In addition, 370 mM sucrose was added to fish synaptosome experiments to maintain iso-osmolarity.

Synaptosomes, suspended in 0.1 mL of binding medium minus BSA, were added to a reaction mixture containing [^{3}H] PbTx-3 and other effectors in 0.9 mL of binding medium in 1.5 mL polypropylene microfuge tubes. After mixing and incubating at the desired temperatures for 1 hr, samples were centrifuged at 15,000 × g for 2 min. Supernatant solutions were sampled for the measurement of free toxin concentrations, and the remainder was aspirated in each case. Pelleted synaptosomes were rapidly washed with 4 drops of a wash medium consisting of 5 mM HEPES (pH 7.4), 163 mM choline chloride, 1.8 mM calcium chloride, 0.8 mM magnesium sulfate, and 1 mg/mL BSA. Pellets were transferred to liquid scintillation vials containing 3 mL of liquid scintillant, and the bound radioactivity was measured using liquid scintillation techniques. Nonspecific binding was measured in the presence of a saturating concentration of unlabeled PbTx-3 and was subtracted from total binding to yield specific binding.

Dissociation Constants and Binding Maxima for Brevetoxin PbTx-3

Brevetoxins bind with high specificity to synaptosomes of fish (*Tilapia sp.*), turtles (*P. scripta*), and rats (Table I). In all cases, the K_d was in the nanomolar

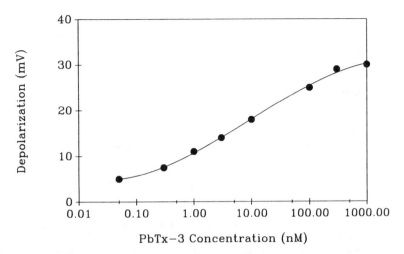

Figure 2. Dose–response curve of membrane depolarization as a function of PbTx-3 concentration (7). Data from a total of 22 axons were pooled; each axon received only one dose. Data are plotted as means of depolarization amplitudes. The solid line is a theoretical 3rd order fit with an ED_{50} of 1.5 nM, maximum observed depolarization of 30 mV, and a Hill's coefficient of 2.

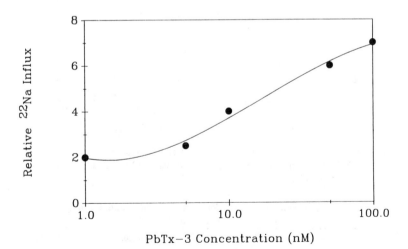

Figure 3. Concentration dependence of the stimulation of $^{22}Na^+$ influx by PbTx-3 (4). Synaptosomes were pre-incubated for 30 min with indicated concentrations of PbTx-3 in the presence of aconitine. Influx is plotted as specific influx, points representing means of triplicate determinations.

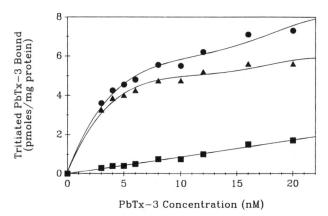

PbTx-3 Concentration (nM)

Figure 4. Binding was measured in rat brain synaptosomes using a rapid centrifugation technique. Total (●), and nonspecific (■) binding of tritiated PbTx-3 were measured, their difference representing specific binding (▲). Rosenthal analysis yields a K_d of 2.6 nM and a B_{max} of 6.0 pmol toxin bound/mg protein.

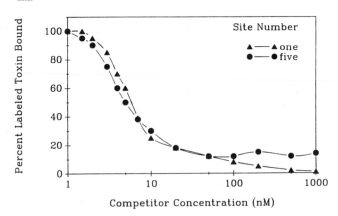

Competitor Concentration (nM)

Figure 5. Comparison of specific displacement of 10 nM tritiated saxitoxin (▲) or 10 nM tritiated PbTx-3 (●) by unlabeled competitor saxitoxin or brevetoxin, respectively, in rat brain synaptosomes. IC_{50} in each case is 5–10 nM.

Table I. Comparison of Dissociation Constant (K_d) and Binding Maximum (B_{max}) in Fish, Turtles, and Rats[a]

Species	K_d (nM)	B_{max} (pmol/mg protein)	Temp. Optimum (°C)	Specific Binding at K_d
Fish	6.1	1.40	23	80%
Turtle	1.5	2.25	4	80%
Rat	2.6	6.80	4	90%

[a] Mean values for K_d and B_{max}, n = 9, 4, 6 for fish, turtles, and rats, respectively.

concentration range, and B_{max} was in the pmol/mg protein range. Specific binding was 80–90% at the dissociation constant concentration in each case. These findings illustrate the general phylogenetic topographic homology of the brevetoxin binding site, and also illustrates that binding maxima increase from fish to turtles and rats.

Specific Displacement of [³H] PbTx-3 by Agonist Brevetoxins

In rat synaptosomes, six of the eight known brevetoxins displace tritiated PbTx-3 from its specific binding site (Figure 6). At 10 nM tritiated PbTx-3, IC_{50} values for the brevetoxins range from 3.5 to 20 nM in the rat system (Table II). Only four brevetoxins could be isolated in sufficient quantities from laboratory cultures, or through chemical modification of another natural toxin, to permit classical competition binding studies. These four toxins are PbTx-2 and PbTx-3, each based on type-1 structural brevetoxin backbone (Figure 1); and PbTx-1 and PbTx-7, toxins based on type-2 structural brevetoxin backbone (Figure 1) with equivalent substituent derivatization to PbTx-2 and PbTx-3 distal to the lactone functionality. Tritiated PbTx-3 is competitively displaced by unlabeled brevetoxins, with K_i values determined by classical graphical methods, ranging from 1.4 to 9.9 nM (*5*).

Regardless of the organism used for synaptosomal preparations, it is apparent to us that the topographic characteristics of the brevetoxin binding site on the VSSC are comparable. Using brevetoxins PbTx-1, -2, and -3, IC_{50} data for specific displacement of tritiated PbTx-3 shows comparable data in each case. The more hydrophobic type-2 brevetoxins are most efficacious in their ability to compete for site 5 binding. It is of interest to note that ciguatoxin is thought to resemble PbTx-1 (*14*). (*See* Frelin et al., this volume.)

Implication of Binding Experiments

Several brevetoxins have been examined for their respective abilities to competitively displace tritiated brevetoxin PbTx-3 from its specific site of action in brain synaptosomes. Analysis of IC_{50} values revealed no marked differences in the displacing abilities between any of the type-1 toxins, and similarly there was no apparent difference between displacing abilities of PbTx-1 or -7, both type-2 toxins. Although some specific details require correlation, a gross comparison indicates that sodium channels in brain are similar in the systems examined. In the system studied most extensively, the rat brain synaptosome, t-test analysis revealed no significant differences between PbTx-2 and PbTx-3 IC_{50}, or between PbTx-1 and PbTx-7 IC_{50}, but statistically significant differences were found between the two classes ($P < 0.01$) (*5*). If the Cheng–Prusoff equation (*15*) is applied:

$$K_i = \frac{IC_{50}}{(1 + \frac{C}{K_d})}$$

where K_i = the inhibition constant, IC_{50} = the inhibitory concentration of competitor toxin required for 50% specific displacement of radioactive toxin, C = the concentration of radioactive toxin, and K_d = the dissociation constant of the radioactive toxin,

then relative affinities of the toxins for the various species receptors can be determined (Table III). Determination of inhibition constants by this equation, however, requires that the compounds of interest interact with only a single receptor subclass. In other words, the inhibition must be clearly and solely competitive in nature. Also implicit is the use of radioactive toxin concentrations at or near the

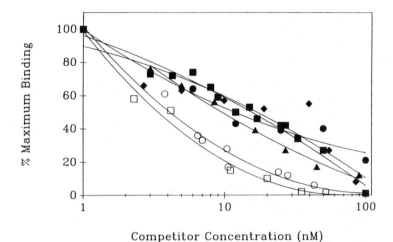

Competitor Concentration (nM)

Figure 6. Effect of brevetoxins on tritiated PbTx-3 binding to rat brain synaptosomes. Incubations, in the presence of 50 μg synaptosomal protein and 16 nM tritiated PbTx-3 with increasing amounts of unlabeled PbTx-1 (□), PbTx-2 (■), PbTx-3 (●), PbTx-5 (▲), PbTx-6 (◆), or PbTx-7 (○) were for 1 hr at 4 °C. Each point represents the mean of three triplicate determinations.

Table II. Specific Displacement of [³H] PbTx-3 from Synaptosome Binding by Unlabeled Brevetoxins, Comparison with LD$_{50}$

	Competitor Toxin Concentration (nM)			
Toxin	Turtle[a] (IC_{50})	Fish[b] (IC_{50})	Rat[c] (IC_{50})	Fish (LD_{50})
PbTx-1	3.0	30	3.5	4.4
PbTx-2	10.3	70	17.0	21.8
PbTx-3	15.0	110	12.0	10.9
PbTx-5	----	---	13.0	42.5
PbTx-6	----	---	32.0	35.0
PbTx-7	----	---	4.1	4.9

Tritiated toxin concentrations: [a]10.0 nM, [b]12.0 nM, [c]10.0 nM.

Table III. Inhibition Constants for Derivative Brevetoxins
Derived from the Cheng–Prusoff[a] Equation

Toxin	K_i (nM)		
	Turtle	Fish	Rat
PbTx-1	0.39	10.10	0.72
PbTx-2	1.34	23.57	3.51
PbTx-3	1.96	37.04	2.47
PbTx-5	----	----	2.68
PbTx-6	----	----	6.60
PbTx-7	----	----	0.85

[a] *See* discussion concerning the limits of applicability for this treatment.

K_d, which necessitates use of high specific activity radioactive toxin (at or above 10 Ci/mmol). Work in progress indicates that all type-1 brevetoxins inhibit tritiated PbTx-3 binding in a purely competitive manner, whereas the type-2 brevetoxins inhibit in a mixed competition manner at higher concentrations (Figure 7).

The comparison of fish bioassays (6) with the calculated effective doses indicate that the two most potent brevetoxins, PbTx-1 and PbTx-7, also are most effective at displacing tritiated probe from its specific site of action. The considerably lower potency of brevetoxins PbTx-5 and PbTx-6 in the rat system suggest that these two toxins may bind with lesser affinity to site 5. In a general sense, this is indicated in Table II, and is summmarized in Table III.

The affinities of each toxin, in each test system, are presumably based on structural considerations: the portion of the toxin molecule which binds to the specific site on the sodium channel must retain sufficient structural integrity to permit binding. Obviously, those toxins which are altered sufficiently so they no longer bind to the site are no longer toxic. This conjecture is supported by past work which indicated that oxidation of the C-42 aldehyde of PbTx-2 to the corresponding carboxylic acid reduced both potency and affinity for the site (11), and that opening of the lactone in ring A destroys all activity (6). Thus, detailed studies of derivative brevetoxins based on the type-1 backbone may lead to increased understanding of the three-dimensional site specificity of brevetoxin binding.

Of significance is the demonstrated increased potency of the type-2 toxins PbTx-1 and PbTx-7 (2, 5, 8) and their increased affinity for the site associated with sodium channels (5). The larger 8- and 9-membered rings (D, E, F) of these toxins may confer a greater binding affinity to synaptosomes. This increased affinity may be a function of increased flexibility of the type-2 backbone (about a 40° bending capability, 14) over the type-1 backbone. The added flexibility and its enhanced propensity to conform to the topography of the channel is an area of potential importance (Figure 1). Preliminary work not reported herein indicates that tritium-labeled PbTx-7 (reduced PbTx-1), PbTx-9 (doubly reduced PbTx-2), and PbTx-10 (doubly reduced PbTx-1) also interact with site 5 associated with the VSSC, with dissociation constants and binding maxima consistent with the data presented for tritiated PbTx-3.

Huang et al. (7) suggested that the brevetoxin binding site lies in the hydrophobic portion of the channel, and since PbTx-1 and PbTx-7 are also the most hydrophobic of the toxins, their potency may also be in part due to solubility considerations. In general, the more hydrophobic the toxin, the higher is its potency and ability to displace tritiated PbTx-3 from its specific binding site. It is our contention that substituent character on each toxin's K (type-1) or J (type-2) ring

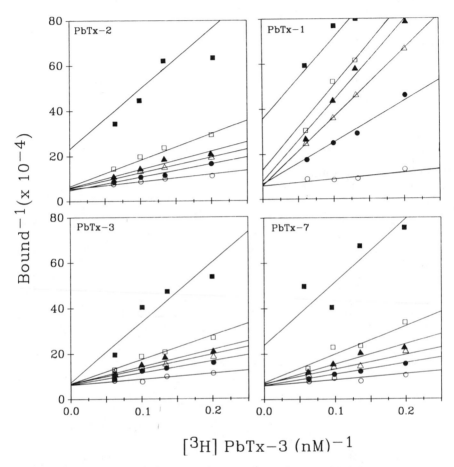

Figure 7. Inhibition of tritiated PbTx-3 specific binding by unlabeled brevetoxins, PbTx-2, PbTx-1, PbTx-3, PbTx-7. Specific binding was measured in intact synaptosomes at 4°C degrees in standard binding medium using four different tritiated toxin concentrations—5.0, 7.5, 10.0, and 15.0 nM (inverse [^{3}H] PbTx-3 abscissa values)—in the presence of unlabeled toxins at 0 (○), 5.0 (●), 7.5 (△), 10.0 (▲), 25.0 (□), 50.0 (■), or 100.0 (data not shown) nM. Points are means of triplicate determinations at each concentration.

determines solubility and hence access to the site. The A-ring in each toxin backbone carries the active portion of the toxin. It is clear to us that both solubility and flexibility characteristics confer differential potency and binding abilities to individual toxins. We feel that naturally occurring toxins, as well as synthetic derivatives, will aid in our understanding of the site-specific potency of toxin from *P. brevis*. Our principal aim is to synthesize derivative toxins with discrete alterations at specific sites without changing their lipid solubilities and hence their access to site 5 associated with voltage-sensitive sodium channels.

Acknowledgments

This work was supported in part by the U.S. Army Medical Research and Development Command, Contract Numbers DAMD17–C–5171 and DAMD17–87–C–7001. Portions of this research were performed in partial fulfillment of the requirements for the M.S.(R.A.E.), Ph.D.(V.L.T.), and M.D. with Research Distinction(A.M.S.) degrees at the University of Miami. The expert technical assistance of Laurie Roszell and Lloyd Schulman is acknowledged.

Literature Cited

1. Steidinger, K. A.; Baden, D. G. In *Dinoflagellates*; Spector, D., Ed.; Academic: New York, 1984; p 201.
2. Shimizu, Y.; Chou, H. N.; Bando, H.; VanDuyne, G.; Clardy, J. C. *J. Am. Chem Soc.* 1986, *108*, 514.
3. Lin, Y. Y.; Risk, M.; Ray, S. M.; Van Engen, D.; Clardy, J.; Golik, J.; James, J. C.; Nakanishi, K. *J. Am. Chem. Soc.* 1981, *103*, 6773.
4. Poli, M. A.; Mende, T. J.; Baden, D. G. *Molec. Pharmacol.* 1986, *30*, 129.
5. Baden, D. G.; Mende, T. J.; Szmant, A. M.; Trainer, V. L.; Edwards, R. A.; Roszell, L. E. *Toxicon* 1988, *26*, 97.
6. Baden, D. G.; Mende, T. J.; Poli, M. A.; Block, R. E. In *Seafood Toxins*; Ragelis, E. P., Ed.; ACS Symposium Series 262, 1984; p. 359.
7. Huang, J. M. C.; Wu, C. H.; Baden, D. G. *J. Pharmacol. Exp. Ther.* 1984, *229*, 615.
8. Catterall, W. A.; Risk, M. A. *Molec. Pharmacol.* 1981, *19*, 345.
9. Catterall, W. A.; Gainer, M. *Toxicon* 1985, *23*, 497.
10. Baden, D. G.; Mende, T. J.; Walling, J.; Schultz, D. R. *Toxicon* 1984, *22*, 783.
11. Poli, M. A. Ph.D. Dissertation, University of Miami, 1985, 129 pp.
12. Dodd, P. R.; Hardy, J. A.; Oakley, A. E.; Edwardson, J. A.; Perry, E. K.; Delaunoy, P. *Brain Res.* 1981, *226*, 107.
13. Bradford, M. M. *Anal. Biochem.* 1976, *72*, 248.
14. Nakanishi, K. *Toxicon* 1985, *23*, 473.
15. Cheng, Y. C.; Prusoff, W. H. *Biochem. Pharmacol.* 1973, *22*, 3099.

RECEIVED June 14, 1989

Chapter 12

Detection, Metabolism, and Pathophysiology of Brevetoxins

Mark A. Poli, Charles B. Templeton, Judith G. Pace, and Harry B. Hines

Pathophysiology Division, U.S. Army Medical Research Institute of Infectious Diseases, Fort Detrick, Frederick, MD 21701–5011

Methods of detection, metabolism, and pathophysiology of the brevetoxins, PbTx-2 and PbTx-3, are summarized. Infrared spectroscopy and innovative chromatographic techniques were examined as methods for detection and structural analysis. Toxicokinetic and metabolic studies for in vivo and in vitro systems demonstrated hepatic metabolism and biliary excretion. An in vivo model of brevetoxin intoxication was developed in conscious tethered rats. Intravenous administration of toxin resulted in a precipitous decrease in body temperature and respiratory rate, as well as signs suggesting central nervous system involvement. A polyclonal antiserum against the brevetoxin polyether backbone was prepared; a radioimmunoassay was developed with a sub-nanogram detection limit. This antiserum, when administered prophylactically, protected rats against the toxic effects of brevetoxin.

Red tides resulting from blooms of the dinoflagellate *Ptychodiscus brevis* in the Gulf of Mexico have elicited a great deal of scientific interest since the first documented event over 100 years ago (*1*). Human intoxications from the ingestion of contaminated shellfish and the impact of massive fish kills on the tourist industry along the Gulf coast of the United States have resulted in a concerted research effort to understand the genesis of red tides and to isolate and characterize the toxins of *P. brevis*.

Research in this area advanced in the 1970's as several groups reported the isolation of potent toxins from *P. brevis* cell cultures (*2–7*). To date, the structures of at least eight active neurotoxins have been elucidated (PbTx-1 through PbTx-8) (*8*). Early studies of toxic fractions indicated diverse pathophysiological effects in vivo as well as in a number of nerve and muscle tissue preparations (reviewed in *9–11*). The site of action of two major brevetoxins, PbTx-2 and PbTx-3, has been shown to be the voltage-sensitive sodium channel (*8,12*). These compounds bind to a specific receptor site on the channel complex where they cause persistent activation, increased Na^+ flux, and subsequent depolarization of excitable cells at resting membrane potentials. At present, the brevetoxins are the only ligands unequivocally demonstrated to act at this site (neurotoxin receptor site 5), although mounting evidence *(13)* suggests the toxins involved in ciguatera intoxication also may bind there.

We are investigating low-molecular-weight toxins of animal, plant, and microbial origin. Our goals are to develop methods to detect these compounds in both environmental and biological samples and to develop prophylaxis and treatment regimens. This chapter summarizes the results of our current investigations of the brevetoxins. Some of these studies will be published elsewhere in greater detail.

Chemical Detection and Stability

A variety of chromatographic and spectrographic techniques have been applied to the study of the brevetoxins. High performance liquid chromatography (HPLC) and thin-layer chromatography (TLC) were instrumental in the initial isolation and purification processes. Mass spectrometry (MS), infrared spectroscopy (IR), circular dichroism (CD), nuclear magnetic resonance spectroscopy (NMR), and X-ray crystallography all played important roles in structure determinations. As research efforts expand to include metabolic studies, these techniques become increasingly important for detection and quantification of exposure as well as structural elucidation of metabolites.

Chromatography. A number of HPLC and TLC methods have been developed for separation and isolation of the brevetoxins. HPLC methods use both C18 reversed-phase and normal-phase silica gel columns *(8, 14, 15)*. Gradient or isocratic elutions are employed and detection usually relies upon ultraviolet (UV) absorption in the 208–215-nm range. Both brevetoxin backbone structures possess a UV absorption maximum at 208 nm, corresponding to the enal moeity *(16,17)*. In addition, the PbTx-1 backbone has an absorption shoulder at 215 nm corresponding to the γ-lactone structure. While UV detection is generally sufficient for isolation and purification, it is not sensitive (>1 ppm) enough to detect trace levels of toxins or metabolites. Excellent separations are achieved by silica gel TLC *(14, 15, 18-20)*. Sensitivity (>1 ppm) remains a problem, but flexibility and ease of use continue to make TLC a popular technique.

Mass Spectrometry. Mass spectrometry holds great promise for low-level toxin detection. Previous studies employed electron impact (EI), desorption chemical ionization (DCI), fast atom bombardment (FAB), and cesium ion liquid secondary ion mass spectrometry (LSIMS) to generate positive or negative ion mass spectra *(15-17, 21-23)*. Firm detection limits have yet to be reported for the brevetoxins. Preliminary results from our laboratory demonstrated that levels as low as 500 ng PbTx-2 or PbTx-3 were detected by using ammonia DCI and scans of 500–1000 amu (unpublished data). We expect significant improvement by manipulation of the DCI conditions and selected monitoring of the molecular ion or the ammonia adduction.

The success of the soft ionization techniques (DCI, FAB, and LSIMS) presents several possibilities for detection of brevetoxins in complex matrices. Positive-ion DCI was used for the analysis of PbTx-3 metabolites generated in vitro by isolated rat hepatocytes (see below). Unmetabolized parent was conclusively identified and metabolites were tentatively identified, pending confirmation by alternate methods (see below).

Chemical Stability and Decontamination. The stability of the brevetoxins is of great interest from the standpoints of detection, metabolism, and safety. PbTx-2 and PbTx-3 have been investigated in our laboratories in order to design rational safety protocols for toxin handling and disposal of contaminated waste (24; R.W. Wannamacher, unpublished data). These compounds were stable for months when stored in the refrigerator either dry or in organic solvents such as ethanol, methanol, acetone, or chloroform. However, toxins with the PbTx-1 backbone have been reported to be unstable in alcohol (15). PbTx-2 and PbTx-3 were unstable at pH values less than 2 and greater than 10, in the presence of 50 ppm chlorine, and at temperatures greater than 300°C. For decontamination of laboratory glassware and surfaces, greater than 99% of the detectable brevetoxin was destroyed by a 10-min exposure to 0.1 N NaOH. Disposable waste can be incinerated if the combustion chamber temperature reaches at least 300°C. Autoclaving was shown to be ineffective for decontamination (24).

Distribution and Metabolic Fate—In Vivo and In Vitro Studies

Of particular interest in brevetoxin research are the diagnosis of intoxication and identification of brevetoxins and their metabolites in biological fluids. We are investigating the distribution and fate of radiolabeled PbTx-3 in rats. Three model systems were used to study the toxicokinetics and metabolism of PbTx-3: 1) rats injected intravenously with a bolus dose of toxin, 2) isolated rat livers perfused with toxin, and 3) isolated rat hepatocytes exposed to the toxin in vitro.

In the first study, male Sprague-Dawley rats (300–350 g) were given an intravenous bolus of [^{3}H]PbTx-3 (9.4 Ci/mmol, 6 μg/kg body weight) via the penile vein (25). The plasma concentration curve (Figure 1) was bi-exponential with a rapid distribution phase (half-life approx. 30 sec) and a slower elimination phase (half-life = 112 min). Toxin clearance (dose administered/area under the plasma concentration curve) in the whole animal was 0.23 ml/min/g liver. In a 325-g rat, hepatic blood flow (Q) is 13.2 ml/min (26), and, assuming hepatic clearance was equal to mean total clearance (Cl), the calculated in vivo extraction ratio (Cl/Q) was 0.55. Within 1 min, 94% of the administered toxin had distributed to the tissues. After 30 min, the liver contained 16%, skeletal muscle 70%, and the gastrointestinal tract 8% of the administered radioactivity. The heart, kidneys, testes, brain, lungs, and spleen each contained less than 1.5%. By 24 hr, radioactivity in skeletal muscle decreased to 20% of the total administered dose. Over the same period, radioactivity remained constant in the liver and increased in the stomach, intestines, and feces, suggesting biliary excretion was an important route of toxin elimination. By day 6, 89% of the total radioactivity had been excreted in the urine and feces in a ratio of 1:5. TLC analysis of urine (Figure 2) and feces indicated that the parent toxin had been metabolized to several more polar compounds.

To further investigate the role of the liver in brevetoxin metabolism, PbTx-3 was studied in the isolated perfused rat liver model (27, 28). Radiolabeled PbTx-3 was added to the reservoir of a recirculating system and allowed to mix thoroughly with the perfusate. Steady-state conditions were reached within 20 min. At steady-state, 55–65% of the delivered PbTx-3 was metabolized and/or extracted by the liver; 26% remained in the effluent perfusate. Under a constant liver perfusion rate of 4 ml/min, the measured clearance rate was 0.11 ml/min/g liver. The calculated extraction ratio of 0.55 was in excellent agreement with the in vivo data. Radioactivity in the bile accounted for 7% of the total radiolabel perfused through the liver. PbTx-3 was metabolized and eliminated into bile as parent toxin plus four more-polar metabolites (Figure 3). Preliminary results of samples stained with 4-(p-nitrobenzyl)-pyridine (29) indicated the most polar metabolite was an epoxide.

In vitro metabolism of [^{3}H]PbTx-3 was studied in isolated rat hepatocytes (25). Hepatocyte monolayers cultured in 6-well plates containing 1 ml modified Williams E medium were incubated with 0.1 μg radiolabeled toxin at 37°C for 24 hr. The

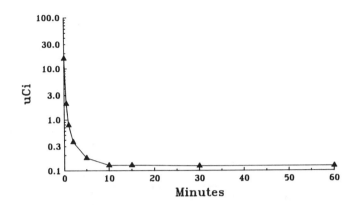

Figure 1. Semilogarithmic plot of brevetoxin (μCi) in plasma over time after an intravenous injection of tritium-labeled PbTx-3.   T 1/2 alpha = 30 sec; T 1/2 beta = 112 min.

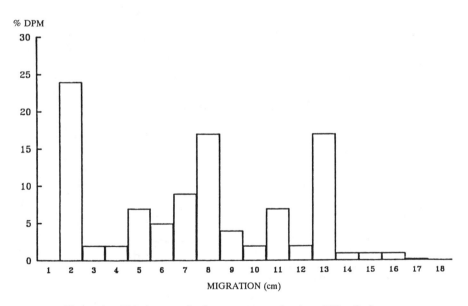

Figure 2.   Thin-layer radiochromatogram of urine (100 μl) from rats injected with labeled PbTx-3.   TLC plates were developed in two sequential solvent systems: chloroform:ethyl acetate:ethanol (50:25:25; 80:10:10).   Radioactive zones were scraped and counted in a liquid scintillation counter.   Native PbTx-3 runs at 13 cm.

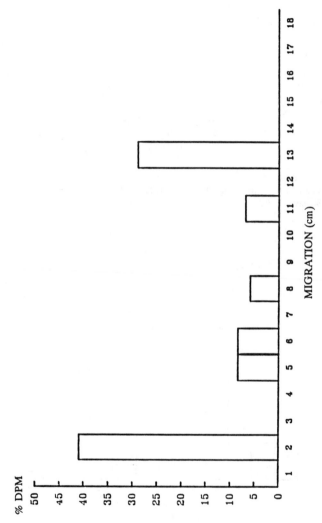

Figure 3. Thin-layer radiochromatogram of bile (1 hr). Aliquots of bile (20 μl) were analyzed by TLC by two sequential solvent systems of chloroform:ethyl acetate:ethanol (50:25:25; 80:10:10). Native PbTx-3 runs at 13 cm.

cell culture medium was sampled periodically during the incubation and analyzed for labeled metabolites via HPLC. The hepatocytes metabolized PbTx-3 to at least three metabolites that exhibited greater polarity than the parent toxin and corresponded closely in relative polarity to the major metabolites obtained in vivo and from the isolated perfused liver. In addition, a fourth peak appeared after approximately 20 min (Figure 4). From preliminary results, it is not clear whether this fourth peak was related to cellular metabolism. The differences in metabolite profiles between this system and the bile produced by the isolated perfused liver are under investigation. Two of the three metabolites generated in vitro cross-reacted strongly with our anti-PbTx antiserum (see below), suggesting conservation of at least a portion of the native backbone structure (data not shown).

By applying an extension of the clearance concept (*30, 31*), in vitro metabolism was used to predict in vivo toxin elimination. Hepatocytes were incubated with 0.5 to 10 μg unlabeled PbTx-3 containing 0.1 μg radiolabeled toxin as tracer. Disappearance of parent compound and the appearance of metabolites were measured by HPLC equipped with a Radiomatic isotope detector. V_{max} (1.6 nmol/min/g liver) and K_m (8.56 μM) were determined by non-linear regression. Intrinsic clearance (V_{max}/K_m) was 0.15 ml/min/g liver and the calculated hepatocyte extraction ratio, 0.46, again was in good agreement with the in vivo data.

These studies represent the first report of the metabolism of brevetoxins by mammalian systems. PbTx-3 was rapidly cleared from the bloodstream and distributed to the liver, muscle, and gastrointestinal tract. Studies with isolated perfused livers and isolated hepatocytes confirmed the liver as a site of metabolism and biliary excretion as an important route of toxin elimination. [^{3}H]PbTx-3 was metabolized to several compounds exhibiting increased polarity, one of which appeared to be an epoxide derivative. Whether this compound corresponds to PbTx-6 (the 27,28 epoxide of PbTx-2), to the corresponding epoxide of PbTx-3, or to another structure is unknown. The structures of these metabolites are currently under investigation.

Data from both in vivo and in vitro systems showed PbTx-3 to have an intermediate extraction ratio, indicating in vivo clearance of PbTx-3 was equally dependent upon liver blood flow and the activity of toxin-metabolizing enzymes. Studies on the effects of varying flow rates and metabolism on the total body clearance of PbTx-3 are planned. Finally, comparison of in vivo metabolism data to those derived from in vitro metabolism in isolated perfused livers and isolated hepatocytes suggested that in vitro systems accurately reflect in vivo metabolic processes and can be used to predict the toxicokinetic parameters of PbTx-3.

Pathophysiological Effects of PbTx-2 in Conscious Rats

The pathophysiological effects of PbTx-2 were examined in the conscious, tethered rat (*32*). This is the model of choice because neurological and behavioral responses can be characterized without interference from anesthetic effects. Male rats (Sprague-Dawley, 350–500 g) were anesthetized with 55 mg/kg pentobarbital and placed on a heated surgical board. A catheter, placed into the carotid artery and advanced until the distal tip resided in the aorta, was used to measure arterial blood pressure and to sample blood for blood gas measurements. Another catheter, placed into the jugular vein and advanced until the distal tip was near the cranial vena cava, was used for experimental infusions. Thermistor probes were implanted into the abdominal cavity and subcutaneously over the sternum. Electrocardiograph (ECG) leads were placed subcutaneously over the ventral and dorsal thorax to obtain a V10 tracing. All lead wires and catheters were tunneled subcutaneously to the dorsal cervical area and passed through a 20-cm steel spring tether. Catheters

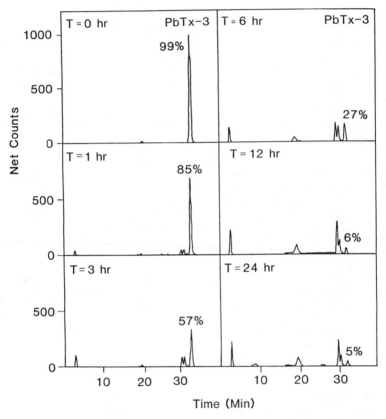

Figure 4.   C18 reverse-phase HPLC radiochromatogram of PbTx-3 and metabolites in hepatocyte media (50 μl) over 24 hr.   Nonlinear gradient of methanol:water; flow rate = 1 ml/min.

were then flushed with heparinized saline and plugged with stainless steel pins. The animals were placed in cages and the tether passed through the wire mesh cage top to allow sampling. After recovery from anesthesia, the rats were able to move freely about the cage while being monitored; sampling was conducted without perturbing the animals.

The initial phase of the study focused on determining a route of toxin administration that would result in sublethal toxic effects. Intravenous, intraarterial, intraperitoneal, and subcutaneous bolus injections were examined. With the exception of the subcutaneous route, bolus injections of greater than 200 μg/kg resulted in total body paralysis, shock, and death within 2–3 min. Onset was so rapid that the animals often entered paralysis during the several seconds required to complete the bolus injection. Doses of 100 μg/kg or greater resulted in rapid-onset cardiac and respiratory paralysis and death within 5–10 min. With lower, non-lethal, intravascular or intraperitoneal doses (≤25 μg/kg), rats showed immediate cardiac and respiratory effects, but compensation occurred within 5–10 min. There were no apparent skeletal muscle contractions in these lower-dose animals. Both subcutaneous and intraperitoneal administration caused sustained effects, but responses varied among animals. However, when rats were slowly infused intravenously at doses of ≤100 μg/kg over 1 hr, animal-to-animal variation decreased markedly and responses were sustained for a longer time. Therefore, to allow sufficient monitoring time and to minimize variations among animals, a 1-hr intravenous infusion was chosen for all subsequent studies.

To characterize the responses to PbTx-2, five dose rates (0, 12.5, 25, 50, and 100 μg/kg/hr in 2 ml saline) were infused into the jugular catheters of rats (four per group). Heart rates, systolic and diastolic arterial blood pressures, pulse pressures, respiratory rates, core and peripheral body temperatures, lead V10 ECGs, and arterial blood gases were monitored. Clinical signs and behaviors were recorded by video camera. After infusion, animals were monitored for 6 hr, by which time most had either died or recovered to near baseline physiological levels.

All rats infused with 100 μg/kg PbTx-2 died within 2 hr. Only one rat infused with 50 μg/kg died during the 6-hr study. All other animals survived. The calculated 6-hr LD_{50} for this study of 20 animals given a 1-hr intravenous infusion of PbTx-2 was 60 μg/kg. Since all of the rats given 50 μg/kg by bolus injection died very early, it is evident that the LD_{50} is lower if the toxin is given intravascularly by bolus injection.

The most dramatic change occurred in the respiratory rates. Upon the beginning of infusion, there was an immediate, precipitous, dose-dependent decrease in respiratory rates (Figure 5). The respiratory rates of the three highest dose groups fell to 20% of baseline. Except for the terminal values of the high-dose group, the blood gas analyses indicated a compensatory response that yielded insignificant changes in pO_2, pCO_2, HCO_3, base excess, pH, and total CO_2 during the entire 6-hr study. The 100 μg/kg group displayed typical terminal hypoventilation: hypercarbia, acidosis, and low oxygen tension (data not shown). Clinically, the animals ventilated very deeply, indicating much larger tidal volumes. All survivors had respiratory rates within the normal range within 6 hr, except the 50 μg/kg group, which recovered to only about 60% of baseline.

Each treatment group showed a dose-dependent decrease in core (Figure 6) and peripheral (not shown) body temperatures. The three highest dose groups decreased 1.5–2.0 degrees centigrade. The decrease in the lowest dose group averaged 0.5 degrees. In all groups, the decrease occurred during the infusion period. In the 50 μg/kg group, survivors showed no significant recovery during the 6-hr study. This simultaneous drop in core and peripheral body temperature may reflect decreased oxygen consumption and reduced heat production.

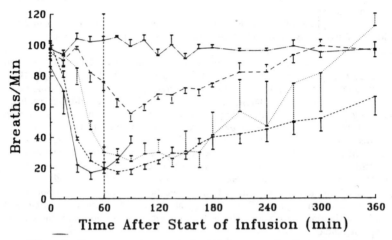

Figure 5. Dose–response curve of respiratory rate for rats given 1-hr infusion of PbTx-2. [_____ = 100 μg/kg; --- = 50 μg/kg; ... = 25 μg/kg; — — — = 12.5 μg/kg; —.— = 0 μg/kg.]

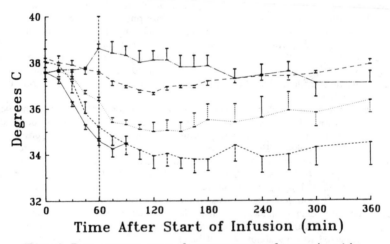

Figure 6. Dose–response curve of core temperature for rats given 1-hr infusion of PbTx-2. [_____ = 100 μg/kg; --- = 50 μg/kg; ... = 25 μg/kg; — — — = 12.5 μg/kg; —.— = 0 μg/kg.]

The lead V10 ECG indicated conduction defects in the Purkinje system of the heart. Numerous arrhythmias were recorded, including premature ventricular depolarizations, paroxysmal ventricular tachycardia, and complete heart block (Figure 7). Heart rates were not significantly altered, however, except in the 100 μg/kg group (data not shown). Systolic and diastolic arterial blood pressures did not change significantly in the higher-dose groups, but pulse pressures increased, suggesting compensation by increased cardiac outputs due to increased stroke volumes. Failure to maintain baseline heart rate and arterial pressure was seen in the 100 μg/kg group just before death, when the animals showed signs of complete cardiorespiratory collapse.

Rats administered PbTx-2 exhibited numerous and varied clinical signs. Consistent in all animals were gasping-like respiratory movements, head-bobbing, depression, and ataxia, all of which could be originating in the brain stem. Time of onset varied, but depression and gasping movements were generally the first clinical manifestations of intoxication. Head-bobbing and ataxia usually began 2-3 hr after infusion. These signs, coupled with the decrease in core temperature, suggested central nervous system involvement, and, more specifically, the brain stem and cerebellum. This conclusion was sustained in later studies when several animals developed head-tilt, a condition indicative of central nervous system involvement. In some animals, an apparently uncontrolled muscular contraction occurred, usually in the hindquarters, which caused the animals to lunge violently across the cage in one coordinated motion. Both hindquarters contracted simultaneously, indicating the initiating impulse originated at the level of the spinal cord or higher. This movement and head-bobbing are indicative of cerebellar involvement. All control animals sustained normal physiological parameters, ECG's, and behavior.

In summary, intravenous infusion of PbTx-2 caused toxic signs indicative of central nervous system involvement, including a precipitous fall in core and peripheral body temperatures and clinical signs of dysfunction that could be originating in the motor cortex, the cerebellum, or the spinal cord. The profound decreases in respiratory and heart rates could be peripherally mediated; these changes were detrimental only in the highest dose group animals. Compensation for the rate changes in the surviving animals occurred to an extent that blood gases and blood pressures were essentially normal.

Radioimmunoassay and Immunoprophylaxis

The ability to treat brevetoxin intoxication is dependent upon diagnosis and quantifying exposure. Currently, neither an effective, specific treatment nor a reliable assay for exposure to the brevetoxins exists. We prepared a specific antiserum against the brevetoxins and evaluated its use in an assay for exposure to the brevetoxins and in the treatment and prophylaxis of intoxication.

A toxin-bovine serum albumin (BSA) conjugate was prepared by succinylation of the alcohol function of PbTx-3, followed by standard carbodiimide coupling of the PbTx-3 hemisuccinate to the lysine residues of BSA (33). Fresh conjugate was prepared every two weeks to avoid decomposition; the molar ratios of toxin:BSA conjugate ranged from 7–14. Aliquots of conjugated toxin containing 0.25–0.50 mg equivalents of toxin in Freund's adjuvant were administered intramuscularly to an adult female goat. Blood samples were taken at weekly intervals, beginning at week 8, and analyzed for anti-PbTx-3 binding activity by standard radioimmunoassay.

Anti-PbTx antibodies appeared rapidly in the goat serum after the initial immunization. By the first bleed at week 8, binding capacity was approximately 0.5 nmol PbTx-3/ml serum, where it remained constant for several weeks. After cessation of boosts at week 10, the serum titers remained elevated for 6 weeks before

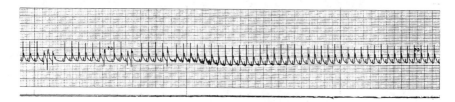

Premature Ventricular Depolarization
No PbTx Antibody Treatment
(Lead V10, 1 cm/mv, 25 mm/sec)

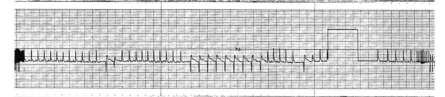

Paroxysmal Ventricular Tachycardia
No PbTx Antibody Treatment
(Lead V10, 1 cm/mv, 25 mm/sec)

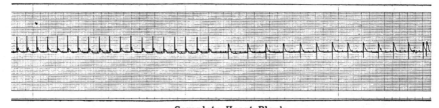

Complete Heart Block
No PbTx Antibody Treatment
(Lead V10, 1 cm/mv, 25 mm/sec)

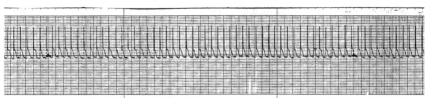

Normal Sinus Rhythm
PbTx Antibody—Treated
(Lead V10, 1 cm/mv, 25 mm/sec)

Figure 7. Electrocardiograms of rats at 30 to 90 min after beginning a
1-hr infusion of PbTx-2 (25 μg/kg). Bottom tracing is a representative
ECG of a normal rat.

falling to very low levels by week 21. Resuming boosts at this time caused an immediate rebound in binding capacity to approximately 1.0 nmol/ml. Ammonium sulfate precipitation of the serum at week 22 resulted in $\geq$85% of the anti-PbTx-3 binding capacity precipitating with the serum IgG fraction. Rosenthal analyses of saturation curves before and after precipitation indicated a single class of high-affinity antibodies with an apparent affinity constant (K_d) of 1 nM (0.5–2.1, n=11) (data not shown).

Standard curves performed under our defined radioimmunoassay conditions ([^{3}H]PbTx-3 = 1 nM, antiserum dilution = 1:2000, assay volume = 1 ml) demonstrated the ability of this antiserum to bind equally to PbTx-2 and PbTx-3, suggesting specificity for the cyclic polyether backbone region of the molecule (Figure 8). The linear portion of the curve indicated a lower detection limit of 0.2–0.5 ng in saline buffer under these conditions. Evaluation of this assay for use with biological fluids and tissue extracts is underway.

After initial experiments demonstrating that the antiserum was capable of completely inhibiting the binding of [^{3}H]PbTx-3 to its receptor site in rat brain membranes (Figure 9), we began studies designed to evaluate potential of the antiserum for prophylaxis and treatment of brevetoxin intoxication (*34*). The tethered rat model was used, and surgical implantations were identical to those described above. Heart rate, core and peripheral body temperatures, lead V10 ECG, and arterial blood pressure were monitored continuously. Respiratory rate was recorded each 5 min for the first 3 hr, then each 15 min until 6 hr.

Conscious rats were pretreated with a 10-min infusion of anti-PbTx antiserum (25 mg/kg total protein, calculated PbTx binding capacity 0.29 nmol) or saxitoxin (control) antibody matched for total protein content. Twenty minutes after completion of the antisera infusion, brevetoxin (25 μg/kg, 28 nmol) was infused over 1 hr. Rats pretreated with control antiserum showed signs consistent with our pathophysiological characterization of brevetoxin intoxication at this dose: decreased respiratory rate, reduction in core and peripheral body temperatures, and cardiac arrhythmias. Arterial blood pressures and heart rates were variable. Rats treated with anti-PbTx antiserum showed no decrease in temperatures (Figure 10), no significant decreases in respiratory (Figure 11) or heart rates, and normal ECG's (Figure 7). Our antiserum was thus capable of blocking the in vivo effects of intravenously administered PbTx-2 under these conditions. Experiments with varying antibody:toxin molar ratios and pretreatment times are planned.

Summary

In summary, a high-affinity antiserum was successfully raised in a goat by immunization with a PbTx-3/BSA conjugate. This antiserum did not differentiate between PbTx-2 and PbTx-3, suggesting specificity for the polyether backbone structure of the molecule. A radioimmunoassay using this serum has been developed. Standard curves indicate a detection limit of 0.2–0.5 ng toxin in phosphate-buffered saline. Development of the assay for use in biological fluids or tissue extracts is underway. At least two of the three PbTx-3 metabolites produced in vitro by isolated rat hepatocytes (see above) cross-reacted strongly with this antiserum. This suggests use of this antiserum as a screen for exposure to PbTx-3 and other brevetoxins sharing the same backbone structure. This antiserum completely inhibited in vitro binding of [^{3}H]PbTx-3 to its receptor site in rat brain tissue, and blocked the in vivo pathophysiological effects of intravenous infusions of PbTx-2 when administered prophylactically. Studies to evaluate its therapeutic potential in the management of brevetoxin intoxication are underway.

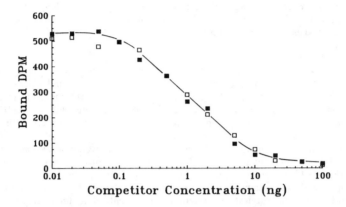

Figure 8. Standard curves for PbTx-2 (□) and PbTx-3 (■) in the brevetoxin radioimmunoassay. Lower detection limits are 0.2 – 0.5 ng in phosphate-buffered saline (PBS). Standard RIA conditions: [^{3}H]PbTx-3 = 1 nM; antiserum dilution = 1:2000; sample vol. = 1 ml; buffer = 0.1 M PBS, pH = 7.4.

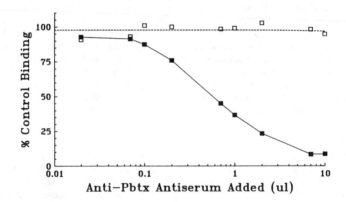

Figure 9. Anti-PbTx antiserum inhibition of [^{3}H]PbTx-3 binding to its receptor site in rat brain membrane preparations. Labeled toxin (0.5 nM in 1 ml PBS) was incubated with rat brain membranes (125 μg total protein) and increasing amounts of anti-PbTx antiserum (-■-) or pre-immune serum (-□-) for 1 hr at 4°C. Membrane-bound radioactivity was then measured in a centrifugation assay as previously described (8).

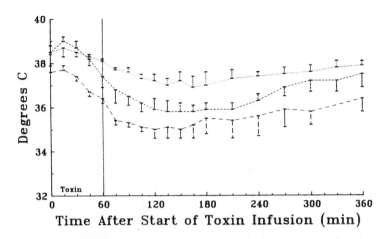

Figure 10. Response curves of core temperature for rats given 1-hr infusion of PbTx-2 (100 μg/kg) followed by a 10-min infusion of 2 ml of saline (———), 6:1 molar ratio of control antibody:toxin (- - -), or 6:1 molar ratio of anti-PbTx antibody:toxin (. . .).

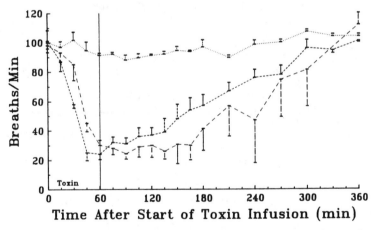

Figure 11. Response curves of respiratory rate for rats given 1-hr infusion of PbTx-2 (100 μg/kg) followed by a 10-min infusion of 2 ml of saline (———), 6:1 molar ratio of control antibody:toxin (- - -), or 6:1 molar ratio of anti-PbTx antibody:toxin (. . .).

Acknowledgments

The work of M.P. was performed during his tenure as a National Research Council Postdoctoral Associate at the United States Army Medical Research Institute of Infectious Diseases in Frederick, Maryland.
All procedures performed in this study conform to the "Guide to the Care and Use of Laboratory Animals" published by the National Institutes of Health, Bethesda, Maryland. All research facilities are accredited by the American Association for Accreditation of Laboratory Animal Care.
The opinions of the authors in no way reflect the opinions of the Department of the Army or the Department of Defense.

Literature Cited

1. Davis, C.C. *Bot. Gaz.* **1947**, *109*, 358–360.
2. Trieff, N.M.; Ramanujam, V.M.S.; Alam, M.; Ray, S.M. *Proc. 1st Intl. Conf. Toxic Dinoflagellate Blooms* **1975**, p 309–321.
3. Alam, M.; Trieff, N.M.; Ray, S.M.; Hudson, J. *J. Pharmaceut. Sci.* **1975**, *64*, 685.
4. Risk, M.; Lin, Y.Y.; MacFarlane, R.D.; Sadagopa-Ramanujam, V.M.; Smith, L.L.; Trieff, N.M. In *Toxic Dinoflagellate Blooms*; Taylor, D.L.; Seliger, H.H., Eds.; Elsevier North Holland: New York, **1979**; p 335–344.
5. Baden, D.G.; Mende, T.J.; Block, R. In *Toxic Dinoflagellate Blooms*; Taylor, D.L.; Seliger, H.H., Eds.; Elsevier North Holland: New York, **1979**; p 327–334.
6. Lin, Y.Y.; Risk, M.; Ray, S.M.; Van Engen, D.; Clardy, J.; Golik, J.; James, J.C.; Nakanishi. K. *J. Am. Chem. Soc.* **1981**, *103*, 6773–6774.
7. Shimizu, Y. *Pure Appl. Chem.* **1982**, *54*, 1973–1980.
8. Poli, M.A.; Mende, T.J.; Baden, D.G. *Mol. Pharmacol.* **1986**, *30*, 129–135.
9. Baden, D.G. *Int. Rev. Cyt.* **1983**, *82*, 99–149.
10. Steidinger, K.A.; Baden, D.G. In *Dinoflagellates*; Specter, D., Ed.; Academic: New York, **1984**; p 201–261.
11. Poli, M.A. Ph.D. Thesis, University of Miami, Florida, 1985.
12. Huang, J.M.C.; Wu, C.H.; Baden, D.G. *J. Pharmacol. Exp. Ther.* **1984**, *229(2)*, 615–621.
13. Lombet, A.; Bidard, J.-N.; Lazdunski, M. FEBS Lett. **1987**, *219(2)*, 355-359.
14. Risk, M.; Lin, Y.Y.; Ramanujam, S.; Smith, L.L.; Ray, S.M.; Trieff, N.M. *J. Chrom. Sci* 1979, *17*, 400–405.
15. Whitefleet-Smith, J.; Boyer, G.L.; Schnoes, H.K. *Toxicon* **1986**, *24*, 1075–1090.
16. Pawlok, J.; Tempesta, M. S.; Golik, J.; Zagorski, M. G.; Lee, M.S.; Nakanishi, K.; Iwashita, T.; Gross, M.L.; Tomer, K.B. *J. Am. Chem. Soc.* **1986**, *109*, 1144–1150.
17. Nakanishi, K. *Toxicon* **1985**, *23*, 474–479.
18. Pace, J.G.; Watts, M.R.; Burrows, E.D.; Dinterman, R.E.; Matson, C.; Hauer, E.C.; Wannemacher, R.W., Jr. *Toxicol. Appl. Pharmacol.* **1985**, *80*, 337–385.
19. Baden, D.G.; Mende, T.J. *Toxicon* **1982**, *20*, 457–461.
20. Baden, D.G.; Mende, T.J.; Lichter, W.; Wellham, L. *Toxicon* **1981**, *19*, 455–462.
21. Chou, H.; Shimizu, Y. *Tetrahed. Lett.* **1982**, *23*, 5521–5524.
22. Golik, J.; James, J.C.; Nakanishi, K. *Tetrahed. Lett.* **1982**, *23*, 2535–2538.
23. Shimizu, Y.; Chou, H.; Bando, H. *J. Am. Chem. Soc.* **1986**, *108*, 514–515.
24. Poli, M.A. *J. Assoc. Off. Analyt. Chemists.* **1988**, *71(5)*, 1000-1002.
25. Poli, M.A. *Toxicon* **1988**, *26*, 36.
26. Dobson, E.L.; Jones, H.B. *Acta Med. Scand.* **1952**, *144* (Suppl. 273), 34.

27. Pace, J.G. *Fund. Appl. Toxicol.* **1986**, *7*, 424–433.
28. Pace, J.G.; Poli, M.A.; Canterbury, W.J.; Matson, C.F. *Pharmacologist* **1987**, *29*, 150.
29. Takitani, S.; Asabe, Y.; Kata, T.; Ueno, Y. *J. Chromatog.* **1979**, *172*, 335–342.
30. Rane, A.; Wilkinson, G.R.; Shand, D.G. *J. Pharmacol. Exp. Ther.* **1977**, *200*, 420–424.
31. Rowland, M. *Eur. J. Pharmacol.* **1972**, *17*, 352–356.
32. Templeton, C.B.; Poli, M.A.; LeClaire, R.D. *Toxicon* **1989**, in press.
33. Baden, D.G.; Mende, T.J.; Walling, J.; Schultz, D.R. *Toxicon* **1984**, *22*, 783–789.
34. Templeton, C.B.; Poli, M.A.; Solow, R. *Toxicon* **1989**, in press.

RECEIVED June 26, 1989

Chapter 13

The Molecular Basis of Ciguatoxin Action

Christian Frelin[1], Monique Durand-Clément[2], Jean-Noël Bidard[1], and Michel Lazdunski[1]

[1]Centre de Biochimie du Centre National de la Recherche Scientifique, Parc Valrose, 06034 Nice Cedex, France
[2]Institut National de la Santé et de la Recherche Médicale, U 303, B.P. 3, 06230 Villefranche-sur-Mer, France

Ciguatoxin is the main toxin involved in ciguatera fish poisoning. It is produced by the dinoflagellate *Gambierdiscus toxicus* and accumulates in large piscivorous fishes. Ciguatoxin increases the membrane permeability to Na^+ ions of excitable tissues by opening the voltage-dependent Na^+ channel. Ciguatoxin binds to the Na^+ channel at a binding site that is different from all other toxin binding sites previously recognized. This action accounts for the neurotoxicity of the compound. Brevetoxins also act specifically on voltage-dependent Na^+ channels; their site of action is identical to the site of action of ciguatoxin.

Ciguatera, a specific endemic affliction of many tropical islands, is due to alimentary consumption of toxic individuals of many species of fishes that are associated with coral reefs (*1*). This food poisoning includes both gastrointestinal and polymorphous neurological symptoms (*2*) with typical gastroenteritis, itching of the skin, peripheral neuropathy, and central nervous system dysfunction (*3*). Symptoms usually appear from a few hours to twelve hours after fish consumption and pass within a few weeks. Some of the neurological disorders may persist for months. The disease is severe but rarely fatal (less than 0.5% mortality); it may affect up to 1% of the individuals in some Pacific island populations. No effective drug is currently known for therapy.

The major toxic compound that is responsible for the illness, ciguatoxin (CTx), was first isolated from the liver of moray eels by Scheuer et al. (*4*). Randall (*5*) suggested that CTx might be synthesized at the base of the food chain, by a benthic alga associated with coral reefs. Yasumoto et al. (*6*) found later that ciguateric reefs in the Gambier islands (Pacific) were populated by large populations of a new dinoflagellate species named *Gambierdiscus toxicus* (*7*) and that fish toxicity in the Gambier islands was correlated to *Gambierdiscus toxicus* population levels. The dinoflagellate is found in an epibenthic association on various macroalgae hosts. Benthic phytoplankton samples were collected and extracted to yield two toxins: CTx and a more polar toxin, maitotoxin (*8*). The two toxins are transmitted to fish through the marine food chain (*2,8,9*). Maitotoxin is mainly found in the gut of herbivorous fishes. CTx, which seems to be chemically more stable than maitotoxin, is mainly found in the liver (*10*) but also in muscles, skin, and bones of large carnivorous fishes (*2,11*). It may even be found in some pelagic species (*12*). Although *Gambierdiscus toxicus* was the first dinoflagellate species to be involved in the genesis of ciguateric toxins, other toxic dinoflagellates have been

0097–6156/90/0418–0192$06.00/0

suspected to be potential sources of ciguateric toxins (*13,14*). Therefore, a number of benthic microorganisms may contribute to the complex syndrome of ciguatera fish poisoning (*15,16*).

Cultures of *Gambierdiscus toxicus* have been obtained in several laboratories (*14,17-19*). These cultures produce large amounts of maitotoxin and low amounts of lipid-soluble CTx-like toxin. However, in most cases, this toxin has not been unequivocally identified as CTx. The only firm evidence that cultures of *Gambierdiscus toxicus* produce CTx was provided by Baden et al. (*20*) who used radioimmunoassays and electrophysiological experiments to characterize the toxin. It is possible that cultured *Gambierdiscus toxicus* produce only trace amounts of CTx and that levels of production comparable to those found in natural populations are dependent on yet undefined environmental parameters.

Chemical characterization of CTx has been difficult because of its very low concentration in fishes (1-20 ppb). The full molecular structure has not yet been elucidated. Data obtained at this time indicate that it is a polar, highly oxygenated polyether compound of molecular weight 1112 and of formula $C_{53}H_{77}NO_{24}$ or $C_{54}H_{78}O_{24}$ (*21*). The purified toxin can be separated into two related forms by alumina chromatography (*22*). Toxicity of the purified toxin is high in mice [0.45 µg/kg, i.p. (*23*)]. CTx is also toxic to mangooses, cats, chicken, and mosquitos (*24*). All these species can be used in toxicological screenings. More recent approaches use anti-ciguatoxin antibodies (*25,26*) or anti-brevetoxin antibodies (*20*).

Brevetoxins (PbTx) are toxins synthesized by another dinoflagellate species (*Ptychodiscus brevis*). Brevetoxins are also of polycyclic etheral structure (*27*). They cause massive fish mortality and human intoxication during "red tides" in the Gulf of Mexico and Florida coast. Toxicity results from ingestion of contaminated shellfish and from inhalation of sea spray aerosols (*28*). Neurotoxic shellfish poisoning syndrome consists of both gastrointestinal and neurological symptoms (*29*). They are very similar to those produced by ciguatoxin.

This chapter summarizes recent work on the molecular basis of the toxic actions of ciguatoxin and brevetoxins. It is shown (i) that the molecular target for these toxins is the voltage-dependent Na^+ channel of excitable tissues and (ii) that ciguatoxin and brevetoxins share a common receptor site on the Na^+ channel.

The Voltage-Dependent Na^+ Channel

The voltage-dependent Na^+ channel is an important protein structure of the plasma membrane of excitable cells, such as neuronal cells, skeletal muscle cells and cardiac cells. It is responsible for the ascending phase of the action potential. The channel consists of a single polypeptide chain of large molecular weight (270 kD)(*30-36*). The structure has been cloned and its amino acid sequence determined (*37, 38*). The purified protein has been reconstituted into lipid bilayers (*39-42*). The cloned protein has been expressed in *Xenopus* oocytes (*38,43,44*). These experiments showed that all the properties of the channel, i.e. the voltage dependence of its opening and closing, its ionic selectivity, and its sensitivity to various neurotoxins, are encoded by the 270 kD protein.

One important point is that the voltage-dependent Na^+ channel is the target for a large variety of neurotoxins. Five groups of neurotoxins have been recognized. Each group corresponds to a distinct binding site on the Na^+ channel and produces well-defined changes in the Na^+ channel response to depolarizing stimuli. These are:

1. The Na^+ channel antagonists tetrodotoxin (TTX) and saxitoxin (*45*). Tetrodotoxin and saxitoxin are important tools to discriminate between different types of Na^+ channels. Some excitable tissues (e.g., neuronal cells and innervated

mammalian skeletal muscle) express Na^+ channels that are highly sensitive to TTX (K_d = 1 nM). Conversely, denervated skeletal muscles and cardiac muscle cells express Na^+ channels that are 100 to 1000 times less sensitive to TTX (46-51). The TTX and saxitoxin binding site is also recognized by μ-conotoxins isolated from the venom of Conus geographus. The μ-conotoxins are single polypeptide chains of 22 amino acids. They specifically inhibit, in a TTX-like fashion, Na^+ channels from innervated skeletal muscles and from related tissues such as the eel electroplax. They are inactive on Na^+ channels from neuronal tissues and on TTX-insensitive Na^+ channels of denervated muscles and heart (52-55).

2. The lipid-soluble toxins (veratridine, batrachotoxin, aconitine, grayanotoxins). These toxins cause persistent activation of Na^+ channels, i.e., their permanent opening and hence membrane depolarization (56-58).

3. The polypeptide neurotoxins from sea anemone (Anemonia sulcata, Anthopleura xanthogrammica, Actinodendron plumosum, Stoichactis giganteus, Radianthus paumotensis) and from the venoms of the scorpions Leiurus quinquestriatus and Androctonus australis. The different sea anemone toxins display large sequence homologies. They are unrelated to scorpion toxins. Some of them are highly toxic to invertebrates, others are more toxic to vertebrate species. Site 3 toxins specifically slow down the inactivation of the Na^+ channel and dramatically prolong the duration of action potentials (59-61). Sea anemone toxins are useful tools to distinguish different isoforms of the Na^+ channel. As a rule, Na^+ channels that have a high affinity for TTX and saxitoxin have a low affinity for the various sea anemone toxins. Conversely, Na^+ channels that have a low affinity for TTX and saxitoxin have a high affinity for sea anemone toxins (62).

4. The polypeptide toxins from the scorpions Centruroides suffusus and Tityus serrulatus. These toxins act by shifting the voltage dependence of the activation of Na^+ channels, thereby inducing a Na^+ channel activity at negative potentials at which Na^+ channels are normally closed (63,64). Site 4 toxins, because of their high affinity for the Na^+ channel, have been efficient tools to elucidate the molecular structure of the Na^+ channel (30,65,66).

5. The pyrethroid insecticides. These molecules transform fast Na^+ channels into slower ones (67-69).

An important point is that two toxins that act on distinct binding sites can synergize each other. This means that interactions between different neurotoxin binding sites exist so that the binding of one toxin to its binding site can influence the binding of a second toxin to its own binding site. Interactions between site 2, which recognizes lipid-soluble toxins, site 3, which recognizes sea anemone toxins, and site 5, which recognizes pyrethroid molecules have been analyzed in great detail (70-73).

The interaction of neurotoxins with the Na^+ channel are usually studied using electrophysiological experiments, $^{22}Na^+$ uptake experiments, and also binding experiments involving radiolabelled toxins.

Evidence that CTx and PbTx Act on the Na^+ Channel

CTx that has been purified from muscles of Gymnothorax javanicus stimulates the release of neurotransmitters such as γ-aminobutyric acid and dopamine from rat brain nerve terminals. It causes a membrane depolarization of mouse neuroblastoma cells and, under appropriate conditions, it creates spontaneous oscillations of the membrane potential and repetitive action potentials. It stimulates the influx of $^{22}Na^+$ in mouse neuroblastoma cells and in rat skeletal muscle cells. All these effects are blocked by tetrodotoxin (74). In frog sartorius muscle CTx causes a Na^+-dependent and TTX-sensitive depolarization (75). In isolated cardiac preparations, CTx produces a positive inotropic effect which is mediated by the release of endogenous catecholamines and by a direct action on Na^+ channels of the cardiac cells (76-78). CTx also produces contraction of guinea pig vas deferens and relaxation of guinea pig taenia caecum (79-81). These actions are believed to be mediated via a release of endogenous neurotransmitters from adrenergic nerves. Detailed voltage-clamp experiments performed on the node of Ranvier indicate that CTx causes spontaneous action potentials by modifying a fraction of Na^+ channels. In the presence of CTx these channels open normally but do not close during long-lasting depolarizations (82).

In squid giant axons, PbTx causes a depolarization of the plasma membrane, repetitive discharges followed by depression of action potentials, and a complete blockade of excitability. This action is antagonized by TTX (83,84). PbTx depolarizes nerve terminals and induces neurotransmitter release (85,86); it depolarizes skeletal muscle cells (87) and increases the frequency of action potentials in crayfish nerve cord (88). PbTx also produces a contraction of the guinea pig ileum (89). All these effects are prevented by TTX.

All these results are clear indications that CTx and PbTx have similar modes of action and that they increase the membrane permability of excitable cells to Na^+ ions by opening voltage-dependent Na^+ channels. This action fully accounts for the toxicity of ciguatoxin and brevetoxins.

Characterization of the Binding Site for CTx

In mouse neuroblastoma cells and in rat skeletal muscle cells, CTx stimulates $^{22}Na^+$ uptake through the voltage-dependent Na^+ channel. The action of CTx is synergistically enhanced by veratridine and batrachotoxin (site 2 toxins), by sea anemone and *Androctonus* toxins (site 3 toxins), and by pyrethroid insecticides (site 5 toxins). It is blocked by TTX (site 1 toxin) (74). CTx has qualitatively similar effects on mouse neuroblastoma cells and on rat skeletal muscle cells. This means that the toxin is unable to discriminate between isoforms of the Na^+ channel that differ in their sensitivity to TTX, saxitoxin, and sea anemone toxins. Binding experiments further show that CTx does not prevent the binding of ^{125}I-labelled *Tityus* toxin (site 4 toxin) to the Na^+ channel in rat brain synaptosomes (74). Finally Lombet et al. (90) reported that CTx stimulates the binding of labelled batrachotoxinin A 20 α-benzoate to rat brain synaptosomes and that this effect is potentiated by sea anemone toxins. These results indicate that CTx binds to a site on the Na^+ channel that is distinct from all previously recognized binding sites for neurotoxins.

Characterization of the Binding Site for Brevetoxins

In mouse neuroblastoma cells, PbTx potentiates the action of veratridine on $^{22}Na^+$ uptake (91). It stimulates the binding of [^{3}H]batrachotoxinin A 20 α-benzoate (90,92) and of labelled *Centruroides suffusus* toxin (site 4 toxin) (93) to rat brain synaptosomes. As for CTx, the action of PbTx on [^{3}H]batrachotoxinin A 20 α-benzoate binding is synergistically enhanced by sea anemone toxins (90). The properties of the PbTx binding to the Na^+ channel were further analyzed using tritiated PbTx-3 (94). [^{3}H]PbTx-3 binding to rat brain synaptosomes is best observed when Na^+ channels have been treated with sea anemone toxins or pyrethroid insecticides, i.e., binding is enhanced by site 3 and site 5 toxins. Under these conditions, [^{3}H]PbTx-3 binds to a single family of sites in rat brain synaptosomes with a

K_d value of 17 nM and a maximum number of binding sites of 6 pmol/mg of protein. Using slightly different experimental conditions, Poli et al. (94) reported a K_d value for [^{3}H]PbTx-3 of 2.9 nM and a maximum binding capacity of 6.8 pmol/mg of protein. Binding is unaffected by TTX or veratridine, meaning that the binding site for PbTx is not site 1 or site 2 (90). These results indicate that PbTx binds to a site on the Na$^+$ channel that is distinct from all previously recognized binding sites for neurotoxins.

Evidence that CTx and PbTx Bind to the Same Receptor Site on the Na$^+$ Channel

Similarities in the actions of CTx and PbTx suggest that the two toxins might share a common receptor site on the Na$^+$ channel. This is readily checked using [^{3}H]PbTx-3 binding experiments. CTx, like PbTx-2, was found to inhibit [^{3}H]PbTx-3 binding to nerve terminals. In the presence of CTx, the apparent affinity of the Na$^+$ channel for [^{3}H]PbTx-3 is decreased and the maximum number of binding sites is not modified (90). These observations suggest that CTx inhibition of [^{3}H]PbTx-3 binding is competitive. Furthermore, CTx does not accelerate the dissociation of [^{3}H]PbTx-3 from its receptor site (90). This is good evidence that the two toxins share a common receptor site on the Na$^+$ channel. This result is consistent with the observation that the symptoms that follow ingestion of food poisoned by CTx or by PbTx are nearly identical. The only difference between the two toxins is potency. In binding experiments, CTx (K_i = 0.14 nM) is 40 times more potent than PbTx-2 (K_i = 5.6 nM) (90). The same difference in potency was noted by comparing the toxicities of the two compounds after intraperitoneal injection to mice. The LD_{50} for CTx is 0.45 μg/kg whereas the LD_{50} for PbTx-2 is 180 μg/kg (23,95). Further evidence that CTx and PbTx are related toxins comes from the observation that CTx recognizes anti-brevetoxin antibodies (20). It will be of interest to know the detailed chemical structure of CTx and to define its relationship with the known structures of brevetoxins.

Conclusion

All these results taken together indicate that ciguatoxin and brevetoxins act specifically on the voltage-dependent Na$^+$ channel of excitable tissues. They also indicate that the two toxins bind to the same receptor site on the Na$^+$ channel and that this receptor site is distinct from the other neurotoxin binding sites on the channel. The action of the two toxins is synergistically enhanced by sea anemone toxins (site 3 toxins) and by pyrethroid insecticides (site 5 toxins). Conversely, CTx and PbTx enhance the effect of the lipid-soluble toxins veratridine and batrachotoxin (site 2 toxins). Finally PbTx enhances the binding of site 4 toxins to the Na$^+$ channel. These findings improve our knowledge of the functioning of the voltage-dependent Na$^+$ channel, of the relationships between the different neurotoxin binding sites on the channel, and of their effect on the gating mechanism. They also provide new analytical techniques for detecting CTx- or PbTx-like compounds in toxic sea food.

Acknowledgments

Thanks are due to Dr. K. Nakanishi for his generous gift of PbTx and to Dr. R. Bagnis and A.M. Legrand for providing the original extracts which served in the purification of CTx.

Literature Cited

1. Banner, A.H. In *Biology and Geology of Coral Reefs*; Jones, O.A.; Endean, R., Eds.; Academic Press, New York, 1976; Vol. III, pp. 177–213.

2. Bagnis, R. *Oceanol. Acta* **1981**, *4*, 375–387.
3. Allsop, J.L.; Martini, L.; Lebris, H.; Pollard, J.; Walsh, J.; Hodgkinson, S. *Rev. Neurol. (Paris)* **1986**, *142*, 590–597.
4. Scheuer, P.J.; Takahashi, W.; Tsutsumi, J.; Yoshida, T. *Science* **1967**, *55*, 1267–1268.
5. Randall, J.E. *Bull. Marine Sci. Gulf and Carib.* **1958**, *8*, 236–267.
6. Yasumoto, T.; Nakajima, I.; Bagnis, R. *Bull. Jap. Soc. Sci. Fisheries* **1977**, *43*, 1021–1026.
7. Adachi, R.; Fukuyo, Y. *Bull. Jap. Soc. Sci. Fisheries* **1979**, *45*, 67–71.
8. Bagnis, R.; Chanteau, S.; Yasumoto, T. *Rev. Int. Med.* **1977**, *45–46*, 29–34.
9. Shimizu, Y.; Shimizu, H.; Scheuer, P.J.; Hokama, Y.; Oyama, M.; Miyahara, J.T. *Bull. Jap. Soc. Sci. Fisheries* **1982**, *48*, 811–813.
10. Yasumoto, Y.; Scheuer, P.J. *Toxicon* **1969**, *7*, 273–276.
11. Vernoux, J.P.; Lahlou, N.; Abbad El Andalousi, S.; Riyeche, R.; Magras, L.P. *Acta Tropica* **1985**, *42*, 225–233.
12. Lewis, R.J.; Endean, R. *Toxicon* **1983**, *21*, 19–24.
13. Nakajima, I.; Oshima, Y.; Yasumoto, T. *Bull. Jap. Soc. Sci. Fisheries* **1981**, *47*, 1029–1033.
14. Tindall, D.M.; Dickey, R.W.; Carlson, R.D.; Morey-Gaines, G. In *Seafood Toxins*; Ragelis, E.P., Ed.; ACS Symposium Series No. 262; American Chemical Society, Washington, DC, **1984**, pp. 224–240.
15. Steidinger, K.A.; Baden D.G. In *Dinoflagellates*; Spector, D. L., Ed.; Academic Press, New York, **1984**; pp. 201–261.
16. Anderson, D.M.; Lobel, P.S. *Biol. Bull.* **1987**, *172*, 89–107.
17. Yasumoto, T.; Nakajima, Y.; Oshima, Y.; Bagnis, R. In *Toxic Dinoflagellate Blooms*; Taylor, D.L.; Seliger H.H., Eds.; Elsevier North Holland, Inc.: New York, **1979**, pp. 65–70.
18. Durand-Clément, M. *Toxicon* **1986**, *24*, 1153–1157.
19. Durand-Clément, M. *Biol. Bull.* **1987**, *172*, 108–121.
20. Baden, D.G.; Mende, T.J.; Brand, L.E. In *Toxic Dinoflagellates*; Anderson, D.M.; White, A.W.; Baden, D.G., Eds.; Elsevier North Holland, Inc.: New York, 1985, pp. 363–368.
21. Tachibana, K.; Nukina, M.; Joh, Y.; Scheuer, P.J. *Biol. Bull.* **1987**, *172*, 122–127.
22. Nukima, M.; Koyanagi, L.M.; Scheuer P.S. *Toxicon* **1984**, *22*, 169–176.
23. Tachibana, K. PhD Thesis, University of Hawaii, 1980.
24. Chungue, E.; Bagnis, R.; Parc, F. *Toxicon* **1984**, *22*, 161–164.
25. Hokama, Y.; Bannee, A.H.; Boylan, D.B. *Toxicon* **1977**, *15*, 317–325.
26. Hokama, Y.; Shirai, L.K.; Iwamoto, L.M.; Kobayashi, M.N.; Goto, C.S.; Nakagawa, L.K. *Biol. Bull.* **1987**, *172*, 144–153.
27. Nakanishi, K. *Toxicon* **1985**, *23*, 473–479.
28. Pierce, R.H. *Toxicon* **1986**, *24*, 955–965.
29. Baden, D.G.; Mende, T.J. *Toxicon* **1982**, *20*, 457–461.
30. Barhanin, J.; Giglio, J.R.; Leopold, P.; Schmid, A.; Sampaio, S.V.; Lazdunski, M. *J. Biol. Chem.* **1982**, *257*, 12553–12558.
31. Hartshorne, R.P.; Catterall, W.A. *Proc. Natl. Acad. Sci. USA* **1981**, *78*, 4620–4624.
32. Hartshorne, R.P.; Catterall, W.A. *J. Biol. Chem.* **1984**, *259*, 1667–1675.
33. Agnew, W.S.; Levinson, S.R.; Brabson, J.S.; Raftery, M.A. *Proc. Natl. Acad. Sci. USA* **1978**, *75*, 2602–2610.
34. Miller, J.A.; Agnew, W.S.; Levinson, S.R. *Biochemistry* **1983**, *22*, 462–470.
35. Barchi, R.L. *Ann. N.Y. Acad. Sci.* **1986**, *479*, 179–185.
36. Norman, R.I.; Schmid, A.; Lombet, A.; Barhanin, J.; Lazdunski, M. *Proc. Natl. Acad. Sci. USA* **1983**, *80*, 4164–4168.

37. Noda, M.; Shimizu, S.; Tanabe, T.; Takai, T.; Kayano, T.; Ikeda, T.; Takahashi, H.; Nakayama, H.; Kanaoka, Y.; Minamino, N.; Kagawa, K.; Matsuo, H.; Raftery, M.A.; Hirose, T.; Inayama, S.; Hayashida, H.; Miyata, T.; Numa, S. *Nature (London)* **1984**, *312*, 121–127.
38. Noda, M.; Ikeda, T.; Kayano, T.; Suzuki, H.; Takeshima, H.; Kurasaki, M.; Takahashi, H.; Numa, S. *Nature (London)* **1986**, *320*, 188–192.
39. Rosenberg, R.L.; Tomiko, S.A.; Agnew, W.S. *Proc. Natl. Acad. Sci. USA* **1984**, *81*, 1239–1243.
40. Hartshorne, R.P.; Keller, B.U.; Talvenheimo, J.A.; Catterall, W.A.; Montal, M. *Proc. Natl. Acad. Sci. USA* **1985**, *82*, 240–244.
41. Hanke, W.; Boheim, G.; Barhanin, J.; Pauron, D.; Lazdunski, M. *The EMBO J.* **1984**, *3*, 509–515.
42. Tamkun, M.M.; Talvenheimo, J.A.; Catterall, W.A. *J. Biol. Chem.* **1984**, *259*, 1676–1688.
43. Stühmer, W.; Methfessel, C.; Sakman, B.; Noda, M.; Numa, S. *Eur. Biophys. J.* **1987**, *14*, 131–138.
44. Suzuki, H.; Beckh, S.; Kubo, H.; Yahagi, N.; Ishida, H, Kayano, T.; Noda, M.; Numa, S. *FEBS Letters* **1988**, *228*, 195–200.
45. Narahashi, T. *Physiol. Rev.* **1974**, *54*, 813–889.
46. Pappone, P.A. *J. Physiol. (London)* **1980**, *306*, 377–410.
47. Lawrence, J.C.; Catterall, W.A. *J. Biol. Chem.* **1981**, *256*, 6213–6222.
48. Renaud, J.F.; Kazazoglou, T.; Lombet, A.; Chicheportiche, R.; Jaimovitch, E.; Romey, G.; Lazdunski, M. *J. Biol. Chem.* **1983**, *258*, 8799–8805.
49. Lombet, A.; Frelin, C.; Renaud, J.F.; Lazdunski, M. *Eur. J. Biochem.* **1982**, *124*, 199–203.
50. Frelin, C.; Vigne, P.; Lazdunski, M. *J. Biol. Chem.* **1983**, *258*, 7256–7259.
51. Sherman, S.J.; Lawrence, J.C.; Messner, D.J.; Jacoby, K.; Catterall, W.A. *J. Biol. Chem.* **1983**, *258*, 2488–2495.
52. Cruz, L.J.; Gray, W.R.; Olivera, B.M.; Zeikus, R.D.; Kerr, L.; Yoshikami, D.; Moczydlowski, E. *J. Biol. Chem.* **1985**, *260*, 9280–9288.
53. Kobayashi, M.; Wu, C.H.; Yoshii, M.; Narahashi, T.; Nakamura, H.; Kobayashi, K.; Ohizumi, Y. *Pflügers Arch* **1986**, *407*, 241–243.
54. Ohizumi, Y.; Nakamura, H.; Kobayashi, J.; Catterall, W.A. *J. Biol. Chem.* **1986**, *261*, 6149–6152.
55. Moczydlowski, E.; Olivera, B.M.; Gray, W.R.; Strichartz, G.R. *Proc. Natl. Acad. Sci. USA* **1986,**,83, 5321–5325.
56. Ulbricht, N. *Erg. Physiol.* **1969**, *61*, 18–71.
57. Khodorov, B.I.; Revenko, S.V. *Neuroscience* **1979**, *4*, 1315–1330.
58. Honerjáger, P.; Frelin, C.; Lazdunski, M. *Naunyn-Schmiedeberg's Arch. Pharmacol.* **1982**, *321*, 123–129.
59. Romey, G.; Abita, J.P.; Schweitz, H.; Wunderer, G.; Lazdunski, M. *Proc. Natl. Acad. Sci. USA* **1976**, *73*, 4055–4059.
60. Schweitz, H.; Vincent, J.P.; Barhanin, J.; Frelin, C.; Linden, G.; Hughes, M.; Lazdunski, M. *Biochemistry* **1981**, *20*, 5245–5252.
61. Schweitz, H.; Bidard, J.N.; Frelin, C.; Pauron, D.; Vijverberg, H.P.M.; Mahasneh, D.M.; Lazdunski, M. *Biochemistry* **1985**, *24*, 3554–3561.
62. Frelin, C.; Vigne, P.; Schweitz, H.; Lazdunski, M. *Mol. Pharmacol.* **1984**, *26*, 70–74.
63. Vijverberg, H.P.M.; Pauron, D.; Lazdunski, M. *Pflügers Arch.* **1984**, *401*, 297–303.
64. Barhanin, J.; Meiri, H.; Romey, G.; Pauron, D.; Lazdunski, M. *Proc. Natl. Acad. Sci. USA* **1985**, *82*, 1842–1846.

65. Barhanin, J.; Pauron, D.; Lombet, A.; Norman, R.I.; Vijverberg, H.P.M.; Giglio, J.R.; Lazdunski, M. *EMBO J.* **1983**, *2*, 915–920.
66. Lombet, A.; Lazdunski, M. *Eur. J. Biochem.* **1984**, *141*, 651–660.
67. Jacques, Y.; Romey, G.; Cavey, M.T.; Kartalovski, B.; Lazdunski, M. *Biochim. Biophys. Acta* **1980**, *600*, 882–897.
68. Vijverberg, H.P.M.; Van Der Zalm, J.; Van Den Bercken, J. *Nature (London)* **1982**, *295*, 601–603.
69. Narahashi, T. *Neurotoxicol.* **1985**, *6*, 3–22.
70. Catterall, W.A. *Annu. Rev. Pharmacol. Toxicol.* **1980**, *20*,15–43.
71. Lazdunski, M.; Renaud, J.F. *Annu. Rev. Physiol.* **1982**, *44*, 463–473.
72. Lazdunski, M.; Frelin, C.; Barhanin, J.; Lombet, A.; Meiri, H.; Pauron, D.; Romey, G.; Schmid, A.; Schweitz, H.; Vigne, P.; Vijverberg, H.P.M. *Ann. N.Y. Acad. Sci.* **1986**, *479*, 204–220.
73. Strichartz, G.; Rando, T.; Wang, G.K. *Ann. Rev. Neurosci.* **1987**, *10*, 237–268.
74. Bidard, J.N.; Vijverberg, H.P.M.; Frelin, C.; Chunge, E.; Legrand, A.M.; Bagnis, R.; Lazdunski, M. *J. Biol. Chem.* **1984**, *259*, 8353–8357.
75. Rayner, M.D. *Fed. Proc.* **1972**, *31*, 1139–1145.
76. Ohshika, H. *Toxicon* **1971**, *9*, 337–343.
77. Miyahara, J.T.; Akau, C.K.; Yasumoto, T. *Res. Commun. Chem. Pathol. Pharmacol.* **1979**, *25*, 177–180.
78. Lewis, R.J.; Endean, R. *Naunyn-Schmiederberg's Arch. Pharmacol* **1986**, *334*, 313–322.
79. Ohizumi, Y.; Shibata, S.; Tachibana, K. *J. Pharmacol. Exptl. Ther.* **1981**, *217*, 475–480.
80. Ohizumi, Y.; Ishida, Y.; Shibata, W. *J. Pharmacol. Exptl. Ther.* **1982**, *221*, 748–752.
81. Lewis, R.J.; Endean, R. *J. Pharmacol. Exptl. Ther.* **1984**, *228*, 756–760.
82. Benoit, E.; Legrand, A.M.; Dubois, J.M. *Toxicon* **1986**, *24*, 357–364.
83. Westerfield, M.; Moore, J.W.; Kim, Y.S.; Padilla; G.M. *Amer. J. Physiol.* **1977**, *232*, C23–C29.
84. Huang, J.C.M.; Wu, C.H. Baden; D.G. *J. Pharmacol. Exptl. Ther.* **1984**, *229*, 615–621.
85. Wu, C.H.; Huang, J.M.C.; Vogel, S.M.; Luke, S.; Atchison, W.D.; Narahashi, T. *Toxicon* **1985**, *23*, 482–487.
86. Gallagher, J.P.; Shinnick-Gallagher, P. *Br. J. Pharmacol.* **1980**, *69*, 373–378.
87. Gallagher, J.P.; Shinnick-Gallagher, P. *Toxicon* **1985**, *23*, 489–496.
88. Parmentier, J.L.; Narahashi, T.; Wilson, W.A.; Trieff, N.M.; Ramanujan, V.M.S.; Risk, M.; Ray, S.M. *Toxicon* **1978**, *16*, 235–244.
89. Grunfeld, Y.; Spiegelstein, M.Y. *Br. J. Pharmacol.* **1974**, *51*, 67–72.
90. Lombet, A.; Bidard, J.N.; Lazdunski, M. *FEBS Letters* **1987**, *219*, 355–359.
91. Catterall, W.A.; Risk, M. *Mol. Pharmacol.* **1981**, *19*, 345–348.
92. Catterall, W.A.; Gainer, M. *Toxicon* **1985**, *23*, 497–504.
93. Sharkey, R.G.; Jover, E.; Couraud, F.; Baden, D.G.; Catterall, W.A. *Mol. Pharmacol.* **1987**, *31*, 273–278.
94. Poli, M.A.; Mende, T.J.; Baden, D.G. *Mol. Pharmacol.* **1986**, *30*, 129–135.
95. Baden, D.G.; Mende, T.J.; Poli, A.M.; Block, R.E. In *Seafood Toxins*; Ragelis, E.P., Ed.; ACS Symposium Series No. 262; American Chemical Society: Washington DC, **1984**; pp. 359–367.

RECEIVED May 15, 1989

PALYTOXIN

Chapter 14

A Perspective on Palytoxin

Gary Strichartz

Anesthesia Research Laboratories, Brigham and Women's Hospital, Harvard Medical School, 75 Francis Street, Boston, MA 02115

As one of the most potent lethal toxins known, palytoxin (PTX) has attracted attention from a broad range of investigators (1). The varieties of physiological and biochemical effects of these non-protein toxins (mw 2659-2677) from the genus (Coelenterata) *palythoa* provide a dizzying choice of toxic mechanisms. Almost all mammalian tissues or cells studied so far respond to PTX (2). The reported effects range from membrane depolarization (3-10), increased Na^+ (11,12) and Ca^{2+} influx (13), K^+ efflux (11,14-16), stimulation of arachidonic acid release [through de-esterification of membrane lipids (2)], stimulation or induction of neurotransmitter (norepinephrine) release (13,17), inhibition of Na^+/K^+-ATPase (10,11,14,18), induction or stimulation of contraction of smooth muscle (3,13,17-19), promotion of actions of tumor-producing compounds (20), inhibition of action of growth-related factors (21) (by down regulation of their specific membrane receptors), and erythrocyte haemolysis (22).

As is our tendency, we molecular pharmacologists are attempting to identify one primary action of PTX that is common to all of these eventual effects. However, that task is confounded by several uncontrolled variables. First, in few of the experiments is a range of PTX concentrations greater than 10^3-fold carefully examined. This approach may be acceptable in those cases where a single receptor of defined affinity can be identified, but it fails in the case of compounds like PTX for which effects are reported from 10^{-12}M [inhibition of binding of epidermal growth factor (21)] to 10^{-6}M [enhancement of Ca^{2+} influx in PC12 cells (13)].

Second, the time-course of the assays are often of different dimensions; depolarizations and ion fluxes are measured in fractions of minutes, while tumor promotion or receptor regulation may take tens of minutes to hours.

Third, the pharmacological specificity is almost never tested by the same criteria in different assays. For example, Na^+ flux and cellular depolarization by PTX are often subjected to tests for inhibition by ouabain, since Habermann and colleagues showed that these responses were sensitive to cardiac glycosides [(14) but not to their aglycones], suggesting an action through the Na^+/K^+ ATPase. But experimentalists testing other effects rarely conduct this assay, although they often propose an increase in Na^+ influx as a possible step in the final effect (e.g., ref. 21). Even when this test is included, the results may appear contradictory; a fast-responding contraction of the *vas deferens* to PTX was abolished by ouabain, whereas a slower responding contraction was potentiated (13,18). Although separate actions of PTX on smooth muscle and on nerve, respectively, are posited to underly these two phases, the details of the mechanisms remain obscure.

One fact does emerge from the assembled literature, despite its disarray. Palytoxin does not appear to act by binding to one single cellular component to trigger a cascade of responses. Original speculations that it might activate voltage-

0097–6156/90/0418–0202$06.00/0

gated Na$^+$ channels seem to be disproven by its insensitivity to that channel's high affinity blocker, tetrodotoxin *(4,10,13)*.

At this time the actions of palytoxin on cellular responses is complex, and the overall response may depend strongly on the toxin's concentration and the time over which the assay is conducted. Parallel conclusions probably hold for the toxic responses of organisms. Different tissues may contribute differentially to the final lethality (or its prevention), dependent on the dose and the course of toxin distribution among the sensitive organs. A rational approach to identifying the lethal actions of PTX requires knowledge of each affected organ's acute and long-term sensitivity and its ability to recover normal function, as well as the biochemical and physiological responses of tissues and individual cells.

Literature Cited

1. Kaul, P.N.; Farmer, M.R.; Ciereszko, L.S. *Proc. West. Pharmacol. Soc.* **1974,** *17,* 294.
2. Levine, L.; Fujiki, H.; Gjika, H.B.; Van Vunakis, H. This volume, chapter 17.
3. Deguchi, T.; Urakawa, N.; Takamatsu, S. *Animal, Plant Microbial Toxins* **1976,** *2,* 379.
4. Dubois, J.M.; Cohen, J.B. *J. Pharmac. Exp. Therap.* **1977,** *201,* 148.
5. Kudo, Y.; Shibata, S. *Br. J. Pharmacol.* **1980,** *71,* 575.
6. Muramatsu, I.; Uemura, D.; Fujiwara, M.; Narahashi, T. *J. Pharmacol. Exp. Therap.* **1984,** *231,* 488.
7. Sauviat, M.P.; Pater, C.; Berton, *J. Toxicon* **1987,** *25,* 695.
8. Tesseraux, I.; Harris, J.B.; Watkins, S.C. In *Toxins as Tools in Neurochemistrv;* Hucho, F., Ovchinnikov, Y.A., Eds.; Walter de Gruyter and Company, 1983; p. 91.
9. Weidmann, S. *Experimentia* **1977,** *33,* 1487.
10. Castle, N.A.; Strichartz, G.R. *Toxicon* **1988,** *10,* 941.
11. Chhatawal, G.S.; Hessler, H.-J.; Habermann, E. *Naunyn-Schmiedeberg's Archs. Pharmac.* **1983,** *319,* 101.
12. Muramatsu, I.; Nishio, M.; Uemura, D. *J. Pharmacobio. Dyn.* **1987,** *10,* S-113.
13. Ohizumi, Y. This volume, chapter 16.
14. Bottinger, H.; Beress, L.; Habermann, E. *Biochim. Biophys. Acta* **1986,** *861,* 165.
15. Chhatwal, G.S.; Hessler, H.J.; Habermann, E. *Naunyn-Schmiedelbergs Arch. Pharmacol.* **1983,** *323,* 261-268.
16. Maramatsu, I.; Nishio, M.; Kigoshi, S.; Uemura, D. *Br. J. Pharmacol.* **1988,** *93,* 811-816.
17. Ishida, Y.; Kajiwara, A.; Takagi, K.; Ohizumi, Y.; Shibata, S. *J. Pharmacol. Exp. Therap.* **1985a,** *232,* 551.
18. Ishida, Y.; Satake, N.; Habon, J.; Kitano, H.; Shibata, S. *J. Pharmacol. Exp. Therap.* **1985b,** *232,* 557.
19. Ito, K.; Karaki, H.; Ishida, Y.; Urakawa, N.; Deguchi, T. Jon. IJ. Pharmacol. **1976,** *26,* 683.
20. Fujiki, H.; Suganuma, M.; Nakayasu, M.; Hakii, H.; Horiuchi, R.; Takayama, S.; Sugimura, T. *Carcinogenesis* **1986,** *7,* 707-710.
21. Wattenberg, E.V.; Fujiki, H.; Rosner, M.R. This volume, chapter 15.
22. Habermann, E.; Ahnert-Hilger, G.; Chhatwal, G.S.; Beress, L. *Biochim. Biophys. Acta* **1981a,** *649,* 481-486.

RECEIVED May 4, 1989

Chapter 15

Mechanism of Palytoxin Action on the Epidermal Growth Factor Receptor

Elizabeth V. Wattenberg[1], Hirota Fujiki[2], and Marsha Rich Rosner[3,4]

[1]Department of Applied Biological Sciences, Massachusetts Institute of Technology, Cambridge, MA 02139
[2]Cancer Prevention Division, National Cancer Center Research Institute, Tsukiji 5-1-1, Chuo-ku, Tokyo 104, Japan
[3]The Ben May Institute and Department of Pharmacological and Physiological Sciences, University of Chicago, Chicago, IL 60637

Palytoxin is a potent marine toxin and mouse skin tumor promoter that is able to act on a wide variety of systems. We have studied the mechanism of palytoxin action in the context of growth control by analyzing the effect of palytoxin on the epidermal growth factor (EGF) receptor in murine fibroblasts. Our results indicate that picomolar levels of palytoxin are able to down modulate the EGF receptor by reducing the number and affinity of EGF binding sites. The mechanism of palytoxin action differs from that of 12-O-tetradecanoylphorbol-13-acetate (TPA) type tumor promoters in several respects, including kinetics, dose-response, and the fact that it is not dependent upon protein kinase C. Further, under our conditions, palytoxin action is sodium-dependent rather than calcium dependent, and palytoxin causes sodium influx with a dose-response that parallels the effects on EGF binding. These results suggest that palytoxin is able to activate a sodium pump or channel, resulting in sodium influx that leads to loss of epidermal growth factor binding sites.

Many environmental toxins interact with specific cellular receptors, including enzymes, ion channels and ion pumps, and thus provide natural tools for the study of cellular signalling pathways. Palytoxin, a compound isolated from the coelenterate of genus *Palythoa*, is one such useful and intriguing compound. The structure of palytoxin was first determined in 1981 independently by Hirata (*1*) and Moore (*2*). As one of the most potent marine toxins known, palytoxin has been studied in a variety of systems ranging from erythrocytes to neurons. As a tumor promoter of the non 12-O-tetradecanoylphorbol-13-acetate (TPA) type, palytoxin can also be studied in the context of a growth control system.

Palytoxin is a relatively large (MW 2681), hydrophilic compound (*1, 2*), unlike the prototypical TPA-type tumor promoters, the phorbol esters. Although palytoxin

---

[4] Address correspondence to this author. Formerly at the Department of Applied Biological Sciences, Massachusetts Institute of Technology, Cambridge, MA 02139.

0097–6156/90/0418–0204$06.00/0
© 1990 American Chemical Society

is a tumor promoter in the two stage mouse skin assay, palytoxin does not activate all of the cellular systems activated by the TPA-type tumor promoters (3). For example, palytoxin does not induce ornithine decarboxylase in mouse skin (4). Most distinctively, palytoxin does not bind to or activate protein kinase C and is therefore classified as a non-TPA-type tumor promoter (3, 4).

On a cellular level it has been shown that palytoxin can act synergistically with TPA-type tumor promoters, further suggesting that these different types of tumor promoters activate different cellular pathways. For example, palytoxin acts synergistically with TPA-type tumor promoters and other activators of protein kinase C to stimulate prostaglandin release (5, 6). Palytoxin also synergizes with TPA-type tumor promoters in the stimulation of histamine release from rat peritoneal mast cells and superoxide release from neutrophils (7, 8). Recent evidence indicates that a variety of growth control systems are regulated by multiple pathways. For example, the induction of ODC and the proto-oncogenes c-myc and c-fos in cell culture can be mediated by protein kinase C-independent as well as protein kinase C-dependent pathways (9–11). Therefore, it is of interest to determine whether palytoxin can modulate a growth control system and, if so, whether palytoxin activates alternate signal transduction pathways.

Evidence from a number of systems suggests that ion flux plays a role in palytoxin action. In a wide range of systems, palytoxin effects are accompanied by a change in intracellular cation levels. For example, the influx of Na^+ and/or Ca^{2+} is associated with palytoxin-stimulated contraction of cardiac and smooth muscle, the release of norepinephrine by rat pheochromocytoma (PC12) cells, and the depolarization of excitable membranes (12–15). Palytoxin also induces K^+ efflux from erythrocytes and thus alters ion flux in a nonexcitable membrane system as well (16–19). In both excitable and nonexcitable membranes, the ultimate action of palytoxin has been shown to be dependent on extracellular cations. The palytoxin-induced effects on smooth muscle and erythroctyes can be inhibited by removing Ca^{2+} from the media, and the palytoxin-induced release of norephinephrine from PC12 cells can be blocked in Na^+ free media (13, 14, 18, 20, 21)

The role of ion flux in signal transduction has been studied extensively, especially in the context of neurological and growth control pathways. In cell growth systems, Na^+ influx and cellular alkalinization appear to correlate with fertilization in sea urchin eggs, and growth factor-induced mitogenesis in cell culture (22, 23). Ca^{2+} appears to be a virtually universal second messenger, acting in regulatory roles as diverse as neurotransmitter release, short term memory, and chemotaxis (24). Ca^{2+} also appears to play a role in growth factor systems. Prior to mitogenesis, epidermal growth factor (EGF) stimulates the influx of extracellular Ca^{2+}, and platelet derived growth factor (PDGF) causes release of Ca^{2+} from intracellular stores (22, 25–27).

One well-characterized model for studying growth control pathways, and in particular the interaction of foreign compounds with growth control systems, is the EGF receptor system. The EGF receptor is subject to transmodulation by a number of agents including other growth factors and tumor promoters. We and others have shown that tumor promoters inhibit EGF binding to a class of high affinity receptors and block EGF-stimulated tyrosine kinase activity (28–30). Since these effects appear to be mediated by protein kinase C, we determined whether the non-TPA-type tumor promoter palytoxin also alters EGF receptors and, if so, whether a similar mechanism is involved. The results indicate that palytoxin, like TPA-type tumor promoters, inhibits EGF binding. However, the mechanism by which palytoxin alters EGF binding differs significantly from that of the TPA-type tumor promoters. In particular, our results suggest a role for Na^+ but not Ca^{2+} in palytoxin action on the EGF receptor.

Experimental

Materials. ^{125}I-EGF was either made by iodinating mouse EGF (Biomedical Technologies Inc.) by the chloramine T method, to a specific activity of approximately 1–2 Ci/μmol, using Na-^{125}I (Amersham) or purchased from New England Nuclear. Phorbol diterpene esters were purchased from Sigma. Palytoxin was isolated from *Palythoa tuberculosa* as previously described (*1*).

Cultures. Swiss 3T3 and A431 cells were grown in a gassed (5.5% CO_2), humidified incubator in Dulbecco's Modified Eagle Media (DME) supplemented with 10% heat inactivated fetal calf serum (FCS).

Quantitation of ^{125}I-EGF binding to high affinity EGF receptors. A431 cells and Swiss 3T3 cells were plated in 24 well dishes at a concentration ranging from 50,000 to 100,000 cells/ml. A431 cells were assayed the day after plating. Swiss 3T3 cells were grown to confluence (appr. 105 cells/well). The medium was removed and replaced with DME containing 0.1% FCS or CR-ITS premix (purchased from Collaborative Research) the night before assaying. The cells were washed in binding medium (DME containing 0.1% ovalbumin and 50 mM Hepes) or incubation medium (130 mM NaCl, 50 mM Hepes, 1.8 mM $CaCl_2$, 0.8 mM $MgSO_4$, 5.5 mM glucose, 0.5 mM boric acid, 0.1% BSA, pH 7.4) and the appropriate agents added in a total volume of 0.4 ml binding medium or incubation medium as indicated in the figure legends and the text. Each variable was tested in triplicate. Cells were then placed on ice and washed with binding media. ^{125}I-EGF (0.05–0.1 nM) in binding media was added for 4–6 hr at 4°C. This concentration of ^{125}I-EGF binds primarily to high affinity EGF receptors, as demonstrated by the 80% reduction in cpm bound after 37°C phorbol dibutyrate (PDBu) treatment of cells that had not been depleted of protein kinase C. Finally, cells were washed, lysed, and quantitated for specific ^{125}I-EGF binding. Data were normalized to the amount of specific ^{125}I-EGF binding to cells that were treated with binding medium or incubation medium alone. In general, the bound cpm ranged from 1000 to 4000 cpm with a standard deviation of 5–10%. The nonspecific EGF binding was 100–250 cpm. The specifics for each experiment presented here are given in the figure legends. In order to generate cells having different levels of protein kinase C, cultures were treated for an additional 72 hr with 0, 20, 200, or 2000 nM PDBu in DME/0.1% FCS as previously described (*31, 32*). The cultures were washed four times with binding medium (DME containing 0.1% ovalbumin) over a period of 2 hr at 37°C prior to incubation with the appropriate agents. Scatchard analysis was done as previously described (*33*).

DNA synthesis assays. DNA synthesis was monitored by incorporation of ^3H-thymidine into TCA precipitable material as described by McCaffrey and Rosner (*32*).

Na^+ Influx Studies. Na^+ influx was monitored according to the procedure of Owen and Villereal (*34*), with some modifications. Cells were seeded onto 60-mm culture dishes, grown, and serum starved as described for the assays above. The cells were washed with incubation media and incubated in 3 ml of the appropriate agent at 37°C. After incubation the cells were rapidly washed in ice cold 0.1 mM $MgCl_2$ and extracted with 5% TCA/0.5% KNO_3 for sodium determination or 0.2% SDS for protein determination. Sodium concentration was measured using a Varian Model 275 Atomic Absorption Spectrophotometer. Protein was determined fluorimetrically.

Results

To determine whether palytoxin could modify EGF receptors, the effect on EGF binding of palytoxin and a TPA-type tumor promoter, PDBu, were compared. Swiss 3T3 cells, a murine fibroblast cell line which is mitogenically responsive to EGF, were treated at 37°C for 15–120 min with 1–11 pM palytoxin or 15–60 min with 2–200 nM PDBu. The cells were then washed and incubated at 4°C with ^{125}I-EGF and cell associated radioactivity was determined. The results indicate that palytoxin, like PDBu, causes inhibition of EGF binding (Figure 1). However, the kinetics of the palytoxin effect differ significantly from that of the PDBu effect. At 11 pM palytoxin, the inhibition of EGF binding begins to plateau by 30–60 min whereas at the lower doses, up to 120 min or longer is required to obtain a similar degree of inhibition. In contrast, PDBu inhibits EGF binding rapidly, and the maximum level of inhibition is strictly dose-dependent. Thus, inhibition of EGF binding by palytoxin occurs at a slower rate than PDBu and has a different dose-dependence.

To ensure that the inhibition of EGF binding by palytoxin was not a consequence of cell toxicity, the effect of palytoxin on DNA synthesis in Swiss 3T3 cells was monitored. When cells were incubated in the presence of palytoxin, 10% fetal calf serum, and ^{3}H-thymidine for 19.5 hr, no depression in the extent of ^{3}H-thymidine incorporation into DNA was detected up to 3.7 pM palytoxin (Table I). Although 11 pM palytoxin was toxic when present for a prolonged period, under the conditions of the assays described above no toxicity was detected (Table I). When cells were incubated in the presence of palytoxin, 0.1% fetal calf serum, and ^{3}H-thymidine, palytoxin did not stimulate significant incorporation of ^{3}H-thymidine into DNA. Thus, although it can modulate the EGF receptor system under these conditions, palytoxin alone does not appear to be mitogenic for Swiss 3T3 cells.

Inhibition of EGF binding by palytoxin could be due to a decrease in receptor affinity, as in the case of TPA-type tumor promoters, and/or a decrease in receptor number. In Swiss 3T3 cells there are two classes of EGF receptors. The dissociation constants for the two EGF receptor classes were determined to be approximately 2×10^{-10} M and 2×10^{-8} M, corresponding to approximately 1×10^4 and 1×10^5 receptor molecules per cell, respectively (*33*). Scatchard analysis revealed that treatment of Swiss 3T3 cells with palytoxin, like PDBu, caused an apparent loss in high-affinity binding (Figure 2). However, in contrast to PDBu, palytoxin also caused a significant (approximately 50%) loss of low affinity EGF binding.

The differences between palytoxin and PDBu with respect to kinetics, temperature dependence, and effect on low affinity binding suggest that these two different types of tumor promoters may be acting through different mechanisms. Further, in contrast to PDBu, the effect of palytoxin is not readily reversible (*33*). To determine where the two pathways differ, we compared the relative ability of palytoxin and PDBu to inhibit EGF binding in protein kinase C depleted cells. Swiss 3T3 cells were depleted of protein kinase C to different extents by exposing confluent quiescent cells to 0, 20, 200, or 2000 nM PDBu for 72 hr. Previous results indicate that this treatment depletes cells of protein kinase C activity in a dose-dependent manner (*31*).

Changing cellular levels of protein kinase C in Swiss 3T3 cells did not significantly affect the ability of palytoxin to inhibit EGF binding (Figure 3). Whereas PDBu action is highly dependent upon cellular levels of protein kinase C, palytoxin inhibits EGF binding to approximately the same extent regardless of the cellular level of protein kinase C. This experiment demonstrates that, unlike the TPA-type tumor promoters, palytoxin modulates EGF receptor properties in an apparently protein kinase C independent manner. Further studies indicated that palytoxin is not inhibiting EGF binding through direct competition with EGF for the receptor,

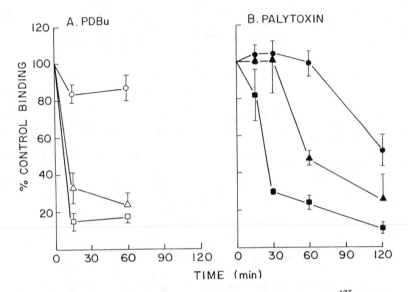

Figure 1.   Effect of tumor promoter treatment on binding of ^{125}I-EGF to Swiss 3T3 cells.   Confluent quiescent Swiss 3T3 cells (A and B) were treated for the indicated times at 37°C with PDBu [A: 2 nM (○), 20 nM (Δ), 200 nM (□)] or palytoxin [B: 1.1 pM (●), 3.7 pM (▲), 11 pM (■)].   Cultures were subsequently washed and assayed for ^{125}I-EGF binding at 4°C as described in the **Experimental** section. The data is expressed as the percent of ^{125}I-EGF binding in cells treated with either PDBu or palytoxin relative to control cells.   The specific binding of ^{125}I-EGF constituting 100% for control cells was appr.   800 cpm for Swiss 3T3 cells.   Each point represents the mean of triplicate samples ± S. D.   (Reproduced with permission from Ref. 33.   Copyright 1987 Cancer Research, Inc.)

Table I. Effect of Palytoxin on DNA Synthesis in Swiss 3T3 Cells

Treatment[a]	^{3}H-Thymidine Incorporated(cpm)[b]	^{3}H-Thymidine Incoroporated(%)
10% FCS[c]	86312 ± 2537	100
10% FCS + PTX (11 pM)[c]	126999 ± 9288	147
10% FCS	82287 ± 626	100
10% FCS + PTX (1 pM)	105780 ± 9735	129
10% FCS + PTX (3.7 pM)	99502 ± 3457	121
10% FCS + PTX (11 pM)	32067 ± 5207	39
0.1% FCS	27631 ± 609	100
0.1% FCS + PTX (3.7 pM)	25915 ± 5238	94
0.1% FCS + PTX (11 pM)	18106 ± 1156	65

[a] Confluent Swiss 3T3 cells were serum-starved by incubation for 48 hr in DME containing 0.1% FCS. Cells were then incubated with the indicated compound at 37°C in the presence of ^{3}H-thymidine, 0.1 or 10% FCS for 19.5 hr, washed, and then assayed for ^{3}H-thymidine incorporation into DNA as described in **Methods**.
[b] The average of duplicate determinations ± range.
[c] Cells were incubated with palytoxin for 2.5 hr at 37°C, washed, and then incubated in the presence of ^{3}H-thymidine and 10% FCS for 19.5 hr.
SOURCE: Reproduced with permission from Ref. 33.
Copyright 1987 Cancer Research, Inc.

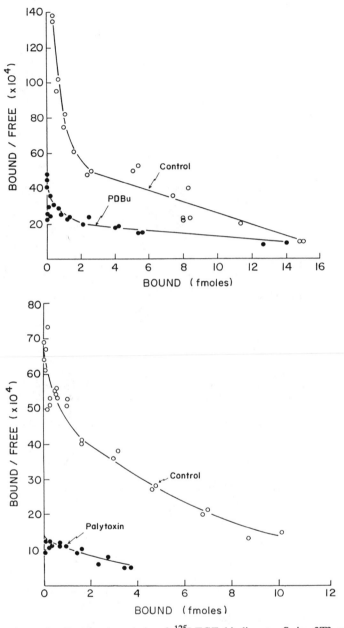

Figure 2. Scatchard analysis of ^{125}I-EGF binding to Swiss 3T3 cells treated with PDBu or palytoxin. Confluent quiescent Swiss 3T3 cells were treated at 37°C with solvent (o) or 200 nM PDBu (•) for 15 min (Upper panel); or with solvent (o) or 11 pM palytoxin (•) for 60 min (Lower panel). Cells were assayed as in Figure 1. (Reproduced with permission from Ref. 33. Copyright 1987 Cancer Research, Inc.)

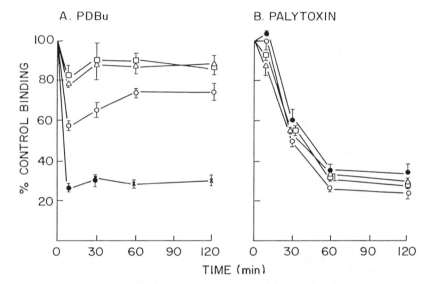

Figure 3.   Relationship between levels of protein kinase C and inhibition of EGF binding to Swiss 3T3 cells by PDBu and palytoxin. Confluent quiescent Swiss 3T3 cells were pretreated for 72 hr with 0.1% FCS plus solvent (•), 20 nM PDBu (o), 200 nM PDBu (Δ), or 2000 nM (□). Cells were then washed and treated with either 200 nM PDBu (A) or 11 pM palytoxin (B) for the times shown and assayed as in Figure 1. Other conditions were as described in the legend to Figure 2. *The data for 200 nM PDBu are composites of two independent experiments. (Reproduced with permission from Ref. 33. Copyright 1987 Cancer Research, Inc.)

and that the mechanism is temperature sensitive and may reflect an energy-dependent process (33).

Since ion flux is implicated in palytoxin action in a number of systems, we investigated the role of Ca^{2+} and Na^2 in palytoxin action on the EGF receptor. For these experiments the cells were incubated in the simpler incubation medium instead of DME (see the discussion of the incubation medium in the **Experimental** section). We have found that there is some variability in the potency of palytoxin from assay to assay, and the use of this borate containing incubation medium has helped to reduce this variability. To determine the effect of Ca^{2+} on palytoxin action, Swiss 3T3 cells were incubated in the presence or absence of palytoxin in either a Ca^{2+} containing medium or a Ca^{2+} deficient medium that also included 100 μM EGTA. The results show that external Ca^{2+} is not necessary for the inhibition of EGF binding by palytoxin (Figure 4), although in some experiments we have noted a slight shift in the dose-response curve. EGF binding does appear to be slightly depressed in the absence of Ca^{2+}, suggesting that external Ca^{2+} may somewhat enhance EGF binding. In concurrence with the result that extracellular Ca^{2+} does not appear to be necessary for palytoxin action on the EGF receptor, we found that palytoxin does not cause Ca^{2+} influx or the release of Ca^{2+} from internal stores at the doses used in these experiments, as measured using the photosensitive protein aqueorin (42). However, palytoxin does appear to cause Ca^{2+} entry at toxic doses, higher than those used in the assays described above.

Na^+ also appears to play a role in palytoxin action in some systems. To determine if there is a Na^+ requirement for palytoxin action on the EGF receptor, Swiss 3T3 cells were assayed for palytoxin activity in Na^+ containing medium versus Na^+ deficient medium. When NaCl is replaced by cholineCl, palytoxin can no longer inhibit EGF binding in Swiss 3T3 cells (Figure 5). By contrast, PDBu is equipotent in both Na^+ containing and Na^+ free media (data not shown).

Because these results suggest that extracellular Na^+ is required for inhibition of EGF binding by palytoxin in these cells, we determined if palytoxin caused Na^+ influx in Swiss 3T3 cells. When Na^+ influx was monitored at an early time point (7 min), it was found that palytoxin causes an influx of Na^+ and that the rate of Na^+ influx is dose dependent (Figure 6). In parallel with its effect on EGF binding, palytoxin at different doses increases intracellular Na^+ to the same final level (42). Although Na^+ influx occurs prior to the inhibition of EGF binding, these results and the apparent Na^+ dependence of the palytoxin effect suggest a role for Na^+ in the action of palytoxin on the EGF receptor.

Discussion

The results discussed here show that the non-TPA-type tumor promoter palytoxin, like the TPA-type tumor promoters, can inhibit EGF binding in Swiss 3T3 cells and A431 cells. However, significant differences in time course, dose response, reversibility, effect on receptor number, dependence upon protein kinase C, and dependence on extracellular Na^+ indicate that the mechanism of action of palytoxin is one that has not been previously described (Figure 7). The effect of TPA-type tumor promoters on EGF binding is rapid, reversible, and strictly dose dependent; the extent of inhibition of EGF binding is directly proportional to the level of protein kinase C, consistent with direct activation of the enzyme by these agents. By contrast, the non-TPA-type tumor promoter palytoxin inhibits EGF binding with slower kinetics and is not readily reversible; the extent of inhibition of EGF binding reaches a maximal level almost independent of the initial dose of palytoxin, consistent with an ion-flux-dependent mechanism. These results indicate that there are at least two

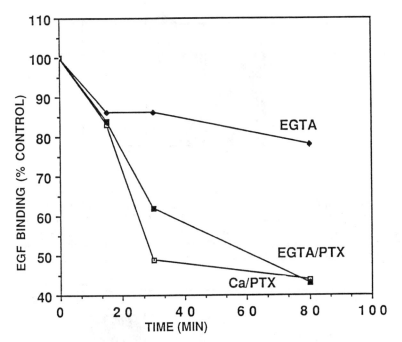

Figure 4. Effect of Ca^{2+} on palytoxin action in Swiss 3T3 cells. Confluent quiescent Swiss 3T3 cells were incubated for the indicated times at 37°C with Ca^{2+}-free incubation media containing 100 μM EGTA (◆), complete incubation media containing 3.7 pM PTX (▫), or Ca^{2+}-free incubation media containing 100 μM EGTA plus 3.7 pM PTX (■). Cells were assayed as in Figure 1. Data expressed as percentage of ^{125}I-EGF binding in cells treated with complete incubation media alone.

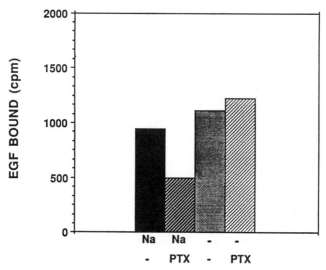

Figure 5. Effect of Na$^+$ on palytoxin action in Swiss 3T3 cells. Confluent quiescent Swiss 3T3 cells were incubated for 120 min at 37°C with incubation media containing media alone (Na/-), incubation media plus 3.7 pM PTX (Na/PTX), Na$^+$-free incubation media containing 130 mM cholineCl (-/-) or Na$^+$-free incubation media containing 130 mM cholineCl plus 3.7 pM palytoxin (-/PTX). Cells were assayed as in Figure 1. Data expressed as specific cpm of ^{125}I-EGF bound per 10^5 cells.

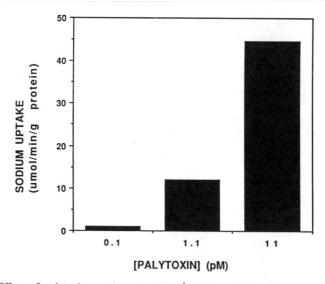

Figure 6. Effect of palytoxin on the rate of Na$^+$ influx in Swiss 3T3 cells. Confluent quiescent Swiss 3T3 cells were incubated for 37°C for 7 min in incubation media containing 0.1 pM PTX, 1.1 pM PTX, or 11 pM PTX. Intracellular Na$^+$ was determined as described in the **Experimental** section. Data points represent the mean of quadruplicate points.

	Palytoxin	PDBu
Kinetics	slow	rapid
Activity at 4 °C	–	+
High Affinity Binding	↓	↓
Low Affinity Binding	↓	—
Reversibility	–	+
Dependence on Protein Kinase C	–	+

Figure 7. Comparison of effects of palytoxin and phorbol dibutyrate on EGF receptor binding.

very different mechanisms by which tumor promoters modulate cellular growth regulatory systems.

The results in our system suggest a role for Na^+ in palytoxin action on the EGF receptor. Palytoxin causes Na^+ influx in Swiss 3T3 cells at a rate which is dose-dependent. Although the effect on Na^+ flux occurs much more rapidly than the effect on the EGF receptor, the results indicate that extracellular Na^+ is required for inhibition of EGF binding by palytoxin and thus suggest a link between the two events. A correlation between Na^+ influx and inhibition of EGF binding is further supported by the observation that, in both cases, a similar endpoint is reached with time almost independent of dose. The Na^+-induced loss of EGF binding may also occur under physiological conditions, since Na^+ influx is an early event in the action of many growth factors.

The mechanism by which palytoxin stimulates ion flux is not yet known. Palytoxin alone does not appear to be an ionophore since the compound does not cause ion flux in artificial liposomes (15, 16). The high potency of palytoxin and the apparent dependence of its action on a cellular component suggest that palytoxin may act through a specific cellular receptor. In various systems, tetrodotoxin does not appear to block the palytoxin-induced monovalent cation flux, diminishing the possibility that palytoxin activates a tetrodotoxin-sensitive Na^+ channel (14, 15, 36). Results obtained with amiloride and more specific Na^+/H^+ antiporter inhibitors in the EGF receptor system (Wattenberg, Cragoe, and Rosner, data not shown) and in the rat erythrocyte system (37) suggest that palytoxin does not cause Na^+ influx by activation of this ion pump. It has been proposed that the Na^+, K^+-ATPase is the cellular receptor for palytoxin based on the observation that oubain can antagonize palytoxin action in some systems and that palytoxin can block ^3H-oubain binding (38, 39). Although these studies suggest that palytoxin may bind to the Na^+, K^+-ATPase, there is no correlation between the sensitivity of a system to oubain and its sensitivity to palytoxin (38); Wattenberg and Rosner, data not shown). In addition, there is no direct evidence that this ion pump mediates palytoxin action. It appears that the sugar moiety of cardiac glycosides is necessary for blocking palytoxin action, but inhibition of the ion pump itself does not affect the activity of the toxin (19, 37). Palytoxin has been shown to inhibit the Na^+, K^+-ATPase in vitro, but at concentrations at least 10,000 times that required to elicit a biological effect such as K^+ release from erythrocytes or inhibition of EGF binding, thus raising into question the significance of these results (40, 41). It remains possible that palytoxin generates or activates a Na^+ channel that has not yet been characterized. The system that we have described should be a useful one for investigating the mechanism of action of palytoxin and determining the biochemical link between ion flux and the modulation of a growth control system.

Acknowledgments

We would like to thank Mitch Villereal and his colleagues for their assistance and helpful discussions. This work was supported by National Cancer Institute Grants CA35541 and CA40407 to M.R.R. and National Institute of Health Toxicology Grant T32–ES07020 to E.V.W.

Literature Cited

1. Uemura, D.; Ueda, K.; Hirata, Y. *Tetrahedron Letters* **1981,** *22*, 2781–2784.
2. Moore, R. E.; Bartolini, G. *J. Am. Chem. Soc.* **1981,** *103*, 2491–2494.
3. Fujiki, H.; Suganuma, M.; Nakayasu, M.; Hakii, H.; Horiuchi, T.; Takayama, S.; Sugimura, T. *Carcinogenesis* **1986,** *7*, 707–710.
4. Fujiki, H.; Suganuma, M.; Tahira, T.; Yoshioka, A.; Nakayasu, M.; Endo, Y.;

Shudo, K.; Takayama, S.; Moore, R. E.; Sugimura, T. *Cellular Interactions by Environmental Tumor Promoters*; Japan Sci. Soc. Press, Tokyo/VNU Science Press, Utrecht, **1984**, pp. 37–45.
5. Levine, L.; Fujiki, H. *Carcinogenesis* **1985**, *6*, 1631–1634.
6. Levine, L.; Xiao, D.; Fujiki, H. *Carcinogenesis* **1986**, *7*, 99–103.
7. Ohuchi, K.; Hirasawa, N.; Takahashi, C.; Watanabe, M.; Tsurufuji, S.; Fujiki, H.; Suganuma, M.; Hakii, H.; Sugimura, T.; Christensen, B. *Biochim. Biophys. Acta.* **1986**, *887*, 94–99.
8. Kano, S.; Izuka, T.; Ishimura, Y.; Fujiki, H.; Sugimura, T. *Biochem Biophys Res Commun.* **1987**, *143*, 672–677.
9. Hovis, J. G.; Stumpo, D. J.; Halsey, D. L.; Blackshear, P. J. *J. Biol. Chem.* **1986**, *261*, 10380–10386.
10. Ran, W.; Dean, M.; Levine, R. A.; Henkle, C.; Campisi, J. *Proc. Natl. Acad. Sci. USA* **1986**, *83*, 8216–8220.
11. Tsuda, T.; Hamamori, Y.; Yamashita, T.; Fukumoto, Y.; Takai, Y. *FEBS Lett.* **1986**, *208*, 39–42.
12. Rayner, M. D.; Sanders, B. J.; Harris, S. M.; Lin, Y. C.; Morton, B. E. *Res. Commun. Chem. Pathol. Pharmacol.* **1975**, *11*, 55–64.
13. Ito, K.; Karaki, H.; Urakawa, N. *Eur. J. Pharmacol.* **1977**, *46*, 9–14.
14. Tatsumi, M.; Takahashi, M.; Ohizumi, Y. *Molecular Pharmacology.* **1984**, *25*, 379–383.
15. Lauffer, L.; Stengelin, S.; Beress, L.; Hucho, F. *Biochim. Biophys. Acta.* **1985**, *818*, 55–60.
16. Ahnert-Hilger, G.; Chhatwal, G. S.; Hessler, H. -J.; Habermann, E. *Biochim. Biophys. Acta.* **1982**, *688*, 485–494.
17. Nagase, H.; Ozaki, H.; Karaki, H.; Urakawa, N. *FEBS Lett.* **1986**, *195*, 125–128.
18. Nagase, H.; Ozaki, H.; Urakawa, N. *FEBS Lett.* **1984**, *178*, 44–46.
19. Ozaki, H.; Nagase, H.; Urakawa, N. *FEBS Lett.* **1984**, *173*, 196–198.
20. Ohizumi, Y.; Shibata, S. *J. Pharmacol. Exp. Ther.* **1980**, *212*, 209–214.
21. Ishida, Y.; Satake, N.; Habon, J.; Kitano, H.; Shibata, S. *J. Pharmacol. Exp. Ther.* **1985**, *232*, 557–560.
22. Rozengurt, E. *Science* **1986**, *234*, 161–166.
23. Bell, J.; Nielsen, L.; Sariban-Sohraby, S.; Benos, D. *Current Topics in Membranes and Transport. Vol. 27.* Academic Press: Orlando, **1986**, pp. 129–162.
24. Rasmussen, H. *New. Eng. J. Med.* **1986**, *314*, 1164–1170.
25. Moolenaar, W. H.; Aerts, R. J.; Tertoolen, L. G.; de Laat, S. W. *J. Biol. Chem.* **1986**, *261*, 279–284.
26. Macara, I. G. *J. Biol. Chem.* **1986**, *261*, 9321–9327.
27. McNeil, P. L.; McKenna, M. P.; Taylor, D. L. *J. Cell Biol.* **1985**, *101*, 372–379.
28. Friedman, B.; Frackelton, A. R.; Ross, A. H.; Connors, J. M.; Fujiki, H.; Sugimura, T.; Rosner, M. R. *Proc. Natl. Acad. Sci. USA* **1984**, *81*, 3034–3038.
29. McCaffrey, P. G.; Friedman, B.; Rosner, M. R. *J. Biol. Chem.* **1984**, *259*, 12502–12507.
30. Foulkes, J. G.; Rosner, M. R. *Molecular Mechanisms of Transmembrane Signalling*; Elsevier Scientific Publishing Co.; Inc.: New York, **1985**, pp. 217–252.
31. Friedman, B.; Rosner, M. R. *J. Cell. Biochem.* **1987**, *34*, 1–11.
32. McCaffrey, P. G.; Rosner, M. R. *Cancer Res.* **1987**, *47*, 1081–1086.
33. Wattenberg, E. V.; Fujiki, H.; Rosner, M. R. *Cancer Research.* **1987**, *47*, 4618–4622.
34. Owen, N. E.; Villereal, M. L. *Cell* **1983**, *32*, 979–985.
35. McCaffrey, P. G.; Rosner, M. R.; Kikkawa, U.; Sekiguchi, K.; Ogita, K.; Ase, K.; Nishizuka, Y. *Biochem. Biophys. Res. Commun.* **1987**, *146*, 140–146.

36. Sauviat, M.; Pater, C.; Berton, J. *Toxicon* **1987**, *25*, 695–704.
37. Ozaki, H.; Nagase, H.; Urakawa, N. *Eur. J. Biochem* **1985**, *152*, 475–480.
38. Habermann, E.; Chhatwal, G. S. *Naunyn Schmiedebergs Arch Pharmacol* **1982**, *319*, 101–107.
39. Bottinger, H.; Beress, L.; Habermann, E. *Biochim. Biophys. Acta.* **1986**, *861*, 165–176.
40. Ishida, Y.; Takagi, K.; Takahashi, M.; Satake, N.; Shibata, S. *J. Biol. Chem.* **1983**, *258*, 7900–7902.
41. Bottinger, H.; Habermann, E. *Naunyn Schmiedebergs Arch Pharmacol* **1984**, *325*, 85–87.
42. Wattenberg, E.V.; McNeil, P.L.; Fujiki, H.; Rosner, M.R. *J. Biol. Chem.* **1989**, *264*, 213–219

RECEIVED June 6, 1989

Chapter 16

Mechanism of Pharmacological Action of Palytoxin

Yasushi Ohizumi

Mitsubishi Kasei Institute of Life Sciences, 11 Minamiooya, Machida, Tokyo 194, Japan

Palytoxin (PTX), isolated from the zanthid *Palythoa* species, caused a rapid contraction followed by a slow phasic contraction of the guinea pig vas deferens. The second component of PTX-induced contraction was markedly inhibited by adrenergic blocking agents, whereas the first component was blocked by ouabain. In pheochromocytoma cells, PTX caused a dose-dependent release of norepinephrine (NE). The NE release induced by lower concentrations of PTX increased proportionately with increasing Na^+ concentrations, but was not modified by tetrodotoxin. However, the NE-releasing action of higher concentrations of PTX was dependent on external Ca^{2+}, but not Na^+. Thus our experimental results suggest that in adrenergic neurons the PTX-induced release of NE by lower concentrations of PTX is brought about by tetrodotoxin-insensitive Na^+ permeability, whereas that induced by higher concentrations is mainly caused by a direct increase of Ca^{2+} influx into smooth muscle cells.

Palytoxin (PTX) is one of the most potent marine toxins known and the lethal dose (LD_{50}) of the toxin in mice is 0.5 µg/kg when injected i.v. The molecular structure of the toxin has been determined fully (*1,2*). PTX causes contractions in smooth muscle (*3*) and has a positive inotropic action in cardiac muscle (*4-6*). PTX also induces membrane depolarization in intestinal smooth (*3*), skeletal (*4*), and heart muscles (*5-7*), myelinated fibers (*8*), spinal cord (*9*), and squid axons (*10*). PTX has been demonstrated to cause NE release from adrenergic neurons (*11,12*). Biochemical studies have indicated that PTX causes a release of K^+ from erythrocytes, which is followed by hemolysis (*13-15*). The PTX-induced release of K^+ from erythrocytes is depressed by ouabain and that the binding of ouabain to the membrane fragments is inhibited by PTX (*15*).

Methods

The vas deferens was removed from male guinea pigs (250–350 g). The preparation of vas deferentia was carried out as described previously (*16*).

Culture of Pheochromocytoma Cells (PC12 Cells). PC12 cells, kindly supplied by Dr. H. Hatanaka of our institute, were maintained in Dulbecco's modified Eagle's medium containing 5% heat-inactivated horse serum.

Assay of Endogenous NE. Endogenous NE release from the guinea pig vas deferens was measured by means of a high pressure-liquid chromatograph with an ODS column and an electrochemical detector as described previously (*16*).

0097–6156/90/0418–0219$06.00/0

Assay of ^{3}H-NE. Experiments to determine ^{3}H-NE released from PC12 cells were performed as described previously (16).

Assay of ^{22}Na and ^{45}Ca Influxes. Experiments to determine ^{22}Na and ^{45}Ca influxes into PC12 cells were done as reported previously (12).

Tissue Na and K Content. Rectangular strips about 20 mg each were made by cutting the vas deferens longitudinally. After 30 min incubation under the control or test conditions, the strip was blotted, weighed, and then digested by incubation in 0.2 mL of a mixture containing equal amounts of HNO_3 (61%) and $HClO_4$ (60%) for 1 hr at 60°C; the remaining tissue was then ashed by an overnight heating at 180°C. The ashed sample was dissolved in cesium solution and the amount of Na or K was determined using an atomic absorption spectrophotometer (Varian, AA-175).

Assay of Ionophore Activity. Ionophoretic activities on rat liver mitochondria and liposomes were performed as described previously (12).

Results

Mechanical Response. PTX (5 x 10^{-10} to 3 x 10^{-8} M) caused a concentration-dependent contraction of the guinea pig vas deferens. The configuration of contractile response indicated the presence of two components, an initial rapid component followed by a second slow component. The first component of the response to PTX was abolished after treatment with Mg^{2+} (10 mM), Ca^{2+}-free medium, or ouabain (10^{-5} M), but remained almost unaffected by phentolamine (10^{-6} M), reserpine, 6-OHDA, atropine, or mecamylamine (10^{-5} M). The second component of the response to PTX was also completely inhibited after the incubation in the high Mg^{2+} or Ca^{2+}-free medium. Phentolamine, reserpine, or 6-OHDA markedly reduced only the second component. The second component also was inhibited by verapamil or low Na^+ medium, but was potentiated after treatment with ouabain (10^{-5} M).

Endogenous NE Release. The release of NE from the vas deferens was markedly increased by treatment with PTX (10^{-8} to 10^{-7} M). The PTX-induced release of NE was abolished by Mg^{2+} (10 mM) or Ca^{2+}-free medium, but was potentiated by ouabain (10^{-5} M).

^{3}H-NE Release. PTX (10^{-9} to 10^{-6} M) caused a concentration-dependent release of ^{3}H-NE from PC12 cells. This releasing action of PTX was markedly inhibited or abolished by Ca^{2+}-free medium, but was not modified by tetrodotoxin. The release of ^{3}H-NE induced by a low concentration (3 x 10^{-8} M) of PTX was abolished in Na^+-free medium and increased with increasing external Na^+ concentrations from 3 to 100 mM. But the release induced by a high concentration (10^{-6} M) was not changed by varying the concentration of external Na^+ from 0 to 100 mM. The release of ^{3}H-NE induced by both concentrations of PTX increased as the external Ca^{2+} concentration was increased from 0 to 3 mM, revealing the dependence on extracellular Ca^{2+}.

^{22}Na and ^{45}Ca Influx. PTX caused a concentration-dependent increase in ^{22}Na and ^{45}Ca influxes into PC12 cells at concentrations of 10^{-10} to 10^{-8} M and 10^{-9} to 10^{-6} M, respectively. The PTX-induced ^{45}Ca influx was markedly inhibited by Co^{2+} but not by verapamil or nifedepine, whereas the PTX-induced ^{22}Na influx was not affected by tetrodotoxin.

Assay of Na and K Content. Figure 1 shows the effects of PTX (10^{-8} M) on the tissue Na and K content of the guinea pig vas deferens. The Na content increased

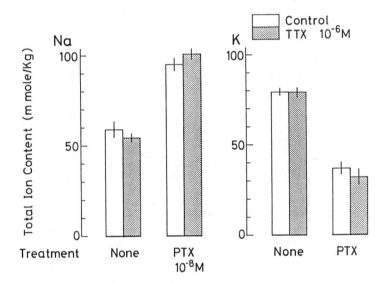

Figure 1. Effects of PTX on the tissue Na and K content of the vas deferens in the presence or absence of tetrodotoxin (TTX). PTX was added 30 min after the application of TTX. The ion content was measured 30 min after the application of PTX.

markedly 30 min after application of PTX, whereas the K content fell. The effect of PTX was not affected by tetrodotoxin (10^{-6} M).

Ionophoretic Activity. PTX, even at high concentrations, had no ionophoretic activity on membranes of mitochondria and liposomes.

Discussion

PTX caused a contraction of the guinea pig vas deferens. The configuration of the contractile response indicates the presence of two components, an initial rapid component followed by a second slow component (11). The second component of PTX-induced contraction, but not the initial component, was markedly inhibited by alpha adrenoceptor blocking agents, reserpine, and 6-OHDA, but was unaffected by cholinergic receptor blocking agents. Both components were abolished by Mg^{2+}-containing or Ca^{2+}-free medium. PTX markedly increased the Na content of the vas deferens. These data support the view that the first component is caused by a direct action of PTX on smooth muscle, possibly due to an increase in Na^+ influx, whereas the second component is the result of an indirect action mediated through NE release from the adrenergic nerve ending of the vas deferens.

The initial component of the PTX-induced contraction is selectively inhibited by ouabain, whereas the second component is rather potentiated by it (17). A similar inhibitory effect of ouabain on the contractile response to PTX is observed in the umbilical artery, which is devoid of adrenergic innervation (18).

Furthermore, it has been found that ouabain, convallatoxin, and cymarin inhibit the PTX-induced contraction of rabbit aortic vascular smooth muscle (19). These observations have also provided evidence for the involvement of Na^+,K^+-ATPase on the contractile effect of PTX in smooth muscle.

PTX caused a release of NE from adrenergic nerve endings of the vas deferens and from PC12 cells. In PC12 cells, the release of NE induced by a low concentration of PTX was not modified by tetrodotoxin, but was abolished in Na^+-free medium. However, the release induced by a high concentration is not affected by variations in the concentration of external Na^+. The release of NE induced by both concentrations of PTX is increased with raised external Ca^{2+} concentration. The above findings suggest that the PTX-induced release of NE by lower concentrations of PTX is primarily brought about by an increased Na^+ permeability, whereas that induced by higher concentrations is mainly caused by a direct increase in Ca^{2+} influx.

Electrophysiological studies on neuronal tissues have indicated that PTX causes a depolarization of the membrane of myelinated fibers (8), spinal cord (9), and squid axons (10). The PTX-induced depolarization is suppressed by removal of Na^+ from the external medium, but only slightly diminished in the presence of tetrodotoxin. PTX causes an increase in the resting Na^+ permeability and changes the current–voltage characteristics of myelinated fibers (8,10). The PTX-poisoned membrane is also permeable to other cations such as Li^+, Cs^+, and NH_4^+ (10). In addition, PTX even at high concentrations does not exhibit any ionophoretic activity. Furthermore, it was reported that PTX triggered the formation of small pores linked with Na^+,K^+-ATPase in resealed ghosts (20). These observations taken together suggest that PTX creates a novel channel in the membrane, thereby causing an increase in cation permeability and depolarization and thus inducing a release of NE from the adrenergic nerve.

Acknowledgments

I am greatly indebted to Prof. Y. Hirata of Meijo University and Dr. D. Uemura of Shizuoka University for kindly supplying palytoxin. I am grateful to Ms. Y. Murakami of this institute for secretarial assistance.

Literature Cited

1. Moore, R.E.; Bartolini, G. *J. Am. Chem. Soc.* **1981**, *103*, 2491.
2. Uemura, D.; Ueda, K.; Hirata, Y.; Naoki, H.; Iwashita, T. *Tetrahedron Lett.* **1981**, *22*, 2781.
3. Ito, K.; Karaki, H.; Urakawa, N. *Eur. J. Pharmacol.* **1977**, *46*, 9.
4. Deguchi, T.; Urakawa, N.; Takamatsu, S. In *Animal, Plant, and Microbial Toxins*; Vol. 2, A. Ohsaka; K. Hayashi; Y. Sawai Eds.; Plenum Publishing Corporation, New York, **1976**; p. 379.
5. Ito, K.; Karaki, H.; Urakawa, N. *Jpn. J. Pharmacol.* **1979**, *29*, 467.
6. Weidmann, S. *Experientia* **1977**, *33*, 1487.
7. Ito, K.; Saruwatari, N.; Mitani, K.; Enomoto, Y. *Naunyn-Schmiedeberg's Arch. Pharmacol.* **1985**, *330*, 67.
8. Dubois, J.M.; Cohen, J.B. *J. Pharmacol. Exp. Ther.* **1977**, *201*, 148.
9. Kudo, Y.; Shibata, S. *Br. J. Pharmacol.* **1980**, *71*, 575.
10. Muramatsu, I.; Uemura, D.; Fujiwara, M.; Narahashi, T. *J. Pharmacol. Exp. Ther.* **1984**, *231*, 488.
11. Ohizumi, Y.; Shibata, S. *J. Pharmacol. Exp. Ther.* **1980**, *214*, 209.
12. Tatsumi, M.; Takahashi, M.; Ohizumi, Y. *Mol. Pharmacol.* **1983**, *25*, 379.
13. Habermann, E.; Ahnert-Hilger, G.; Chhatwal, G.S.; Beress, L. *Biochim. Biophys. Acta.* **1981**, *649*, 481.
14. Ahnert-Hilger, G.; Chhatwal, G.S.; Hessler, H.J.; Habermann, E. *Biochim. Biophys. Acta.* **1982**, *688*, 486.
15. Habermann, E.; Chhatwal, G.S. *Naunyn-Schmiedeberq's Arch. Pharmacol.* **1982**, *319*, 101.
16. Ohizumi, Y.; Kajiwara, A.; Yasumoto, T. *J. Pharmacol. Exp. Ther.* **1983**, *227*, 199.
17. Ishida, Y.; Kajiwara, A.; Takagi, K.; Ohizumi, Y.; Shibata, S. *J. Pharmacol. Exp. Ther.* **1985**, *232*, 551.
18. Ishida, Y.; Sataka, N.; Habon, J.; Kitano, H.; Shibata, S. *J. Pharmacol. Exp. Ther.* **1985**, *232*, 557.
19. Ozaki, H.; Nagase, H.; Urakawa, N. *J. Pharmacol. Exp. Ther.* **1984**, *231*, 153.
20. Chhatwal, G.S.; Hessler, H.-J.; Haberman, E. *Naunyn-Schmiedeberg's Arch. Pharmacol.* **1983**, *323*, 261.

RECEIVED June 8, 1989

Chapter 17

Production of Antibodies and Development of a Radioimmunoassay for Palytoxin

Lawrence Levine[1], Hirota Fujiki[2], Hilda B. Gjika[1], and Helen Van Vunakis[1]

[1]Department of Biochemistry, Brandeis University, Waltham, MA 02254
[2]Cancer Prevention Division, National Cancer Center Research Institute, Chuo-ku, Tokyo 104, Japan

Palytoxin (a) stimulates arachidonic acid metabolism in cells in culture, (b) hemolyzes rat erythrocytes, and (c) is lethal to mice when administered intraperitoneally. Serum from rabbits, immunized with a conjugate of palytoxin and bovine albumin via palytoxin's amino function, neutralized all of these biologic activities. The titer of the serum required for the neutralization of palytoxin's stimulation of arachidonic acid metabolism increased with each course of immunization and was reduced to less than 99% by precipitation of the rabbit γ-immunoglobulin with a goat anti-rabbit γ-immunoglobulin. Palytoxin, labelled with [^{125}I] Bolton-Hunter Reagent on its terminal amino group, bound specifically to these antibodies. The extent of binding also increased progressively with repeated immunizations. A radioimmunoassay was developed. For 50% inhibition of binding, 0.27 pmol of unlabelled palytoxin was required. Maitotoxin, teleocidin, okadaic acid, debromoaplysiatoxin, and TPA, when tested at 10- to 100-fold higher concentrations than palytoxin, did not affect binding.

The palytoxins synthesized by several *Palythoa* species are potent biologically active substances (*1*) with molecular weights of 2659 to 2677, depending on the species of coelenterate from which they originate. Their long, partially unsaturated aliphatic chains contain interspaced cyclic ethers, 40–42 hydroxy, and two amide groups (*2*, *3*). The chain starts with a hydroxy group and terminates with an amino group (*2*, *3*).

Palytoxin is hemolytic (*4*) and is an extremely potent toxin (*1*). We have shown that in rat liver cells palytoxin stimulates de-esterification of cellular lipids to liberate arachidonic acid (*5*). These rat liver cells metabolize this increased arachidonic acid via the cyclooxygenase pathway to produce prostaglandin (PG) I_2 and lesser amounts of PGE_2 and $PGF_{2\alpha}$. Palytoxin acts on many cells in culture to stimulate the production of cyclooxygenase metabolites (Table I). Clearly, the myriad pharmacological effects of the arachidonic acid metabolites must be considered in any explanation of the many clinical manifestations of palytoxin's toxicity.

0097–6156/90/0418–0224$06.00/0

Table I. Cells in Which Arachidonic Acid Metabolism Is Stimulated
by Palytoxin to Produce Cyclooxygenase Products

Rat liver cells
Bovine aorta smooth muscle cells
Bovine aorta endothelial cells
Porcine aorta endothelial cells
Rat keratinocytes
Squirrel monkey aorta smooth muscle cells
Rat peritoneal macrophages (6)
Mouse calvaria: osteoclasts and/or osteoblasts (7)

To prepare a conjugate for immunization of rabbits (8), an amide bond was
formed between the terminal amino group of palytoxin and carboxyl groups on
bovine albumin (Figure 1). The immune sera neutralized palytoxin's stimulation of
arachidonic acid metabolism by several cell types in culture with the titre increasing
as a function of immunization (Figure 2). The neutralizing activity in the serum
was essentially eliminated after precipitation of the rabbit γ-immunoglobulin (IgG)
with a goat anti-rabbit IgG. The immune serum also neutralized palytoxin's lytic
activity toward rat erythrocytes.

The LD_{50} of palytoxin in female Swiss Albino mice 24 hours following intraperitoneal injection is 5×10^{-4} mg/kg (8). The immune sera also neutralized
palytoxin's lethal effects. As shown in Figure 3, 11/12 mice were killed by palytoxin
(1×10^{-3} mg/kg), whereas 0/12 and 0/11 mice were killed by palytoxin when
injected intraperitoneally in the presence of the immune serum. None of the protected mice showed any signs of distress.

Palytoxin, radiolabelled by reaction of its terminal NH_2 group with [^{125}I]
Bolton-Hunter reagent, bound to the anti-palytoxin, and a radioimmunoassay was
developed (9). The serologic specificity with respect to several toxins and/or tumor
promoters as well as the palytoxins isolated from two *Palythoa* species is shown in
Table II. The anti-palytoxin did not distinguish the homologous palytoxin of *P. cari-
baeorum* from the heterologous palytoxin of *P. tuberculosa* whose structure differs
only in the hemiketal ring at position C55 (2). Maitotoxin, present in some
species of dinoflagellates (10); teleocidin, isolated from *Streptomyces* and similar to
the toxin found in the blue-green alga *Lyngbya majuscula* (11); okadaic acid, a
polyether found in some species of sponge (12); debromoaplysiatoxin, found in
some species of blue-green algae (13); and 12-O-tetradecanoylphorbol-13-acetate, a
tumor promoter isolated from the plant *Croton tiglium*, did not inhibit at the levels
tested. This lack of serologic activity by these toxins and tumor promoters suggests
that this immune system will be specific, at least with respect to these toxins, for
detection of palytoxin-like material in the food chain.

Palytoxin's properties of stimulating arachidonic acid metabolism were decreased
> 90% after exposure to 0.1 N HCl for 60 min at 37°C and decreased about 75%
after exposure to 0.01 N HCl for 60 min at 50°C, but these biologic properties
were stable to boiling in H_2O for 60 min. Palytoxin's serologic activities were stable
to these treatments.

The palytoxin–anti-palytoxin reaction is unique in that its binding increases
with increasing temperature (Figure 4). The apparent association constant of the
palytoxin to anti-palytoxin was 4.9×10^9 M^{-1} at 0°C and 1.1×10^{10} M^{-1} at 35°C,
suggesting that H_2O must be displaced from some of palytoxin's epitopes before
they can bind to their antibody combining sites.

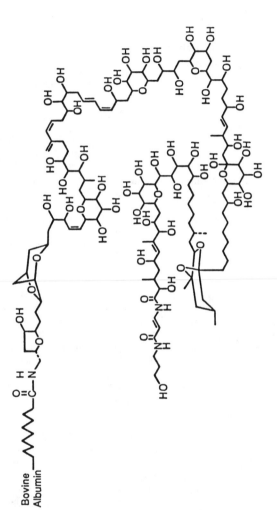

Palytoxin Immunogen

Figure 1. Structure of palytoxin, isolated from *P. tuberculosa* (*3*), immunogen (bovine albumin). The structures of the palytoxins from different *Palythoa* species vary in the hemiketal ring at position C55 (*2, 3*).

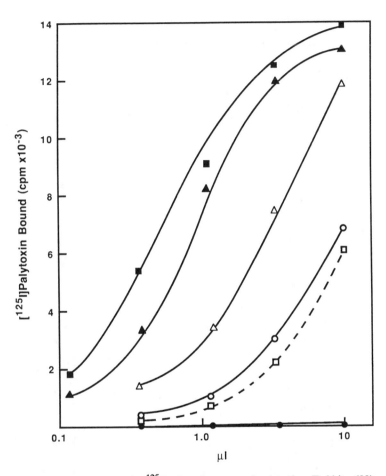

Figure 2. Binding of [^{125}I]palytoxin to anti-palytoxin (Rabbit 633) before (●) and after primary (○), first boost (△), second boost (▲), and third boost (■ , Rabbit 633D-24) of immunization and after absorption of IgG in Rabbit 633D-24 with goat anti-rabbit IgG (□).

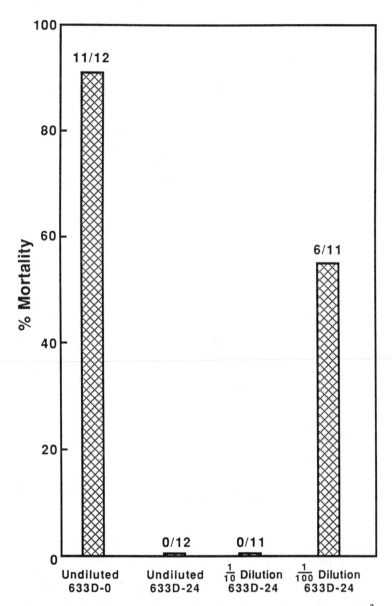

Figure 3. Mortality of mice 24 hr after i.p. injection of 1×10^{-3} mg/kg palytoxin in the presence of pre-immune serum from Rabbit 633; undiluted immune serum, 10% immune serum, and 1% immune serum. Numbers above bars represent the number of animals dead and the number tested. (Reproduced with permission from Ref. 8. Copyright 1987 Pergamon Press.)

Table II. Serologic Specificity of [^{125}I]Palytoxin–anti-Palytoxin
Immune System

Inhibitor	IC_{50}[a]
Palytoxin (*P. tuberculosa*)	0.27[b]
Palytoxin (*P. caribaeorum*)	0.30[b]
Maitotoxin	c
Teleocidin	d
Okadaic acid	e
12-*O*-tetradecanoylphorbol-13-acetate	f
Debromoaplysiatoxin	g

SOURCE: Reproduced with permission from Ref. 9. Copyright 1988,
Pergamon Press PLC.
[a]pmol required for 50% inhibition of binding.
[b]Values calculated from dose-response curves.
[c]0 Inhibition with 2.9 nmol
[d]0 Inhibition with 22.8 nmol
[e]0 Inhibition with 10.0 nmol
[f]0 Inhibition with 16.8 nmol
[g]0 Inhibition with 14.7 nmol

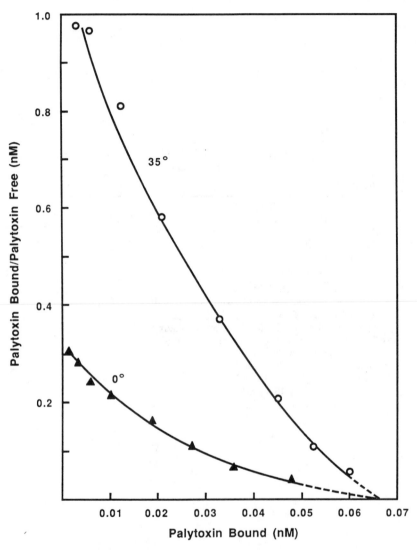

Figure 4. Specific binding of palytoxin to anti-palytoxin (Rabbit 633D-24, 1/25,000) at 35 ° C (○) and 0 ° C (▲).

Acknowledgments

This work was supported by Grant GM 27256 from the National Institutes of Health and Grant DA 02507 from the National Institute of Drug Abuse. L.L. is an American Cancer Society Research Professor of Biochemistry (Award PRP-21). H.V.V. is the recipient of a Research Career Award (5K6-AI-2372) from the National Institute of Allergy and Infectious Disease. We thank Dr. Y. Hirata of Meijo University for generous gifts of palytoxin isolated from *Palythoa tuberculosa*. We thank Dr. T. Yasumoto, Tohoku University, Sendai, Japan, for the maitotoxin preparation. We thank also Jeffrey A. Bessette and Nancy Worth for their technical assistance and Inez Zimmerman for preparation of the manuscript.

Literature Cited

1. Kaul, P. N.; Farmer, M. R.; Ciereszko, L. S. *Proc. West. Pharmacol. Soc.* **1974**, *17*, 294.
2. Moore, R. E.; Bartolini, G. *J. J. Am. Chem. Soc.* **1981**, *103*, 2491.
3. Uemura, D.; Ueda, K.; Hirata, Y.; Naoki, H.; Iwashita, T. *Tetrahedron Lett.* **1981**, *22*, 2781.
4. Habermann, E.; Ahnert-Hilger, G.; Chhatwal, G. S.; Béress, L. *Biochim. Biophys. Acta* **1981**, *649*, 481.
5. Levine, L.; Fujiki, H. *Carcinogenesis* **1985**, *11*, 1631.
6. Ohuchi, K.; Watanabe, M.; Yoshizawa, K.; Tsurufuji, S.; Fujiki, H.; Suganuma, M.; Sugimura, T.; Levine, L. *Biochim. Biophys. Acta* **1985**, 42–47.
7. Lazzaro, M.; Tashjian, A. H., Jr.; Fujiki, H.; Levine, L. *Endocrinology* **1987**, *120*, 1338.
8. Levine, L.; Fujiki, H.; Gjika, H. B.; Van Vunakis, H. *Toxicon* **1987**, *25*, 1273.
9. Levine, L.; Fujiki, H.; Gjika, H. B.; Van Vunakis, H. *Toxicon* **1988**, *26*, 1115.
10. Ohizumi, Y.; Yasumoto, T. *J. Physiol.* **1983**, *337*, 711.
11. Fujiki, H.; Sugimura, T. In *Advances in Cancer Research* **1987**, *49*, 223.
12. Tachibana, K.; Scheuer, P. J.; Tsukitani, Y.; Kikuchi, H.; Van Engen, D.; Clardy, J.; Gopichand, Y.; Schmitz, F. J. *J. Am. Chem. Soc.* **1981**, *103*, 2469.
13. Fujiki, H.; Suganuma, M.; Tahira, T.; Yoshioka, A.; Nakayasu, M.; Endo, Y.; Shudo, K.; Takayama, S.; Moore, R. E.; Sugimura, T. In *Cellular Interactions by Environmental Tumor Promoters*; Fujiki, H.; Hecker, E.; Moore, R. E.; Sugimura, T.; Weinstein, I. B., Eds.; Published jointly by Japan Sci. Soc. Press: Tokyo, and VNU Science Press: Utrecht, 1984; p 37.

RECEIVED May 30, 1989

Chapter 18

New Tumor Promoters from Marine Natural Products

Hirota Fujiki[1], Masami Suganuma[1], Hiroko Suguri[1], Shigeru Yoshizawa[1], Kanji Takagi[1], Michie Nakayasu[2], Makoto Ojika[3], Kiyoyuki Yamada[3], Takeshi Yasumoto[4], Richard E. Moore[5], and Takashi Sugimura[6]

[1]Cancer Prevention Division, National Cancer Center Research Institute, Tsukiji 5-1-1, Chuo-ku, Tokyo 104, Japan
[2]Common Laboratory, National Cancer Center Research Institute, Tsukiji 5-1-1, Chuo-ku, Tokyo 104, Japan
[3]Faculty of Science, Nagoya University, Nagoya 464, Japan
[4]Department of Food Chemistry, Faculty of Agriculture, Tohoku University, Tsutsumidori Amamiya, Sendai 980, Japan
[5]Department of Chemistry, University of Hawaii, Honolulu, HI 96822
[6]National Cancer Center Research Institute, Tsukiji 5-1-1, Chuo-ku, Tokyo 104, Japan

Tumor promoters induce tumor formation from initiated cells. We have found new tumor promoters that are structurally different from 12-O-tetradecanoylphorbol-13-acetate (TPA). These new tumor promoters include lyngbyatoxin A, debromoaplysiatoxin, aplysiatoxin, bromoaplysiatoxin, oscillatoxin A and anhydrodebromoaplysiatoxin isolated from marine blue-green algae, and dibromoaplysiatoxin synthesized chemically; palytoxin isolated from a marine coelenterate; and okadaic acid isolated from a black sponge. These new tumor promoters are classified as TPA-type and non-TPA type tumor promoters on the basis of their abilities to bind to the phorbol ester receptor. Lyngbyatoxin A and aplysiatoxins belong to the TPA-type, and palytoxin and okadaic acid are non-TPA type tumor promoters. TPA-type tumor promoters activate protein kinase C, which serves as the phorbol ester receptor, while non-TPA type tumor promoters do not activate protein kinase C in vitro. Therefore, these two types of tumor promoters provide direct evidence for divergent mechanisms of action in two-stage carcinogenesis experiments on mouse skin.

The process of chemical carcinogenesis consists of two stages, initiation and promotion (1). Initiation is caused by a single application of a small amount of a carcinogen, which induces irreversible genetic damage to DNA. Application of the carcinogen, 7,12-dimethylbenz(a)anthracene (DMBA), for example, to the skin of the back of a mouse was reported to induce mutation of an oncogene, so-called activation of the *ras* gene (2). Agents that then promote carcinogenesis from initiated cells are called tumor promoters (3). Treatment with DMBA, followed by repeated applications of a tumor promoter, results in a high percentage of tumor-bearing mice, whereas treatment with an initiator alone or tumor promoter alone does not produce any tumors in mouse skin. TPA or phorbol-myristate-acetate (PMA) is the classical tumor promoter, isolated from croton oil, the oil of seeds of *Croton tiglium* L., belonging to the Euphorbiaceae *(4,5)* (Figure 1).

We have so far found 20 so-called new tumor promoters, because they are structurally different from phorbol esters *(6)*. Naturally occurring tumor promoters can be classified according to their sources: *Actinomycetes*, terrestrial plants, and marine organisms (Table I) *(7)*. This review deals with tumor promoters isolated from marine organisms, such as lyngbyatoxin A, also named teleocidin A-1, aplysiatoxins, palytoxin, and okadaic acid (Figure 1). It is noteworthy that lyngbyatoxin A, debromoaplysiatoxin, aplysiatoxin, and okadaic acid have the same potent tumor promoting activities as TPA in two-stage carcinogenesis experiments *(8,9,10)*.

TPA-Type Tumor Promoters: Teleocidin Class

Teleocidin isolated from *Streptomyces mediocidicus* is a mixture of two isomers of teleocidin A and four isomers of teleocidin B (Figure 1 and Table I) *(11,12)*. (-)-Indolactam-V is a biosynthetic intermediate of teleocidins A and B *(13)*. Des-O-methylolivoretin C is a regioisomer of teleocidin B-1 (Table I) *(14)*.

Lyngbyatoxin A was isolated from a Hawaiian shallow-water variety of *Lyngbya majuscula*, which grows at Kahala Beach on the island of Oahu *(15)*. In 1979, Cardellina et al. determined the structure of lyngbyatoxin A (Figure 1). Because lyngbyatoxin A was structurally similar to teleocidin B, and was a highly inflammatory and vesicatory substance like teleocidin B, we thought that lyngbyatoxin A might be as active as teleocidin B in various biological and biochemical tests and in a two-stage carcinogenesis experiment on mouse skin. However, we found that lyngbyatoxin A is identical to one of the two isomers of teleocidin A, teleocidin A-1, which corresponds to (19R)-teleocidin A *(16)*. Lyngbyatoxin A was biologically and biochemically active like all isomers of teleocidins A and B. Table II shows that lyngbyatoxin A had approximately the same potency as TPA in an irritant test on mouse ear, in inductions of ornithine decarboxylase (ODC) in mouse skin and adhesion of human promyelocytic leukemia (HL-60) cells, in inhibition of the specific binding of ^{3}H-TPA to a mouse skin particulate fraction, and in activation of protein kinase C in vitro. An irritant test on mouse ear shows the inflammatory effect. Induction of ODC in mouse skin reflects the induction of polyamine biosynthesis associated with cell proliferation. Induction of cell adhesion of HL-60 cells is a measure of changes of the cell surface induced by a tumor promoter. Inhibition of the specific binding of ^{3}H-TPA to a mouse skin particulate fraction measures the phorbol ester receptor binding of a compound. In a two-stage carcinogenesis experiment, lyngbyatoxin A had the same potent tumor promoting activity as TPA, as shown in Table II, as did teleocidin A-2 and the four teleocidin B isomers (data not shown). The two-stage carcinogenesis experiment was carried out by a single application of 100 μg DMBA, followed by repeated applications of the test compounds in the amounts shown in Table II, twice a week, until week 30. Tumor promoting activity was determined as the percentage of tumor-bearing mice in the group treated with DMBA plus the test compound (Table II) *(17)*.

TPA-Type Tumor Promoters: Aplysiatoxin Class

Aplysiatoxin and debromoaplysiatoxin were isolated from another variety of blue-green alga at Kailua Beach on the windward side of Oahu found to be a causative agent of swimmer's itch *(18)*. Aplysiatoxin and debromoaplysiatoxin were also isolated from blue-green alga in Okinawa *(19)*. In addition to aplysiatoxin and debromoaplysiatoxin, bromoaplysiatoxin, oscillatoxin A, and anhydrodebromoaplysiatoxin were also obtained from a mixture of blue-green algae *(20)*. Dibromoaplysiatoxin, which contains three bromine atoms, is a chemically brominated derivative of

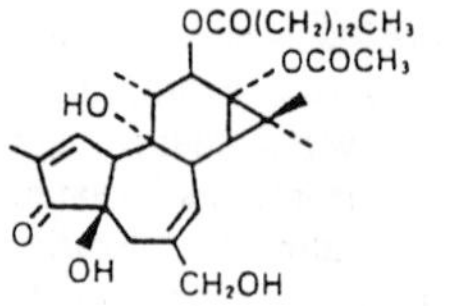

TPA

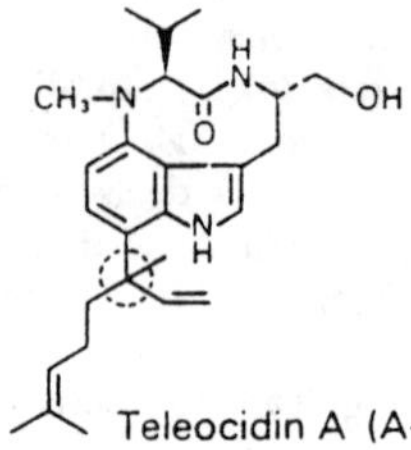

Teleocidin A (A-1, A-2)
Lyngbyatoxin A

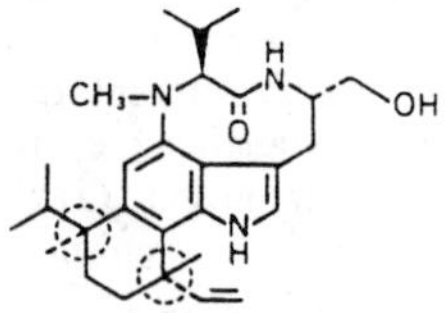

Teleocidin B
(B-1, B-2, B-3, B-4)

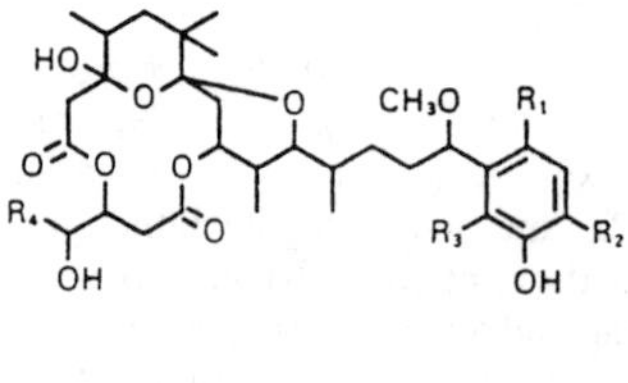

	R_1	R_2	R_3	R_4
Debromoaplysiatoxin	H	H	H	CH_3
Aplysiatoxin	Br	H	H	CH_3
Bromoaplysiatoxin	Br	Br	H	CH_3
Dibromoaplysiatoxin	Br	Br	Br	CH_3
Oscillatoxin A	H	H	H	H

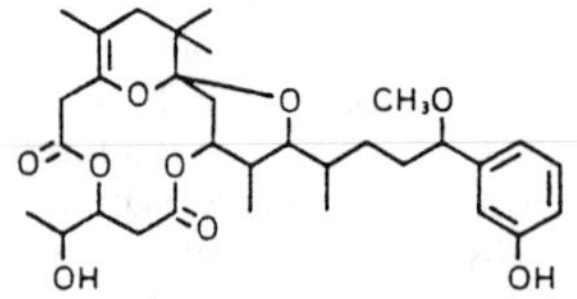

Anhydrodebromoaplysiatoxin

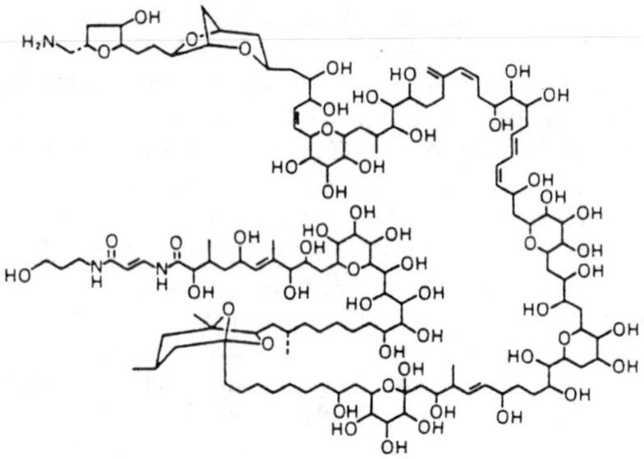

Palytoxin

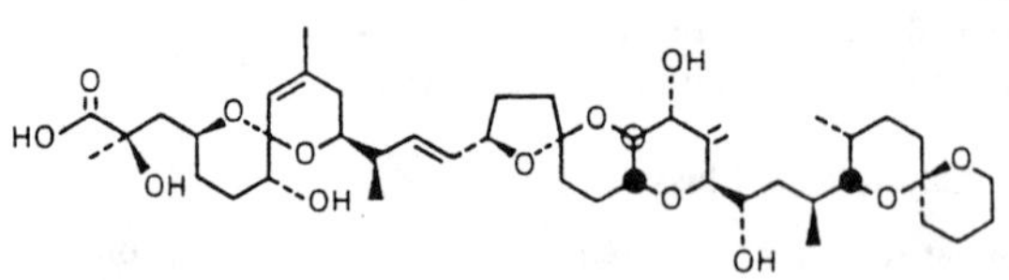

Okadaic acid

Figure 1. Structures of TPA and new tumor promoters from marine organisms

Table I.   New Tumor Promoters Classified According to Their Sources

1. From *Actinomycetes*	3. From marine organisms
Teleocidin A-1, A-2	Lyngbyatoxin A (Teleocidin A-1)
Teleocidin B-1, B-2, B-3, B-4	Debromoaplysiatoxin
(-)-Indolactam-V	Aplysiatoxin
Des-*O*-methylolivoretin C	Bromoaplysiatoxin
	Dibromoaplysiatoxin
	Oscillatoxin A
2. From terrestrial plants	Anhydrodebromoaplysiatoxin
	Palytoxin
Thapsigargin	Okadaic acid

Table II. Effects of New Tumor Promoters Derived from Marine Organisms

Tumor promoter	Irritant test (0.1 nmol)	Induction of ODC (nmol CO_2/mg protein/30 min)	Adhesion of HL-60 cells (ED_{50} ng/mL)	Inhibition of specific binding of ^3H-TPA (ED_{50} nM)	Activation of protein kinase C (pmol/min/1.0 µg compound)	Tumor-bearing mice in week 30 (%)
Lyngbyatoxin A	+++	3.31[a]	1.2	8.0	3.3	86.6[g]
Debromoaplysiatoxin	+++	5.52[b]	128.0	6.8	4.1	71.4[h]
Aplysiatoxin	++++	5.06[b]	1.2	6.6	4.2	73.3[h]
Bromoaplysiatoxin	++	6.38[c]	0.8	6.6	5.0	57.1[h]
Dibromoaplysiatoxin	—	1.21[d]	18.3	3300	5.2	66.7[h]
Oscillatoxin A	++	3.85[c]	78.0	10.0	5.9	53.3[h]
Anhydrodebromo-aplysiatoxin	—	2.02[d]	600.0	530	3.4	40.0[i]
Palytoxin	+	0.[e]	>0.2	>10,000	0	62.5[j]
Okadaic acid	+	1.38[f]	>20.0	>100,000	0	80.0[k]
TPA	+++	3.02[b]	1.5	1.4	6.0	92.7[h]

[a] 11 nmol of lyngbyatoxin A; [b] 3 nmol; [c] 9 nmol; [d] 90 nmol; [e] up to 3 nmol of palytoxin; [f] 12 nmol of okadaic acid.
Dose per application: [g] 6.9 nmol; [h] 4 nmol; [i] 20 nmol; [j] 0.2 nmol; [k] 12 nmol.

debromoaplysiatoxin (21). The effects of these compounds on the activities tested are summarized in Table II. The results for ODC induction and tumor promoting activity were obtained with optimal concentrations of the compounds. Like lyngbyatoxin A, the aplysiatoxin class tumor promoters bind to phorbol ester receptors in the cell membrane and activate protein kinase C in vitro (Table II). Interestingly, all compounds of the aplysiatoxin class were found to be tumor promoters with varying potencies of tumor promoting activity (Table II).

However, it is important to describe the potency of anhydrodebromoaplysiatoxin regarding several biological effects in more detail. Anhydrodebromoaplysiatoxin was reported to be an inactive compound formed by a facile, acid-catalyzed dehydration of debromoaplysiatoxin (22). Table II shows that 0.1 nmol of anhydrodebromoaplysiatoxin was negative in irritant test on mouse ear and our earlier paper reported that 3 nmol of anhydrodebromoaplysiatoxin did not induce any ODC activity (10). However, 90 nmol of the compound induced 2.02 nmol CO_2/mg protein as shown in Table II. Additional results of several experiments, such as induction of HL-60 cell adhesion and activation of protein kinase C, suggested the presence of weak biological activity in anhydrodebromoaplysiatoxin, which was also supported by the evidence of inhibition of metabolic cooperation in the Chinese hamster V79 cell system found by Trosko and his associates (23). The two-stage carcinogenesis experiments with DMBA plus anhydrodebromoaplysiatoxin using two different doses, 4 nmol and 20 nmol, revealed that the percentages of tumor promoting activity were 0% and 40% in week 30, respectively. It was therefore concluded that anhydrodebromoaplysiatoxin is a derivative associated with a weak activity. Furthermore, experiments with ^{3}H-TPA, ^{3}H-lyngbyatoxin A, and ^{3}H-debromoaplysiatoxin indicated three classes of tumor promoters that bind to the same phorbol ester receptors in cell membranes in the same way (24). Since lyngbyatoxin A and the aplysiatoxin class exert tumor promoting activity through the same pathway as TPA, we called them TPA-type tumor promoters (6).

Non-TPA Type Tumor Promoters: Palytoxin and Okadaic Acid Class

In contrast to the TPA-type tumor promoters, palytoxin, thapsigargin, and okadaic acid are classified as non-TPA type tumor promoters, which do not bind to phorbol ester receptors, or activate protein kinase C in vitro (Table II) (6,25-27). In this chapter, thapsigargin is not discussed, because it is derived from terrestrial plants.

Palytoxin was isolated from a marine coelenterate of the genus *Palythoa* (Figure 1) (28,29). Okadaic acid was isolated from a black sponge, *Halichondria okadai* (Figure 1) (30). Okadaic acid is a polyether derivative of a C_{38} fatty acid. As Table II shows, palytoxin and okadaic acid both caused irritation of mouse ear. Palytoxin did not induce ODC in mouse skin or HL-60 cell adhesion, whereas okadaic acid induced ODC in mouse skin, but not HL-60 cell adhesion (Table II) (23,31). Like the TPA-type tumor promoters, palytoxin and okadaic acid stimulated prostaglandin E_2 production from ^{3}H-arachidonic acid-prelabeled macrophages and arachidonic acid metabolism by rat liver cells in culture (data not shown) (32,33). Palytoxin was shown to be the strongest prostaglandin inducer found to date. It is noteworthy that palytoxin and okadaic acid exert various effects and tumor promoting activities; however, they may not exert their effects by the same pathway. Okadaic acid acts on cells in a different way from TPA-type tumor promoters, such as TPA, lyngbyatoxin A, and aplysiatoxin, but its tumor promoting activity is as strong as that of TPA-type tumor promoters (27). Therefore, the mechanism of action of okadaic acid is of interest.

We synthesized ^{3}H-okadaic acid chemically and demonstrated its specific binding to the particulate and cytosolic fractions of mouse skin. The specific binding of ^{3}H-okadaic acid to the particulate fraction was not inhibited by TPA, lyngbyatoxin

A, aplysiatoxin, or palytoxin. Therefore, we proposed that okadaic acid binds to its own receptor and induces various biological effects, including tumor promoting activity, through this receptor (27). We are now investigating the nature of its receptor.

In addition to okadaic acid, dinophysistoxin-1 (i.e., 35-methylokadaic acid), 7-O-palmitoyl-okadaic acid, and pectenotoxin 2 are reported to be diarrhetic toxins from shellfish (34). Application of 1 μg of dinophysistoxin-1 to mouse ear caused as strong irritation as the same dose of okadaic acid. Interestingly, the potencies of these compounds in the irritant test on mouse ear correlated well with their potencies as diarrhetic shellfish poisons. Dinophysistoxin-1 induced ODC activity as strongly as okadaic acid. Recently, we found that dinophysistoxin-1 is also a new non-TPA type tumor promoter with as high activity as okadaic acid (35).

Common Effects of Tumor Promoters

We have shown that TPA-type tumor promoters, such as lyngbyatoxin A and aplysiatoxins exert tumor promoting activities on mouse skin through different mechanisms from those of non-TPA type tumor promoters such as palytoxin, okadaic acid, and dinophysistoxin-1.

Furthermore, we found that the two types of tumor promoters induced common biological effects, such as irritation of mouse ear, and stimulation of prostaglandin E_2 production and of arachidonic acid metabolism in rat macrophages. These common effects seem to be the most essential biological activities in tumor promotion (6).

This chapter reports the presence of various kinds of tumor promoters in marine organisms. The problem of the biological functions of these tumor promoters remains to be investigated. Solution of this problem will throw light on the problems of tumor promotion and tumor development. We are also investigating inhibitors of tumor promotion derived from marine organisms (36). This work shows that marine natural products provide useful tools for understanding chemical carcinogenesis.

Acknowledgments

This work was supported in part by Grants-in-Aid for Cancer Research from the Ministry of Education, Science and Culture, a grant for the Program for a Comprehensive 10-Year Strategy for Cancer Control from the Ministry of Health and Welfare of Japan, grants from the Foundation for Promotion of Cancer Research, the Princess Takamatsu Cancer Research Foundation, and the Smoking Research Foundation., and by Grant CA 12623 from the National Institutes of Health.

Literature Cited

1. Berenblum, I. *Cancer Res.* **1941**, *1*, 44.
2. Balmain, A.; Ramsden, M.; Bowden, G.T.; Smith, J. *Nature* **1984**, *307*, 658.
3. Boutwell, R.K. In *Carcinogenesis: Mechanisms of Tumor Promotion and Cocarcinogenesis*; Slaga, T.J., Sivak, A., Boutwell, R.K., Eds.; Raven Press: New York, 1978: Vol. 2, p 49.
4. Hecker, E. *Methods Cancer Res.* **1971**, *6*, 439.
5. Van Duuren, B.L. *Prog. Exp. Tumor Res.* **1969**, *11*, 31.
6. Fujiki, H.; Sugimura, T. *Adv. Cancer Res.* **1987**, *49*, 223.
7. Fujiki, H.; Suganuma, M.; Hirota, M.; Yoshizawa, S.; Suguri, H.; Suttajit, M.; Wongchai, V.; Sugimura, T. In *Current Status of Cancer Research in Asia, The*

Middle East and Other Countries; Wada, T., Aoki, K., Yachi, A., Eds.; The University of Nagoya Press: Nagoya, 1987; p 215.

8. Fujiki, H.; Suganuma, M.; Hakii, H.; Bartolini, G.; Moore, R.E.; Takayama, S.; Sugimura, T. *J. Cancer Res. Clin. Oncol.* 1984, *108*, 174.

9. Moore, R.E. In *Cellular Interactions by Environmental Tumor Promoters*; Fujiki, H., Hecker, E., Moore, R.E., Sugimura, T., Weinstein, I.B., Eds; Jpn. Sci. Soc. Press, Tokyo/VNU Science Press: Utrecht, 1984: p 49.

10. Suganuma, M.; Fujiki, H.; Tahira, T.; Cheuk, C.; Moore, R.E.; Sugimura, T. *Carcinogenesis* 1984, *5*, 315.

11. Fujiki, H.; Suganuma, M.; Tahira, T.; Yoshioka, A.; Nakayasu, M.; Endo, Y.; Shudo, K.; Takayama, S.; Moore, R.E.; Sugimura, T. In *Cellular Interactions by Environmental Tumor Promoters*; Fujiki, H., Hecker, E., Moore, R.E., Sugimura, T., Weinstein, I.B., Eds; Jps. Sci. Soc. Press, Tokyo/VNU Science Press: Utrecht, 1984: p 37.

12. Fujiki, H.; Sugimura, T. *Cancer Surveys* 1983, *2*, 539.

13. Fujiki, H.; Suganuma, M.; Hakii, H.; Nakayasu, M.; Endo, Y.; Shudo, K.; Irie, K.; Koshimizu, K.; Sugimura, T. *Proc. Jpn. Acad.* 1985, *61*, 45.

14. Ninomiya, M.; Fujiki, H.; Paik, N.S.; Hakii, H.; Suganuma, M.; Hitotsuyanagi, Y.; Aimi, N.; Sakai, S.; Endo, Y.; Shudo, K.; Sugimura, T. *Jpn. J. Cancer Res. (Gann)* 1986, *77*, 222.

15. Cardellina, J.H.; Marner, F.-J.II; Moore, R.E. *Science* 1979, *204*, 193.

16. Sakai, S.; Hitotsuyanagi, Y.; Aimi, N.; Fujiki, H.; Suganuma, M.; Sugimura, T.; Endo, Y.; Shudo, K. *Tetrahedron Lett.* 1986, *27*, 5219.

17. Fujiki, H.; Suganuma, M.; Matsukura, N.; Sugimura, T.; Takayama, S. *Carcinogenesis* 1982, *3*, 895.

18. Moore, R.E. *Pure Appl. Chem.* 1982, *54*, 1919.

19. Fujiki, H.; Ikegami, K.; Hakii, H.; Suganuma, M.; Yamaizumi, Z.; Yamazato, K.; Moore, R.E.; Sugimura, T. *Jpn. J. Cancer Res. (Gann)* 1985, *76*, 257.

20. Mynderse, J.S.; Moore, R.E. *J. Org. Chem.* 1978, *43*, 2301.

21. Moore, R.E.; Blackman, A.J.; Cheuk, C.E.; Mynderse, J.S.; Matsumoto, G.K.; Clardy, J.; Woodard, R.W.; Craig, J.C. *J. Org. Chem.* 1984, *49*, 2484.

22. Kato, Y.; Scheuer, P.J. *Pure Appl. Chem.* 1976, *48*, 29.

23. Jone, C.; Erickson, L.; Trosko, J.E.; Chang, C.C. *Cell Biology and Toxicology* 1987, *3*, 1.

24. Moore, R.E.; Patterson, G.M.; Entzeroth, M.; Morimoto, H.; Suganuma, M.; Hakii, H.; Fujiki, H.; Sugimura, T. *Carcinogenesis* 1986, *7*, 641.

25. Fujiki, H.; Suganuma, M.; Nakayasu, M.; Hakii, H.; Horiuchi, T.; Takayama, S.; Sugimura, T. *Carcinogenesis* 1986, *7*, 707.

26. Hakii, H.; Fujiki, H.; Suganuma, M.; Nakayasu, M.; Tahira, T.; Sugimura, T.; Scheuer, P.J.; Christensen, S.B. *J. Cancer Res. Clin. Oncol.* 1986, 111, 177.

27. Suganuma, M.; Fujiki, H.; Suguri, H.; Yoshizawa, S.; Hirota, M.; Nakayasu, M.; Ojika, M.; Wakamatsu, K.; Yamada, K.; Sugimura, T. *Proc. Natl. Acad. Sci. U.S.A.* 1988, *85*, 1768.

28. Moore, R.E.; Bartolini, G. *J. Am. Chem. Soc.* 1981, *103*, 2491.

29. Uemura, D.; Ueda, K.; Hirata, Y.; Naoki, H.; Iwashita, T. *Tetrahedron Lett.* 1981, *22*, 2781.

30. Tachibana, K.; Scheuer, P.J.; Tsukitani, Y.; Kikuchi, H.; Van Engen D.; Clardy, J.; Gopichand, Y.; Schmitz, F.J. *J. Am. Chem. Soc.* 1981, *103*, 2469.

31. Fujiki, H.; Suganuma, M.; Suguri, H.; Yoshizawa, S.; Ojika, M.; Wakamatsu, K.; Yamada, K.; Sugimura, T. *Proc. Jpn. Acad.* 1987, *63*, 51.

32. Ohuchi, K.; Watanabe, M.; Yoshizawa, K.; Tsurufuji, S.; Fujiki, H.; Suganuma, M.; Sugimura, T.; Levine, L. *Biochim. Biophys. Acta* 1985, *834*, 42.

33. Levine, L.; Fujiki, H. *Carcinogenesis* 1985, *6*, 1631.

34. Murata, M.; Shimatani, M.; Sugitani, H.; Oshima, Y.; Yasumoto, T. *Bull. Jpn. Soc. Fish* **1982**, *48*, 549.
35. Fujiki, H.; Suganuma, M.; Suguri, M.; Yoshizawa, S.; Takagi, K.; Uda, N.; Wakamatsu, K.; Yamada, K.; Murata, M.; Yasumoto, T.; Sugimura, T. *Jpn. J. Cancer Res. (Gann)* **1988**, *79*, 1089.
36. Fujiki, H., Moore, R.E. *Jpn. J. Cancer Res. (Gann)* **1987**, *78*, 875.

RECEIVED June 12, 1989

Chapter 19

Pharmacological and Toxicological Studies of Palytoxin

James A. Vick and Joseph Wiles

U.S. Food and Drug Administration, 200 C Street, SW, Washington, DC 20204

Palytoxin is one of the most potent coronary vasoconstrictors known, producing death by the iv routewithin minutes by diminishing the supply of oxygen to the myocardium. Tests in animals with a number of vasodilators showed that papaverine and isosorbide dinitrate are effective antidotes, but must be injected directly into the ventricle of the heart because of the speed of action of the toxin. Sublethal doses of this toxin given intragastrically and intravenously produce elevated plasma cortisol concentrations and protect against subsequent lethal intravenous challenge. This protection is lost with time; however, a second sublethal injection of palytoxin, after circulating plasma cortisol have decreased and prior to lethal challenge, once again partially increases blood steroid concentrations and reestablishes protection. Pretreatment with hydrocortisone also provided partial protection against lethal doses of the toxin. The one hour required for protection to develop is probably the time required for desensitization of the endothelial wall to vasoconstrictors. Adrenalectomized animals do not develop any degree of protection unless treated with exogenous hydrocortisone, substantiating the involvement of steroids. The possibility of an additional, additive immune mechanism in the response of the animals to palytoxin is also suggested. These studies have provided a foundation for the understanding of how palytoxin works as well as for the development of a therapeutic regime for treating palytoxin poisoning in man.

Palytoxin is a potent toxin extracted from a soft coral sea anemone, *Palythoa vestitus*, found in certain South Pacific islands. Interesting legends concerning a "deadly seaweed" or a "pool of death," which the natives of Hawaii called Limu-make-o-Hana, led scientists to make inquiries about the material found in these waters. They found that the toxic effects were not caused by a seaweed, but by the coral associated with it. Further investigation indicated that the toxin was actually of marine origin and derived from a sea anemone. Initial isolation and testing (*1, 2*), indicated that this substance was indeed highly toxic and that the empirical formula was $C_{30}H_{53}NO_{14}$. The stereochemistry of palytoxin was studied by Cha et al. (*3*) and the actual configuration established by Uemura et al.(*4*).

Earlier studies by Wiles et al. (*5*) in which palytoxin was administered by various routes showed that this material was extremely toxic to rabbits, dogs, and monkeys. The effect of route of administration on toxicity varies in that intravenous (iv), intramuscular (im), and subcutaneous (sc) toxicity is high, yet intrarectal (ir) or oral (po) palytoxin is relatively ineffective. It was also observed that palytoxin caused marked irritation and tissue injury when applied topically to the skin or eyes of animals.

In additional studies, Wiles et al. (5) found that rats and rabbits given palytoxin by either the ir or po route appear to be protected against a subsequent iv or im injection. This tolerance, or immunity, appears to be related to the time between initial exposure and subsequent challenge.

This chapter is concerned therefore with studies of the mechanism of action of palytoxin, how it produces its lethal effects, as well as attempts to develop an antidote for this form of poisoning.

Methods

The palytoxin used was obtained from Moore and Scheuer, University of Hawaii. The crude toxin was purified and analyzed by Sephadex columns, reconstituted in 50% ethanol and water, stored in a refrigerator, and bioassayed for potency at least once a week during use. All dilutions of the palytoxin were made with glass distilled water.

The toxicity of palytoxin in six unanesthetized animal species was established as well as the toxicity of the toxin when administered by various routes. Tables I and II show the doses and numbers of animals used in this phase of the study.

Antidotal Therapy Studies. Thirty-seven dogs were anesthetized with sodium pentobarbital and were given 0.01 $\mu g/kg$ of palytoxin iv into the right femoral vein and treated at onset of cardiac symptoms with either papaverine (10–50 mg) or isosorbide dinitrate (0.5–5.0 mg). (In initial studies, 70 dogs given 0.005–0.015 $\mu g/kg$ of this particular batch of palytoxin iv all died.) Criteria for administration of therapy were a sharp drop in arterial pressure, an increase in the height of the QRS segment of the EKG, a decrease in heart rate, and the development of difficult breathing. The therapeutic injections were made directly into the left ventricle of the heart with a 4.5-in., 18-gauge needle due to the speed with which the palytoxin acts in producing its lethal effects and thus the inadequacy of iv or im therapy which cannot reach the primary site of palytoxin toxicity, the myocardial tissue.

Protection Studies. Animals were given a sublethal dose of palytoxin followed at various time intervals by a lethal dose. Control and treatment data for each route of administration and species studied are given in the appropriate table (Tables III, IV, V, and VI).

Rats (Table III) were given a dose of 5 $\mu g/kg$ of palytoxin either ir or po followed at 1–72 hr by 0.25 $\mu g/kg$ im. A second group of rats were given palytoxin po, 10 $\mu g/kg$, followed by an iv dose of 0.20 $\mu g/kg$.

One group of rabbits (Table IV) was given 2.5 $\mu g/kg$ po, followed at 1–72 hr by 0.025 $\mu g/kg$ iv. A dose of 0.5 $\mu g/kg$ was instilled into one eye intraocular (io) of each rabbit in another group, and this was followed at 2–48 hr by 0.032 $\mu g/kg$ palytoxin iv. The concentration of the aqueous solution of palytoxin administered to these animals was 5–10 $\mu g/mL$. The challenge dose was administered through the marginal ear vein of the rabbits, the femoral vein of the rats, or into the thigh muscle of either species as indicated in the appropriate tables.

Adult mongrel dogs (Table V) and rhesus monkeys (Table VI) were anesthetized with sodium pentobarbital (30 mg/kg) and given sublethal dose of palytoxin followed by challenge (lethal) doses. Arterial and venous blood pressure was monitored continuously using Statham strain gauges, recording on an E&M physiograph. Respiratory rate, electrocardiogram (EKG) and heart rate were continuously monitored using needle-tip electrodes placed in either side of the chest wall.

Table I. Intravenous Toxicity of Palytoxin in Several Animal Species

Species	No. of Animals	24-hr LD_{50} ($\mu g/kg$)
Rabbit	20	0.025 (0.024–0.026)[a]
Dog	20	0.033 (0.026–0.041)
Monkey	10	0.078 (0.60–0.090)
Rat	50	0.089 (0.080–0.098)
Guinea Pig	20	0.11 (0.070–0.170)
Mouse	100	0.45 (0.33–0.62)

[a]95% confidence limits

Table II. Toxicity of Palytoxin Administered by Various Routes in the Rat

Route	No. of Animals	24-hr LD_{50} ($\mu g/kg$)
Intravenous (iv)	20	0.089 (0.80–0.098)[a]
Intramuscular (im)	20	0.24 (0.21–0.28)
Intratracheal	20	0.36 (0.23–0.55)
Subcutaneous (sc)	20	0.40 (0.29–0.54)
Intraperitoneal	20	0.71 (0.45–1.12)
Intragastric (po)	20	>40.0
Intrarectal (ir)	20	>10.0

[a]95% confidence limits

Table III. Toxicity of Palytoxin in Rats When an Initial Sublethal Dose is Followed by a Lethal Dose

		Challenge Injection			
Route	Dose (µg/kg)	Time After Initial Injection (hr)	Route	Dose (µg/kg)	Mortality Fraction (within 48 hr after challenge)
im (control)	0.25	–	–	–	6/12
ir	5.0	No challenge (control)		–	0/8
	5.0	4	im	0.25	0/8
	5.0	24	im	0.25	0/8
	5.0	48	im	0.25	3/8
po	5.0	No challenge (control)		–	0/8
	5.0	1	im	0.25	7/8
	5.0	4	im	0.25	7/8
	5.0	24	im	0.25	5/8
	5.0	48	im	0.25	3/8
	5.0	72	im	0.25	8/8
iv (control)	0.20	–	–	–	6/6
po	10.0	No challenge (control)		–	0/6
	10.0	1	iv	0.20	5/5
	10.0	4	iv	0.20	5/5
	10.0	24	iv	0.20	1/5

Table IV. Toxicity of Palytoxin in Rabbits When an Initial Sublethal Dose is Followed by a Lethal Dose

		Challenge Injection			
Route	Dose (µg/kg)	Time After Initial Injection (hr)	Route	Dose (µg/kg)	Mortality Fraction (within 48 hr after challenge)
iv (control)	0.025	–	–	–	6/12
po	2.5	No challenge (control)		–	0/6
	2.5	1	iv	0.025	3/6
	2.5	4	iv	0.025	4/6
	2.5	24	iv	0.025	2/6
	2.5	48	iv	0.025	0/6
	2.5	72	iv	0.025	6/6
iv (control)	0.032	–	–	–	6/6
Intraocular	0.5	No challenge (control)		–	0/6
	0.5	2	iv	0.032	6/6
	0.5	4	iv	0.032	6/6
	0.5	24	iv	0.032	3/6
	0.5	48	iv	0.032	4/6

Table V. Toxicity of Palytoxin in Dogs (Normal) When an Initial Sublethal Dose is Followed by a Lethal Dose

Initial Injection		Challenge Injection			Mortality Fraction (within 48 hr after challenge)
Route	Dose (μg/kg)	Time After Initial Injection (hr)	Route	Dose (μg/kg)	
iv (control)	0.05	–	–	–	6/6
po	2.5	No challenge (control)		–	0/2
	2.5	4	iv	0.05	3/4
	2.5	24	iv	0.05	2/3
	2.5	48	iv	0.05	0/2
	2.5	72	iv	0.05	2/2
	2.5	7 days	iv	0.05	1/1
iv	0.1	No challenge (control)		–	0/2
	0.01	Immediately	iv	0.05	4/4
	0.01	2	iv	0.05	4/5
	0.01	4	iv	0.05	2/2
	0.01	24	iv	0.05	4/4
	0.01	48	iv	0.05	1/4
	0.01	72	iv	0.05	0/8
	0.01	7 days	iv	0.05	4/8

Table VI. Toxicity of Palytoxin in Monkeys (Normal) When an Initial Sublethal Dose is Followed by a Lethal Dose

Initial Injection		Challenge Injection			Mortality Fraction (within 48 hr after challenge)
Route	Dose (μg/kg)	Time After Initial Injection (hr)	Route	Dose (μg/kg)	
iv	0.01	No challenge (control)		–	1/2
	0.005	No challenge (control)		–	0/4
	0.01	72	iv	0.05	1/4
	0.005	72	iv	0.05	0/4

The dogs were given 2.5 μg/kg of palytoxin (po) or 0.001 μg/kg (iv) followed in 2 hr to 7 days by an LD_{100} dose of 0.05 μg/kg (iv). Certain dogs were given two doses of 0.001 μg/kg (iv) 7 days apart and challenged with 0.05 μg/kg (iv) 24 hr or 15 days after the second sublethal dose.

In a series of range-finding studies in monkeys, the lethal dose of palytoxin was established at 0.05 μg/kg (iv). One of two monkeys survived a dose of 0.01 μg/kg and four of four survived a dose of 0.005 μg/kg. Therefore, and as in the dog (challenge) study, the monkeys were given either 0.01 or 0.005 μg/kg, followed in 72 hr by a dose of 0.05 μg/kg (iv).

Blood samples for determination of hydrocortisone concentrations were taken from a group of 10 dogs prior to and at 2, 24, 48, and 72 hr and 15 days after receiving a sublethal dose of 0.001 μg/kg of palytoxin iv. Concentrations were measured using standard techniques as described by Sweat (6) and Peterson et al. (7).

Adrenalectomized mongrel dogs and rhesus monkeys were studied 2 weeks following surgery. (Steroids were given to the animals only in the immediate postoperative period.) They were anesthetized and prepared for recording of physiological data as previously described. The dogs were given an initial dose of 0.001 μg/kg of palytoxin iv followed in 72 hr by a dose of 0.05 μg/kg iv. The monkeys were given palytoxin 0.005 μg/kg iv followed in 72 hr by an iv dose of 0.05 μg/kg. An additional 6 dogs and 4 monkeys (adrenalectomized) were given 50 mg/kg of hydrocortisone sodium succinate iv into the femoral vein 1 hr prior to challenge with the lethal dose (0.05 μg/kg) of palytoxin. Six normal dogs and 4 normal monkeys were given one dose of 50 mg/kg hydrocortisone 1 hr prior to receiving 0.05 μg/kg palytoxin to further assess the role of steroids in palytoxin poisoning.

The concentration of the aqueous solution of palytoxin given to the dogs and monkeys was 5.6 μg/mL, and was injected directly into the exposed femoral vein of each animal.

The animals given intragastric palytoxin were fasted for 24 hr prior to testing. All animals were observed for 72 hr or until death.

Results

The iv toxicity of palytoxin is shown in Table I. Rabbits and dogs appear to be the sensitive to palytoxin; rats and guinea pigs appear less sensitive. Table II shows the comparative toxicity of palytoxin administered by several routes - again iv palytoxin is extremely toxic while ir and po are relatively without toxic effects.

Antidotal Therapy Studies. One of the most challenging considerations in treating poisoning by palytoxin was that the average time to death in the dogs was 3.5 min. Attempts at iv therapy were ineffective in that stagnation of venous blood flow occurred so rapidly that antidotes were pooled in the venous circulation. Direct injections of either papaverine or isosorbide dinitrate into the left ventricular of the heart were made in attempts to reverse the cardiotoxic effects of the toxin. Additional doses of either agent were required at 1–2-hr intervals over a 12-hr period to maintain normal cardiovascular function. In general, dogs responded more favorably to the administration of isosorbide dinitrate than to papaverine. Of the dogs treated in this fashion, 49% survived (18/37). The effects of a lethal dose of palytoxin on EKG, heart rate, respiration, and blood pressure of dogs are shown in Figure 1. Figure 2 shows the ability of a vasodilator to alter the otherwise lethal course of events of this compound. In those animals which expired, death appeared due to profound coronary vasoconstriction and subsequent cardiac failure.

Protection Studies. The responses of rats and rabbits given a sublethal dose of palytoxin followed by lethal challenges are shown in Tables III and IV.

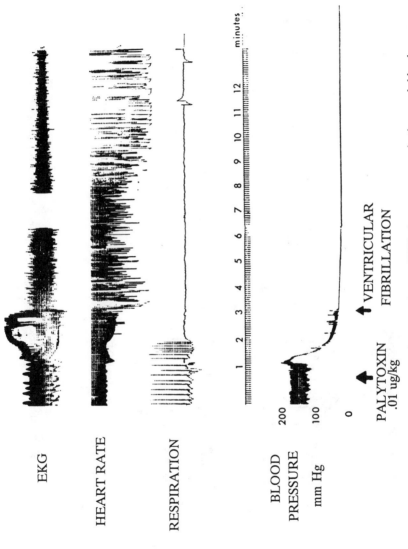

Figure 1. Effect of a lethal dose of palytoxin on EKG, heart rate, respiration, and blood pressure. (Reproduced from Ref. 10.)

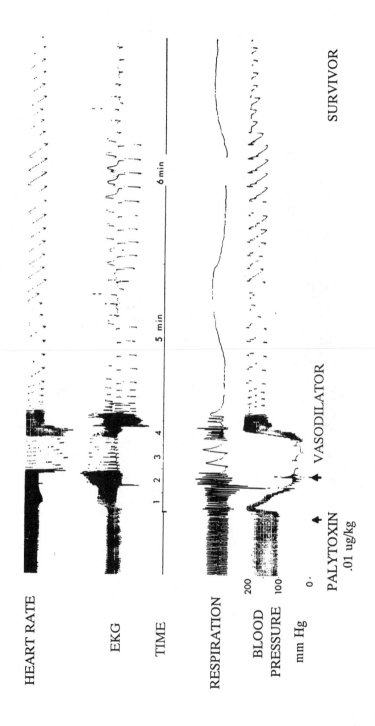

Figure 2. Effect of treatment with isosorbide dinitrate in reversing the otherwise lethal effect of palytoxin. (Reproduced from Ref. 10.)

Rats showed an increase tolerance to im palytoxin at 4 and at 24 hr when the initial dose had been given ir (0/8 vs. 6/12 mortalities). An initial oral po dose also provided some protection against lethal iv challenge, 1/5 vs 6/6 mortalities at 24 hr, however, after these times protection was lost.

Rabbits showed a similar trend to that seen in the rats when the initial dose was administered by either the po (0/6 vs. 6/12 at 48 hr) or the io (3/6 vs. 6/6 at 24 hr).

Anesthetized dogs and monkeys responded similarly to palytoxin (Tables V and VI). In the dog, maximum protection was seen against lethal challenge at 48 (1/4 vs. 6/6) and at 74 (0/8 vs. 6/6) hr after the initial sublethal iv doses. At 7 days after initial dosing only 4 of 8 dogs survived lethal challenge, however, a second sublethal injection administered 7 days after the initial sublethal dose but prior to the lethal challenge reestablished the ability of the animals to withstand a lethal dose of palytoxin (6/6 survived) at 24 hr and (4/5 survived) at 15 days (Figure 3). Monkeys given sublethal iv doses of palytoxin also showed some protection at 48 hr, 0/4 vs. 3/3 mortality.

The protection afforded intact dogs by a sublethal injection of palytoxin appeared to be in part associated with an increased blood steroid concentration in that the postinjection period during which there was the greatest protection was also the period when the highest blood steroid concentrations were recorded (Figure 3). Blood steroids increased from a control of 4.2 μg/100 mL to 18.7 μg/100 mL at 72 hr, decreasing to 7.0 μg/100 mL at 7 days. A second sublethal injection at 7 days increased steroid levels to 23.0 μg/100 mL and provided maximum protection against a lethal dose of palytoxin. Of note is the observation that steroid concentrations decreased to 9.8 μg/100 mL 15 days after the second sublethal dose of palytoxin yet 80% of the animals still survived lethal challenge.

Adrenalectomized dogs and monkeys given sublethal doses of palytoxin iv did not develop any protection against lethal challenge (Table VII), nor did the blood steroid levels increase at any time during the study. However, the administration of hydrocortisone 1 hr prior to challenge did protect both adrenalectomized dogs and monkeys against the otherwise lethal dose of palytoxin. In addition, 3 of 6 normal dogs and 2 of 4 normal monkeys given hydrocortisone 1 hr prior to lethal challenge with palytoxin survived (Figure 4).

Discussion

Palytoxin is probably one of the most potent toxins known to humans. Intravenous LD_{50} values in the six species that have been studied are consistently less than 0.5 μg/kg. In addition, palytoxin possesses a speed of action and other pharmacologic properties that are markedly different from those exhibited by other toxic materials. For example, when injected iv or sc, palytoxin is extremely toxic; yet when given po or ir, it is relatively non-toxic. It is also very interesting that the doses of palytoxin required to kill are somewhat different in anesthetized vs. unanesthetized animals.

The signs noted in intact, unanesthetized animals following iv administration of palytoxin vary somewhat in the different species. Rats, guinea pigs, rabbits, and mice become drowsy and inactive. Prostration, dyspnea, and convulsions often occur just prior to death. No other signs were observed, probably due to the speed with which palytoxin acts. Monkeys become drowsy, weak, and ataxic; then collapse and die. Vomiting occurs. The drowsiness seen in monkeys is not seen in dogs. Instead, dogs normally defecate, vomit, become weak, ataxic, collapse, and die within 5 to 10 min. If death is delayed, a shocklike state is seen. Body temperature falls and extensive hemorrhage into the gastrointestinal tract occurs. Bloody vomitus

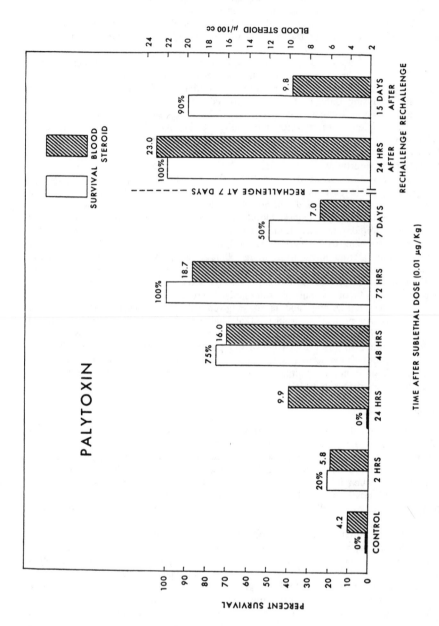

Figure 3. Effect of repeated injections of palytoxin on blood steroids and survival. (Reproduced from Ref. 10.)

Table VII. Toxicity of Palytoxin in Adrenalectomized Dogs and Monkeys When an Initial Sublethal Dose is Followed by a Lethal Dose

Species	Initial Injection		Challenge Injection			Mortality (within 48 hr after challenge)
	Route	Dose (µg/kg)	Time After Initial (hr)	Route	Dose (µg/kg)	
Dog	iv	0.001	No challenge (control)		–	4/10
	iv	0.001	72	iv	0.05	6/6
	iv	0.001	72	iv	0.05	0/6[a]
	–	No	–	iv	0.05	3/6[b]
Monkey	iv	0.005	No challenge (control)		–	0/2
	iv	0.005	72	iv	0.05	2/2
	iv	0.005	72	iv	0.05	0/4[a]
	–	No	–	iv	0.05	2/4[b]

[a] 50 mg/kg iv hydrocortisone 1 hr prior to challenge. Adrenalectomized animals.
[b] 60 mg/kg iv hydrocortisone 1 hr prior to challenge. Animals not adrenalectomized.

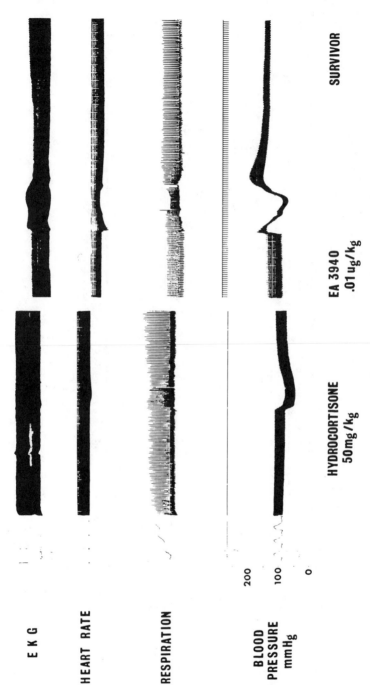

Figure 4. Effect of pretreatment with hydrocortisone (50 mg/kg) on response of animal to a lethal injection of palytoxin (0.01 μg/kg).

and diarrhea are often noted. The massive hemorrhage observed in the gut of the dog is not prominent in other species. Pooling of blood in the hepatosplanchnic bed has been previously reported in dogs given a variety of toxins and may be a species-specific effect (8).

Early death in dogs and monkeys given palytoxin appears to be due to profound coronary vasoconstriction and rapid cardiac failure. Careful analysis of EKG recordings indicates that this toxin produces intense myocardial ischemia which results in either ventricular fibrillation or cardiac arrest. Palytoxin appears to act directly on the smooth muscle of the vessel walls to produce this marked coronary vasoconstriction. This constriction is undoubtedly widespread throughout the body, yet it is primarily the reduction in flow to the heart that produces the irreversible damage and subsequent death. It is also important to note that palytoxin produces vasoconstriction in a dose considerably less than the doses reported for other potent vasoconstrictors. Goodman and Gilman (9) report that angiotensin II produces noticeable changes in vessel tension at $1-10$ μg/kg; this is at least 100 times the dose of palytoxin that not only produced changes in vessel tension but resulted in death of the animal.

The direct injection of potent vasodilatory agents such as papaverine or isosorbide dinitrate, into the ventricles of the heart reverses the action of palytoxin in approximately one-half of the animals. These extreme measures are required because palytoxin kills quickly. Antidotes injected into the venous circulation were not able to reach the heart because the stagnation of venous blood occurs so rapidly that antidotes are simply pooled on the venous side of the circulation and never reach the heart. In these studies isosorbide dinitrate appeared to be approximately twice as effective as papaverine in reversing the toxic effects of palytoxin.

Pretreatment with hydrocortisone protected 50% of the animals against a lethal challenge with palytoxin. This is undoubtedly related to the reported action of this steroid in protecting the vascular tree. The observation that a minimum of 1 hr is required between injection of hydrocortisone and the development of protection indicates that this may be the time required for desensitization of the endothelial wall to vasoconstrictors or, to some other unknown action of steroids in the body. Additional data shows that increased protection against lethal challenge is associated with an increase in steroid concentration which is the result of prior sublethal palytoxin administration. In addition, protection is partially lost as the concentration of the steroid decreases, yet a second sublethal dose of palytoxin given when protection has decreased will once again increase blood steroid concentration and reestablish protection against a lethal dose of toxin. It follows, therefore, that adrenalectomized animals which cannot develop increased plasma steroid levels are not protected against palytoxin. The possibility exists that the immune mechanism may also be involved in this phenomenon in that repeated sublethal doses of palytoxin do not maintain the same high concentrations of blood steroids, yet protection against lethal challenge remains almost maximal.

The cause of death from palytoxin can be attributed to several factors. Early death is probably due to intense spasm of coronary vascular smooth muscle, resulting in marked reduction in cardiac output. This may be an early manifestation of what later becomes a general necrotizing effect of palytoxin on blood vessels, which ultimately causes cell destruction throughout the body. The delayed deaths may be due to a less profound yet widespread decrease in blood flow and oxygen supply, causing ischemia and, ultimately, anoxia to major organ systems. This would account for the gradual reduction in cardiac function and respiration seen in the later animal deaths. Kidney damage due to accumulation of metabolites could also be a factor which contributes to death from palytoxin.

These studies have provided a foundation for the understanding of the possible mechanism by which palytoxin kills and for the development of a therapeutic regimen which might have application to humans.

Literature Cited

1. Scheuer, P.J. *Program Chem. Org. Nat. Prod.* **1964**, *22*, 265–283.
2. Mosher, H.S.; Fuhrman, F.A.; Buckwald, H.D.; Fisher, H.G. *Science* **1964**, *144*, 1100–1110.
3. Cha, J.K.; Christ, W.J.; Finan, J.; Fujioka, H.; Kishi, Y.; Klein, L.; Ko, S.; Leder, J.; McWhorter, W.; Pfaff, K.; Yonaga, M. *J. Am Chem. Soc.* **1982**, *104*, 7369–7371.
4. Uemura, D.; Hirata, Y.; Iwahita, T.; Naoki, H. *Tetrahydron* **1985**, *41*, 1007–1017.
5. Wiles, J.S.; Vick, J.A.; Christenson, M.K. *Toxicon* **1974**, *12*, 427–433.
6. Sweat, M.L. *Anal. Chem.* **1954**, *26*, 773.
7. Peterson, R.E.; Karrer, A.; Guerra, S.L. *Anal. Chem* **1957**, *29*, 144–149.
8. Vick, J.A. *Mil. Med.* **1973**, *138*, 279–285.
9. Douglass, W. W. In *The Pharmacological Basis of Therapeutics*, 3rd ed., Goodman, L.; Gilman, A., Eds.; The MacMillian Company: New York, 1965; p. 653.
10. Vick, J.; Wiles, J. *Toxicol. Appl. Pharmacol.* **1975**, *34*, 214.

RECEIVED June 30, 1989

PEPTIDE   TOXINS

Chapter 20

Conotoxins: Targeted Peptide Ligands from Snail Venoms

Baldomero M. Olivera[1], David R. Hillyard[2,3], Jean Rivier[4], Scott Woodward[3], William R. Gray[1], Gloria Corpuz[5], and Lourdes J. Cruz[1,5]

[1]Department of Biology, University of Utah, Salt Lake City, UT 84112
[2]Department of Pathology, University of Utah, Salt Lake City, UT 84112
[3]Howard Hughes Medical Institute, University of Utah, Salt Lake City, UT 84112
[4]The Salk Institute, La Jolla, CA 92037
[5]Marine Science Institute, University of the Philippines, Quezon City, 1101, Philippines

Several hundred species of *Conus* snails produce a wide array of small (12-30 AA) and mostly tightly disulfide bonded peptide ligands with high affinity for a diverse set of receptor and ion channel types. These include nicotinic acetylcholine receptors, neuronal calcium channels, muscle sodium channels, vasopressin receptors, and *N*-methyl-D-aspartate (NMDA) receptors. Some general features of the structure, function, and evolution of biologically active peptides isolated from *Conus* venom are presented.

The venoms of the predatory marine gastropods in the genus *Conus* (the "cone snails") contain peptide toxins which bind to, and inhibit, the function of key macromolecules in the neuromuscular system (*1-4*). Two novel features of the peptides derived from *Conus* make them of particular interest: (1) The *Conus* peptides are almost always under 30 amino acids in length (and therefore can be chemically synthesized). Nevertheless, they display high affinities for their receptor targets. (2) In *Conus* venoms, a great variety of such small biologically active peptides are found. We expect that they will target a wide spectrum of different receptor and ion channel types.

In this chapter, our aim is to give an overview of the work on conotoxins that has been done to date and to present a biological rationale for why *Conus* venoms have the features above, and are therefore likely to yield an increasingly large array of well-characterized peptides that will be used for probing the nervous system in years to come. Enough work has been done to this point to understand in broad outline the strategy cone snails have adopted to be able to paralyze their diverse prey. We discuss the venoms and toxins of *Conus* at several levels: their general biological context, the underlying toxinological and biochemical strategy, and, finally, initial evidence elucidating probable genetic mechanisms which have allowed the cone snails to generate such a wide variety of diverse toxins.

It is not our intention to present a comprehensive review of the work on *Conus* toxins here; the reader is referred to recent articles already in the literature (*1-5*). We have focused largely on fish-hunting *Conus* toxins, particularly the ω-conotoxins, their targets, and their relationship to some toxins from *Conus textile*, a molluscivorous *Conus*. These examples illustrate general features of *Conus* toxins.

Conus Venoms: General Considerations

The cone snails are predatory, venomous molluscs which use a common general strategy to capture prey (*1, 5-7*). All 300–500 species of *Conus* have a specialized venom apparatus, diagrammed in Figure 1 (*8*). A venom paralytic to the prey is produced in a venom duct and injected through a disposable, harpoon-like tooth (Figure 2). Paralysis of the prey can be remarkably rapid; in the case of certain piscivorous cone species, the fish prey is immobilized in less than one second.

The venom is typically a milky fluid if extracted directly from the venom duct. However, injected venom is usually clear (some venoms have a detergent-like substance and appear foamy). Our recent evidence indicates that extensive processing may take place between extrusion from the duct and injection into prey (C. Hopkins and G. Zafaralla, unpublished results). If the venom collected from the duct is examined under a microscope, a large fraction of the protein is found in the form of granules (*see* Figure 3a). These granules are insoluble and can be sedimented into a pellet; when purified from the soluble peptides in this way, the granules are biologically inactive upon injection into test animals. However, the granules can be activated by treatment with certain proteases (*9, 10*).

Although the activation is inefficient (probably because inappropriate proteases are used), the recovery of any toxicity at all suggests that the granules contain the precursors to at least some of the toxins. Although the process of activation is poorly understood, molecular genetic data described below provide direct evidence that the toxins are synthesized as propeptides, and are then processed to give the final peptide. Cytological data suggest that once venom has been transferred from the venom duct into the proboscis, final processing of peptides may take place. Staining of proboscis sections reveals putative secretory granules which could conceivably be packets of enzymes necessary for the rapid processing of the venom (Figure 3b,c). However, no biochemical data to support this hypothesis have been reported to date.

Conus Species: Feeding Types, Taxonomic Considerations

Compared to other venomous animal groups, the cones as a genus have succeeded in using a rather diverse spectrum of prey comprising at least three different phyla (*11*). *Conus* species prey on various types of marine worms (Figure 4a), other marine snails (Figure 4b), and, most remarkably, a number of cones are the only snails known which can overpower and devour vertebrate prey (Figure 5). However, individually, most *Conus* species are fairly specialized; each cone will usually envenomate a narrow spectrum of prey. It has been reported that some worm-hunting cones feed exclusively on a single polychaete prey species.

The little comparative information available suggests that in terms of their venom composition, *Conus* species do not simply fall into three convenient classes, the worm-hunting, snail-hunting, and fish-hunting *Conus*. Among the snail-hunting cones, for example, rather different sets of toxins have been isolated from the two species that have been examined (i.e., *Conus textile* and *Conus marmoreus*). Nearly all taxonomists agree that this large genus should be split into smaller groups, but no scheme has been generally accepted for dividing the several hundred *Conus* species into compact generic groups. On morphological grounds, it is clear that some groups of species are closely related to each other. Among the snail-hunting cones, for example, the well known *textile* cones, mostly assigned to the subgenus *Cylinder* (Montfort, 1810), are quite distinct from the group of snail-hunting species

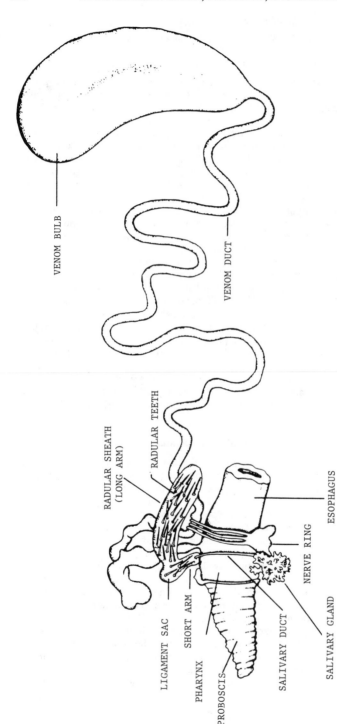

Figure 1. Diagram of the venom duct of *Conus*. The venom is produced in the venom duct, apparently expelled from the duct into the proboscis by contraction of the venom bulb. Simultaneously, a harpoon-like tooth is transferred from the radula sac to the proboscis. When injection takes place, the venom is pushed through the hollow tooth and flows into the prey through a hole at the tip of the tooth. Typically, fish-hunting cones will strike at a fish only once and grasp the tooth after injection has occurred, effectively harpooning their prey while injecting the paralytic venom. In contrast, snail-hunting cones will usually sting their prey several times before total paralysis occurs. (Reprinted with permission from the Second Revised Edition of Ref. 8. Copyright 1988 Darwin Press, Inc.)

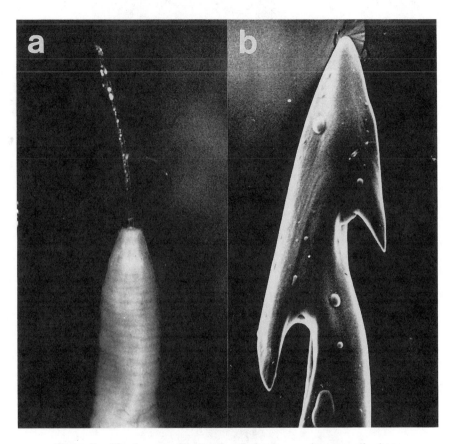

Figure 2.   The harpoon-like tooth of *Conus*.   a. An unusual photograph of a radular tooth at the tip of the proboscis of *Conus purpurascens*. Normally, the tooth would not be ejected from the proboscis until the prey had been harpooned.   Photograph by Alex Kerstitch.   b. A scanning electron micrograph of the tip of the radular tooth of *Conus purpurascens*, showing its harpoon-like form.

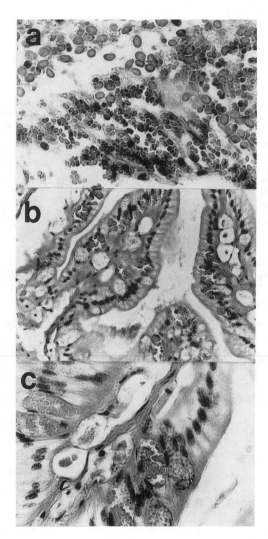

Figure 3. (a) A cross section of the venom duct of *Conus californicus*. This section shows two sizes classes of granules present in the venom duct of this species. Magnification, 640×. (b) A section of the proboscis, showing the villi in the interior wall of the proboscis. The nuclei are stained blackish-blue. Compartments containing what appear to be secretory granules which are stained red; these appear as grape-like clusters. Empty compartments which may have recently expelled their secretory granules are also seen. Magnification, 250×. (c) A closeup of the presumptive secretory granules in a *Conus californicus* proboscis. Magnification, 640×. All of the slides sections shown were stained using a standard hemotoxylin-eosin stain.

Figure 4. Worm-hunting and snail-hunting *Conus.* (a) The vermivorous species *Conus brunneus* about to sting its polychaete worm prey. (b) The molluscivorous species *Conus dalli* stinging the snail prey, *Columbella.* Both *Conus* species were collected in the Gulf of California. Photographs by Alex Kerstitch.

Figure 5. *(See p. 262.)* Fish-hunting *Conus.* This sequence shows *Conus purpurascens* from the Gulf of California feeding on the goby *Elacatinus.* (a) The snail is stinging the fish, causing paralysis. (b,c) The paralyzed fish is swallowed by the distensible stomach (rostrum) of the snail. After 1-2 hours, only the scales and bones of the fish, as well as the harpoon-like tooth used to inject venom will be regurgitated by the snail. Photographs by Alex Kerstitch.

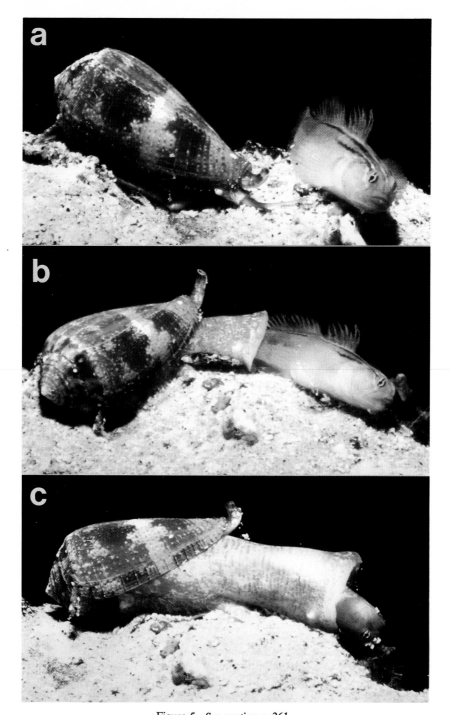

Figure 5. *See caption p. 261.*

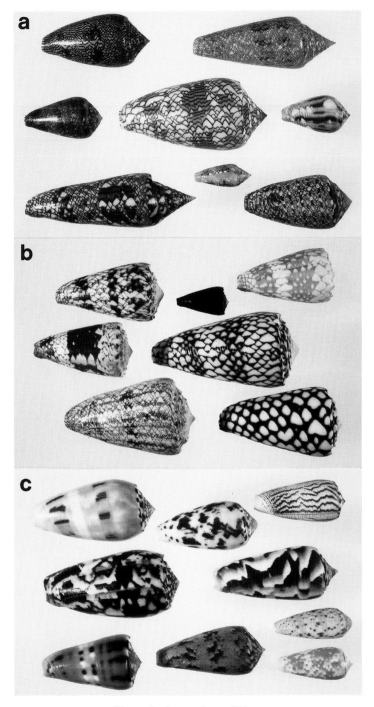

Figure 6. *See caption p. 264.*

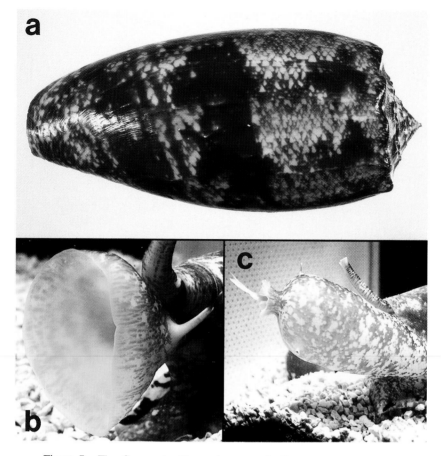

Figure 7. The Geography Cone, the most deadly *Conus* species. **(a)** The shell of *Conus geographus*. **(b)** *Conus geographus* hunting fish. This species uses a "net" strategy; it is nocturnal and it stalks sleeping fish, capturing them by luring them by using its distended rostrum as a net, which is closed quickly when a fish wanders in. **(c)** The fish is generally stung only after it has been engulfed. Photographs by K. S. Matz.

Figure 6. **(a)** *(See p. 263.) The Conus textile* group of species. Shown in this figure are a series of snail-hunting *Conus* species that are clearly related to *Conus textile*. They are all characterized by tent markings, and a series of lines or dark patches that run along the entire length of the shell, although the relative size of the tent markings and the shape of the shells vary significantly. Top row, from left: *Conus eutrios*, Zanzibar; *Conus gloriamaris*, Panglao Island, Philippines, 100 m depth. Center row: *Conus dalli*, Gulf of California; *Conus textile* (75 mm), Cebu Island, Philippines; *Conus rectifer*, Sulu Sea. Bottom row: *Conus bengalensis*, Bay of Bengal, trawled in deep water; *Conus legatus*, Kwajelein; *Conus natalis*, South Africa. *Continued on p. 265.*

Figure 6. (b) *Continued from p. 264.* Snail-hunting cones related to the Marble cone, *Conus marmoreus*. These species are snail-hunting, but are not closely related to the *Conus textile* group shown in Figure 6a. All of them have coronate shoulders and clear tent markings. Top row: *Conus nicobaricus*, Zamboanga, Philippines; *Conus nigrescens*, Samoa; *Conus vidua*, Cuyo Island, Philippines. Center row: *Conus nocturnus*, Philippines; *Conus bandanus* (80 mm), Hawaii. Bottom row: *Conus araneosus*, Indian Ocean; *Conus marmoreus*, Sulu Sea. (c) Fish-hunting cone snails related to *Conus striatus*. Shown are a number of piscivorous species which typically strike their prey by extending their proboscis and harpooning the prey using a hook and line strategy for catching fish. Top row: *Conus chusaki*, Andaman Sea; *Conus floccatus*, Marshall Islands; *Conus magus*, Cebu Island, Philippines. Center row: *Conus striatus* (74.5 mm), Cebu Island, Philippines; *Conus gubernator*, Indian Ocean. Bottom row: *Conus barthelemyi*, Mauritius Island, Indian Ocean; *Conus kinoshitai*, Bohol Island, Philippines; *Conus circumcisus*, Mactan Island, Philippines, 100 m depth; *Conus timorensis*, Indian Ocean. Photographs by K. S. Matz.

related to *Conus marmoreus* (these are shown in Figures 6a and 6b). Similar groupings of related species can also be made for fish-hunting and worm-hunting *Conus*. Thus, Figure 6c shows a group of fish-hunting cones, the *Conus magus-striatus* group. Other piscivorous species such as *Conus geographus* are less closely related to the species shown in Figure 6c.

To date, a preliminary comparison of toxins and venoms have been made within a single venom, between venoms of closely related species (such as *Conus magus* and *Conus striatus*, both in Figure 6c), and between species that hunt the same general prey but are not closely related morphologically (i.e., molluscivorous species in Figures 6a and 6b). Finally, comparisons between venoms of species not closely related on morphological grounds, and whose venoms are not targeted for prey in the same phyla (such as between species in Figures 6a and 6c) can also be made. Although only a preliminary data base has been obtained so far, a certain number of generalizations can already be made about the comparative pharmacology, toxicology, and biochemistry of the toxins in these venoms.

Toxicology of *Conus* Venom

Although all *Conus* species are venomous, the major toxins in any given venom are probably designed primarily toward blocking the neuromuscular system of the prey, which may be a fish, mollusc, or marine worm. Most dangerous to humans are the fish-hunting species, particularly *Conus geographus*, which has reportedly caused over 20 human fatalities (2). Stinging cases in humans have resulted from careless handling by divers and collectors. Normally snails merely retract into their shells when disturbed; however, cones will sting defensively if their shells are broken or scraped, or if they are trapped, i.e., when placed in pockets by divers. Symptoms observed in humans (12) after a *Conus geographus* sting include numbness at the site of stinging which spreads to upper parts of the limb and to the rest of the body. Blurring of vision, impaired speech, and muscle paralysis precede death. Generalized numbness and muscular weakness persist for a day or two in survivors (13). Toxicological experiments with rats and guinea pigs (14) showed marked respiratory depression and respiratory failure accompanied by blood pressure fluctuations upon iv injection of venom from *C. geographus*. Artificial respiration saves experimental animals; death is believed to be due to asphyxiation by paralysis of diaphragm muscles (15). The LD_{50} of freeze-dried *Conus geographus* venom in mice was 0.82 mg/Kg on iv injection and 1.3 mg/Kg upon ip injection (14).

Conus geographus (shown in Figure 7) is clearly the most dangerous species to humans. Other large piscivorous *Conus* species, such as *Conus striatus*, appear to be less prone to sting, and in addition, some of the paralytic toxins can be shown to have narrower phylogenetic specificity. Thus, although most of the toxins of *Conus geographus* are effective in paralyzing all vertebrates, two of the major toxins of *C. striatus* paralyze fish, but not mammals (*see* below). Other than *Conus geographus*, two molluscivorous species have been reported to kill humans, although the accuracy of these reports has been questioned based on tests of venoms collected from dissected snails. However, it should be emphasized that since *Conus* venom injected by the snail may be more highly processed than venom collected from dissected ducts, until more definitive information is available, all *Conus* species should be handled with caution.

Major Venom Components Cause Paralysis of Prey

When a cone snail envenomates its prey, the latter is invariably paralyzed. In all cases, the paralytic toxins in the venom ("conotoxins") appear to be small peptides, most commonly with 3 disulfide bonds (although conotoxins with 2 or 4 S-S bonds are present). The mechanism by which paralysis is effected is best understood for the fish-hunting cone snails, which will be discussed below.

In the venom of *C. geographus* and other fish-hunting species, the conotoxins isolated so far can be divided into three major classes (*1-4*): ω-**conotoxins** which block neuronal calcium channels at the presynaptic terminus of the neuromuscular junction, α-**conotoxins** which inhibit the acetylcholine receptor at the postsynaptic terminus, and μ-**conotoxins** which block Na channels on the muscle membrane.

Between 6 and 10 homologous peptides have been extensively characterized for each toxin class. Although ω- and α-conotoxins have been isolated from several fish-hunting *Conus* species, μ-conotoxins have so far been isolated only from *C. geographus* venom.

The biological targets of paralytic conotoxins are presumably ion channels and receptors in the fish neuromuscular system. Amphibian, avian, and mammalian systems may also be affected by ω-, α-, and μ-conotoxins. However, specific conotoxins vary in their phylogenetic spectrum and detailed physiological effects in vivo.

ω-Conotoxins

The ω-conotoxins are high affinity ligands for voltage-sensitive Ca channels in neuronal tissue. The first ω-conotoxin, GVIA, was isolated from *C. geographus* as a component which caused shaking in mice when injected intracranially but produced no obvious effect when injected intraperitoneally (*16*). In fish, frogs, and chicks, ω-conotoxin GVIA causes paralysis and death. An operational assay for detecting ω-conotoxins is paralysis and death upon ip injection of fish, and the characteristic shaking syndrome upon ic injection of mice. The possibility remains that voltage-sensitive Ca channel blockers with different specificity are present which do not cause the "shaker syndrome" elicited by the main series of ω-conotoxins.

Sequence data for natural ω-conotoxins has provided considerable insight into the range of allowable structural variation. Each species of *Conus* has proven to have its own distinctive set of ω-conotoxins; no two species share the same sequence. Sequence comparison of amino acids in GVIA with ω-conotoxins isolated from several *Conus* species (*see* Table I; an asterisk indicates that the α-carboxyl is known to be aminated) shows absolute conservation of only a glycine residue in the first loop in addition to the six cysteine residues involved in disulfide bonding. ω-Conotoxin GVIA, 27 amino acids long, is 46% homologous to ω-conotoxin MVIIA from *C. magus*; the homology of ω-conotoxin GVIA and MVIIA to ω-conotoxins from other species generally ranges from 40 to 60%. We have argued elsewhere (*1*) that the most divergent sequences are not necessarily found in the venoms from the most unrelated *Conus* species (as judged by morphological characteristics, etc.). Curiously, within a single venom, that of *Conus geographus*, two ω-conotoxins have been found which are surprisingly unrelated to each other (i.e., GVIA and GVIIA in Table I). The presence of two divergent ω-conotoxins in the same venom may signify a specialization for different targets (whether this means voltage-sensitive Ca channels in different kinds of fish, or two different subtypes within the same fish prey is not established at this point).

ω-Conotoxin Receptors

In general, ω-conotoxin receptors are neuronal voltage-sensitive Ca channels (*17*); voltage-sensitive Ca channels that are non-neuronal are neither high affinity binding sites nor physiological targets for ω-conotoxins. Although ω-conotoxins have been widely used to study presynaptic voltage-sensitive Ca channels, not all synaptic Ca channels are targets for ω-conotoxins, and some ω-conotoxin receptor targets are not synaptic.

Table I. Calcium Channel Inhibitors: ω–Conotoxins

Conotoxin	Sequence
GVIA	C K S P G S S C S P T S Y N C C R S C N P Y T K R C Y *
GVIB	C K S P G S S C C S Y N C C R S C N P Y T K R C Y G
GVIC	C K S P G S S C C S P T S Y N C C R S C N P Y T K R C
GVIIA	C K S P G T P C C S R G M R D C C T S C L L Y S N K C R R Y
GVIIB	C K S P G T P C C S R G M R D C C T S C L S Y S N K C R R Y
MVIIA	C K G K G A K C S R L M Y D C C T G S C R S G K C *
MVIIB	C K G K G A S C H R T S Y D C C T G S C N R G K C
Other amino acids found in new ω-conotoxins	. R L S - Q - - R V S - - I - - S - Y - K - - N - - - - G
	L P G K R Y G T
Consensus	C - - - - G - - C - - - - - - C C - - - - C - - - - - - C
Disulfide Bridges	C - - - - - C - - - - C C - - - C - - - - - - C

In vertebrates, the spectrum of ω-conotoxin targets is dependent not only on the species of animal being studied, but on the ω-conotoxin being used. In chicks and frogs all synapses tested are almost completely inhibited by ω-conotoxin GVIA, while in rodents, ω-conotoxin GVIA does not inhibit the neuromuscular synapse, and only a fraction of CNS synapses tested are blocked. In contrast, ω-conotoxin MVIIA acts much more reversibly and with a much reduced affinity for many amphibian, and some mammalian, synapses.

A second generalization that has emerged from both electrophysiological and biochemical data is that in most neuronal tissue, there are several ω-conotoxin receptor subtypes. Electrophysiological data have shown that within a single cell, as many as three different Ca-channel types can occur. In embryonic chick dorsal root ganglion cells, at least two of these were high affinity targets (*18*). Biochemical data (using cross-linking and toxin displacement assays) also reveal multiple targets; how the biochemically defined subtypes correspond to the different electrophysiological types remains to be established.

Cross-linking data reveal multiple specifically cross-linked receptor species (*19*). There appear to be two general classes of cross-linked targets. One is easily cross-linked using a bivalent cross-linker [such as disuccinimidylsuberate (DSS)]; these typically have a denatured molecular weight of 135–150 K. The second class is not easily cross-linked using DSS but requires photoactivatable cross-linking derivatives of ω-conotoxin GVIA – the high molecular weight targets in this class exhibit apparent molecular weights of $\sim$ 220 K and $\sim$ 300 K. In mammalian brain, the latter are the predominant receptors; in chick brain all target classes can be cross-linked.

Our data so far suggest that in the mammalian systems examined (primarily mouse and rat), the ω-conotoxin targets being studied are *not* the primary presynaptic voltage-sensitive Ca channel mediating excitatory input at synaptic junctions. ω-Conotoxin does not cause paralysis and death when injected into mice. Rather, upon ic injection it causes an uncontrollable shaking syndrome, implying that blocking some inhibitory CNS circuit may be the physiological target eliciting the in vivo symptoms.

In the periphery, there are no obvious biological effects from ω-conotoxin when injected into an adult rodent. However, when injected into a neonatal rodent, a slow depression in respiration takes place suggesting that inhibition of some ω-conotoxin binding site in the periphery may result in a dysfunction in respiratory control (J. M. McIntosh, unpublished results).

Thus, when a family of conotoxins such as the ω-conotoxins are used to study a set of homologous receptor targets which are homologous to the actual physiological targets (vertebrate voltage-sensitive Ca channels in this case), a complex matrix of effects is observed. It would be simpler to summarize the results if sweeping generalizations could be made; instead, the strength of toxin/receptor interaction apparently is a rather unpredictable function of both the ω-conotoxin homolog used and the particular receptor target examined. Nevertheless, this complexity and unpredictability is potentially enormously useful, particularly for defining Ca channel subtypes in complex tissues such as the mammalian CNS.

μ-Conotoxins

The presence of toxins in *C. geographus* venom which block the response of vertebrate skeletal muscle to direct electrical stimulation was first detected by Endean et al. (*14*). A toxic component which reversibly blocked the generation of action potentials in mouse diaphragm was subsequently purified and partially characterized by Spence et al. (20). Since then, seven homologous peptides now referred to as μ-conotoxins have been isolated from *C. geographus* and sequenced (21, 22). The toxins cause paralysis in fish and mice on ip injection. So far toxins of this type (*see* Table II) have not yet been found in other fish-hunting *Conus* species.

Tests of μ-conotoxin GIIIA on frog muscle preparation (cutaneous pectoralis) demonstrated block of Na channels activated by veratridine (21). Studies on single Na channels from rat muscle inserted into planar lipid bilayers and activated by batrachotoxin (21) showed that GIIIA as a blocking agent behaves similarly to the guanidinium toxins, tetrodotoxin and saxitoxin: a) block occurs only when the toxin is introduced on the extracellular side of the channel, b) the toxin induces the appearance of long-lived discrete blocked states in current records, and c) blocking activity is voltage dependent. Indeed μ-conotoxins (GIIIA and GIIIB) have been shown to compete with ^{3}H-Lys-tetrodotoxin and ^{3}H-saxitoxin for Site 1 but not with ^{3}H-batrachotoxin A 20-α-benzoate for Site 2 or [125]I-*Leiurus* scorpion toxin for Site 3 of the Na channel (23, 24).

A distinctive characteristic of μ-conotoxins is their much greater tissue specificity compared to the guanidinium toxins. Selectivity of certain toxins in crude *C. geographus* venom for muscle versus nerve was first suggested by Endean's work (14) and then by Spence's work with purified toxin (20). This was verified with GIIIA on frog motor nerve and mouse phrenic nerve diaphragm preparations (21). In binding competition studies of GIIIB with ^{3}H-saxitoxin, $K_{0.5}$ values of 60 nM and 35 nM were obtained with rat skeletal homogenates and T-tubular membranes whereas only 20% block was obtained with 10 μM toxin in membranes of rat cervical ganglion (24). Comparison of μ-conotoxin effects on single Na channel currents from rat muscle and rat brain synaptosomes in planar lipid bilayers, revealed a thousand-fold greater affinity for muscle versus neuronal Na channels suggesting that the difference is due to the intrinsic properties of the channels.

All μ-conotoxins so far isolated have 22 amino acid residues and 3 disulfide bonds. As shown in Table II, the three predominant forms have three hydroxyproline residues. Four minor variants of GIIIA and GIIIB which lack hydroxylation of either Pro[6] and Pro[7] have also been found in the venom (21). The disulfide bonding pattern is not yet known. Nevertheless, GIIIA has recently been synthesized and the 3 disulfide bridges formed by air oxidation (25). The active form was purified by HPLC and shown to be identical to native conotoxin GIIIA from *C. geographus*. In order to use μ-conotoxins as a sensitive probe for Na channels, radioactive derivatives have been prepared. With ^{3}H-monopropionyl-GIIIA prepared from native toxin and eel electroplax membranes, Yanagawa et al. (26) verified competition of μ-conotoxin with saxitoxin and tetrodotoxin but not with other neurotoxins and local anesthetics which are known to interact with other sites of Na channels.

Cruz et al. (25) prepared a ^{125}I-derivative of μ-conotoxin GIIIA. To accomplish this, the Bolton-Hunter reagent was used to derivatize the toxin with a (4-hydroxyphenyl)propionyl moiety (4 HP). Several 4 HP derivatives obtained were biologically active and one of them was radioiodinated. Specific binding of the ^{125}I-3-(4-hydroxyphenyl)-propionyl-GIIIA to eel electroplax membranes was abolished by competition with tetrodotoxin. More importantly, the radioiodinated derivatives can be cross-linked to the channels giving a specifically labeled band with the appropriate denatured molecular weight (> 200 K). The use of different radioiodinated derivatives of μ-conotoxins has a great potential in defining Site 1 of Na channels.

α-Conotoxins

Peptides in the α-conotoxin family are inhibitors of nicotinic acetylcholine receptors. They were first isolated from *C. geographus* venom as components which cause paralysis in mice and fish when injected intraperitoneally (27). Early physiological experiments (28) indicated that α-conotoxins GI, GII, and GIA (*see* Table III) all act at the muscle end plate region. Mini end-plate potentials and end plate potentials evoked in response to nerve stimulation are inhibited in the presence of α-conotoxins in the nM to μM range. α-Conotoxin GI was subsequently shown to compete with *d*-tubocurarine and α-bungarotoxin for the acetylcholine receptor (29).

Like the other paralytic toxins from *Conus* venom, α-conotoxins are small and very tightly folded, structural features which may be advantageous for rapid paralysis of prey (1). α-Conotoxins are typically 13 to 15 amino acids long with two disulfide bridges (*see* Table III). In addition to the five α-conotoxins shown, two new α-conotoxins (SIA and SIB) from *C. striatus* have recently been isolated, sequenced, and chemically synthesized. SIA is very unusual because it is 19 amino acid residues long and it contains 6 cysteine residues, three of which are contiguous near the amino terminus (C. Ramilo et al., unpublished results).

The degree of homology among α-conotoxins is high. Seven amino acids including the four cysteine residues involved in disulfide bonding are absolutely conserved and substitutions at four other residues are conservative. The greatest variability is at position 2 which can be occupied by Glu, Ile, Tyr, and Cys (Table III). Although the toxins have comparable paralytic and lethal activities in fish, the most potent in mice is MI from *C. magus* followed by α-conotoxin GI from *C. geographus*. The α-conotoxins purified from *C. striatus* venom are much less effective in mice; the activity of SI is < 1/50 that of MI. Binding competition of GI, MI, and SI with ^{125}I-α-bungarotoxin showed similar activities for sites on electroplax membranes from *Torpedo* (30). The inability of SI to paralyze mice may be due to the replacement of Arg^9-His^{10} in GI or Lys^{10}-Asn^{11} in MI by Pro^9-Lys^{10} in SI. Such correlation of sequence differences with phylogenetic specificities illustrates the potential use of α-conotoxins as biochemical probes for structural comparison of ACh receptors in different organisms.

Radioiodinated derivatives have been prepared to define more closely the target site of α-conotoxins on the acetylcholine receptor (R. Myers, unpublished data). In membrane preparations from *Torpedo* electroplax, photoactivatable azidosalicylate derivatives of α-conotoxin GIA preferentially label the β and γ subunits of the acetylcholine receptor. However, when the photoactivatable derivative is cross-linked to detergent solubilized acetylcholine receptor (AChR), only the γ subunit is labeled. Since snake α-neurotoxins mainly bind to the α subunits of AChR and α-conotoxins compete directly with α-bungarotoxin, the cross-linking results above are both intriguing and problematic.

Other Peptides in Conus Venoms

The ω-, μ-, and α-conotoxins are the best characterized of the peptides isolated from *Conus* venoms so far. However, a large number of other peptides are found in these venoms. These comprise both paralytic toxins to immobilize the prey of the cone snail, and other biologically active peptides which are not themselves directly paralytic. Only the briefest overview of these peptide components will be presented here.

In the fish-hunting cone snail venoms, α- and ω-conotoxins are ubiquitously distributed. As noted above, μ-conotoxins have only been found in one species, *Conus geographus*. In addition to these three well-characterized classes, however, a fourth class of paralytic conotoxins has been found. In contrast to the α-, μ-, and ω-conotoxins, these peptides cause a spastic paralysis in mice when injected ip. These components have been purified from *Conus magus* and *Conus striatus* venom (M. McIntosh and G. C. Zafaralla, unpublished data), and when applied to the amphibian neuromuscular junction, cause trains of action potentials upon stimulation of the nerve, eventually leading to a block in neuromuscular transmission (F. Abogadie and D. Yoshikami, unpublished data). Two peptides which cause this spastic paralysis have been purified to homogeneity. They have proven to be blocked at the N-terminus, and for this reason, sequence information is still incomplete.

Another potentially paralytic conotoxin was recently described; this was a peptide purified from *Conus geographus* venom, which like μ-conotoxin appeared to target to voltage-sensitive Na channels. However, the structure of "conotoxin GS" [nomenclature of Yanagawa et al. (*31*)] was less homologous to μ-conotoxins than to the ω-conotoxins, which are Ca channel blockers. The same peptide was purified and characterized using a different assay, the induction of highly aberrant behavior upon ic injection of mice (L. J. Cruz, unpublished data).

From the snail-hunting and worm-hunting cone venoms, a large number of peptides have been purified which are paralytic to the prey. In certain cases, these peptides also have activity in vertebrate systems. Since the detailed physiological mechanisms through which these peptides act have not been clarified, we will only summarize the generalizations from the work on toxins from molluscivorous and vermivorous cone venoms. Over 25 peptides have been purified and sequenced; all are cysteine-rich (as are the α-, ω-, and μ-conotoxins from piscivorous *Conus*). In addition, two post-translational modifications are found in many of these peptides, hydroxyproline and γ-carboxyglutamate. The peptides are generally similar in their properties to the conotoxins that have already been well characterized with two exceptions. In contrast to the conotoxins from the fish-hunting cone snails, which are all highly positively charged and hydrophilic, many of the peptides from snail-hunting and worm-hunting cones have a net negative charge and many are significantly more hydrophobic. Included in this set of peptides are the most tightly cross-linked peptides known; one peptide which has been characterized has 3 disulfide bonds although only 12 amino acids in length.

In addition to the paralytic conotoxins, which are produced for paralysis of prey, there are a number of peptides found in *Conus* venoms which do not appear to be directly paralytic to prey, although they have demonstrable biological activity. Two classes of these peptides have been well characterized to date. The first class are the conopressins, which are vasopressin analogs found in *Conus* venoms (*32*). As shown in Table IV, the homology to vasopressins is very strong. Two different conopressins which have been isolated are conopressin-G from *Conus geographus*, and conopressin-S from *Conus striatus*. Both of these peptides are biologically active when injected intracerebrally; however, no tests to determine pressor activity (or other biological activities of neurohypophyseal enzymes) have yet been reported. A second class of biologically active peptides which do not appear to be strongly paralytic are the conantokins. The first of these purified from *Conus geographus* was originally called the sleeper peptide (*33, 34*); we have since determined that a family of related peptides can be defined which induce a sleep-like state in young mice. For this series of *Conus* peptides we have coined the term *conantokin*, from *antokin*, the Pilipino word for sleepy. The structure of conantokin-G from *Conus geographus* venom is shown on Table IV. Other conantokins have been found in fish-hunting *Conus* venoms, and the peptide from *Conus tulipa* has been sequenced and synthesized (*35*). Conantokins recently have been demonstrated to block the NMDA receptor in a noncompetitive manner (*35*; E. E. Mena, unpublished data).

Table II. Sodium Channel Inhibitors: μ-Conotoxins

μ-Conotoxin	Sequence
GIIIA	R D C C T *P P* K K C K D R Q C K *P* Q R C C A *
GIIIB	R D C C T *P P* R K C K D R R C K *P* M K C C A
GIIIC	R D C C T *P P* K K C K D R R C K *P* L K C C A

Table III. Acetylcholine Receptor Inhibitors: α-Conotoxins

α–Conotoxin	Sequence
GI	E C C N P A C G R H Y S C *
GIA	E C C N P A C G R H Y S C G K *
GII	E C C H P A C G K H F S C *
MI	G R C C H P A C G K N Y S C *
SI	I C C N P A C G P K Y S C *
Amino acids found in other α-conotoxins	- Y - - - - - - - - - - D - - T S C S C
Consensus	- - C C - P A C G - - - - C - -

Disulfide bridges	C C - - - C - - - - - C

Table IV. Non-Paralytic Biologically Active Peptides

Peptide	Sequence
Arg-conopressin-S	C I I R N C P R G *
Lys-conopressin-G	C F I R N C P K G *
Conantokin GV	G E γ γ L Q γ N Q γ L I R γ K S N *

A large number of other non-paralytic but biologically active peptides have been purified and sequenced. In most cases however, the detailed mechanism of action has not yet been elucidated. Clearly, *Conus* venoms will be a rich source of such peptides that can be used to probe various receptor targets, particularly in the central nervous system.

The Generation of Diverse Toxins: Genetic Mechanisms and Evolution

Cone snails have generated a remarkably wide array of toxins. A variety of data suggest that this genus may use novel mechanisms for evolving diverse cysteine-rich peptides.

Structural comparisons between toxins that have been sequenced so far reveal two types of related peptide structures. First, there are the truly homologous structures ("homologs"), which are related both structurally and functionally (e.g., the ω-conotoxin series from fish-hunting cone venoms). However, a second class of related structures has emerged, peptides which are related in sequence, particularly at the cysteine residues, but which appear to have an entirely different receptor target specificity. This can occur within a single venom, or in different venoms; examples of all these relationships are shown in Table V. Since these are peptides which are structurally (and almost certainly evolutionarily) related to each other, but not functionally related, they are the opposite of analogs. We have coined the term "disanalogs" for this second relationship. Therefore, structurally related conotoxins can either be homologs or if not functionally related, "disanalogs".

As more and more peptides from snail-hunting and worm-hunting cones are sequenced, an increasing number of disanalogs are being discovered. A comparison of ω-conotoxin MVIIA from *Conus magus* venom (*1*) and the King Kong peptide from *Conus textile* venom (*36*) serves as an example. These two snails have entirely different phylogenetic specificity; *Conus magus* preys exclusively on fish, *Conus textile* exclusively on gastropod molluscs. Even correcting for divergent phylogenetic specificity, the two peptides show no similarity in mechanism. While ω-conotoxin MVIIA causes inhibition of calcium currents, no such inhibition can be demonstrated using the King Kong peptide from *Conus textile* even in molluscan systems. Although ω-conotoxin MVIIA causes paralysis of fish (the natural prey of *Conus magus*), the King Kong peptide induces a rhythmic peristaltic movement in snails (the natural prey of *Conus textile*). Thus, there is no congruence of function in the two disanalogs, irrespective of the assay. However, the structural similarity is striking as shown in Table V. Thus, although the peptides are structurally homologous, they are functionally divergent.

A possible origin for such disanalogs in *Conus* venoms has been suggested by recent cloning data. An unexpected result obtained when attempting to clone the King Kong peptide was the discovery of three different transcripts encoding a "King-Kong" disanalog family in *Conus textile* venom.

The primary translation products predicted from the three transcript sequences are a propeptide family containing a relatively long (~ 50 AA) N-terminal domain which is cleaved off to yield a final 26–27 amino acid Cys-rich conotoxin. A strong sequence homology is found between the large N-terminal excised regions of all three propeptides (S. Woodward and D. Hillyard, unpublished data). In addition, all six of the Cys residues in the mature King-Kong peptide are conserved. However, the amino acids between the Cys residues are not conserved at all. Thus, the three *Conus textile* propeptides show constant and hypervariable regions: the hypervariable regions are the four regions in the final peptide toxins between the Cys residues (*see* Table V). while the large excised N-terminal region and the six Cys residues of the final peptide are constant. The three disanalog sequences

Table V. Structural Similarity of "Disanalogs" from *C. textile* and *C. magus*

Species	Peptide	Sequence
C. textile	KK-O	W C K Q S G E M C N L L D Q N C C D G Y C I V L V C T
C. magus	ω-MVIIA	C K G K G A K C S R L M Y D C C T G S C R S G K C *
Absolutely conserved residues	Includes other KingKong disanalogs	C - - - - - C - - - - - - C C - - - C - - - - C

The four hypervariable regions are indicated by the bars. The constant regions in this disanalog family include a ~ 50 AA N-terminal region (not shown) and the 6 cysteine residues.

predicted by cloning result from the mollusc-hunting *Conus textile* venom are even more unrelated to each other than they are to ω-conotoxin MVIIA from a fishhunting snail.

We suggest that the constant region of the three *Conus textile* disanalogs (the disulfides in the final toxin, and the N-terminal excised region) functions to form the biologically active disulfide-bonded configuration in the three disanalogs. In order to evolve a ligand with new specificity from a tightly disulfide-bonded polypeptide, the necessity for folding into a specific disulfide bonding configuration presents a significant constraint to amino acid variation. Mutations which might contribute to new specificity could simultaneously destabilize or randomize the disulfide configuration, leading to significant loss of activity. However, if the disulfide bonding were largely determined by sequences outside the final peptide, these could be conserved and the spectrum of amino acid changes tolerated in inter-Cys (i.e., hypervariable) regions to give biologically useful variants would be much greater. Thus, conserving a structural framework around which variation can be introduced may be the most efficient route to generating peptides with new specificity, while minimizing the probability of losing biological activity. In this scenario, the pathway for generating diverse toxins is to conserve the disulfide structural framework, but to change the amino acid sequences between the disulfide bonds, which presumably would be the actual contact points with the target receptors. Indeed, the different nucleotide sequences which code for some of the hypervariable regions are essentially random with respect to each other and show no detectable homology. This suggests that rather than a gradual accumulation of point mutations, unusual genetic mechanisms such as cassette switching may be employed for varying the inter-Cys regions. By combining a conserved structural framework (determined by the conserved N-terminal regions and the Cys residues), with a cassette replacement mechanism for the hypervariable propeptide regions, the cone snails are, in effect. uncoupling peptide folding from target specificity. Conotoxins with new target specificity can therefore be generated without perturbing determinants to form specific disulfide bonds.

The strategy proposed for conotoxins is not unlike the generation of antibody molecules; these, too, have conserved and hypervariable regions, which correspond, respectively, to a structural framework and residues that determine ligand specificity. For both antibodies and conotoxin propeptides, the biological purpose for evolving molecules with sharply defined conserved and hypervariable sequence motifs is the generation of structurally conservative ligands of diverse target specificity.

Conclusions

The cone snails constitute one of the largest groups of animals that use venoms as their primary weapon for capturing prey (300–500 species). Furthermore, they have successfully envenomated a much wider variety of prey than has been observed for most venomous animal genera.

The analysis of toxins in *Conus* venoms reveals a novel strategy of venom design. The primary paralytic molecules in the venom are the conotoxins, exceptionally small, tightly disulfide cross-linked peptides made in a venom duct. The conotoxins are synthesized as propeptides and processing takes place to generate the final peptide toxins, typically 12–30 AA long, most commonly with 3 disulfide bonds. Suggestive evidence has been obtained that the entire propeptide is necessary for folding the final peptide toxin into a specific disulfide-bonded configuration. Each class of toxins binds to a specific receptor target in the prey, altering function. There are multiple classes of toxins in a single *Conus* venom. Toxins which block voltage-sensitive Ca channels, muscle Na channels, the nicotinic acetylcholine receptor at the neuromuscular junction, the vasopressin receptor, and the NMDA receptor are found in venom of a single fish-hunting *Conus* species.

Although functionally diverse sets of conotoxins have been characterized from *Conus* venom, these are surprisingly conservative structurally, particularly with respect to the placement of Cys residues in the peptide. Cloning experiments have revealed that a family of functionally diverse but structurally conserved conotoxins can be generated in the same venom. The details of sequence indicate that a cassette replacement mechanism best explains the generation of such structurally related but functionally divergent toxins. Thus, *Conus* venoms contain targeted peptide ligands in which general structural patterns emerge, indicative of channeled evolutionary pathways for the generation of toxins with differing specificity.

Acknowledgments

The work described in this paper was primarily supported by grants GM 22737 from the National Institutes of General Medical Sciences, N00014-80-K-0178 from the Office of Naval Research, and the Marine Science Institute, Quezon City, Philippines. Support was also provided by grants AM 2674 (J.R.), The Howard Hughes Medical Institute (D.R.H. and S.R.W.), GM 34913 (W.R.G.), the International Foundation for Science, Stockholm, Sweden (L.J.C.), and the National Research Council of the Philippines (L. J. C.).

Literature Cited

1. Olivera, B. M.; Gray, W. R.; Zeikus, R.; McIntosh, J. M.; Varga, J.; Rivier, J.; de Santos, V.; Cruz, L. J. *Science* **1985**, *230*, 1338-43.
2. Cruz, L. J.; Gray, W. R.; Yoshikami, D.; Olivera, B. M. *J. Toxicol-Toxin Rev.* **1985**, *4*, 107-32.
3. Olivera, B. M.; Gray, W. R.; Cruz, L. J. In *Marine Toxins and Venoms*; Tu, A. T., Ed.; Handbook of Natural Toxins; Marcel Dekker, Inc.: New York and Basel, 1988; Vol. 3, pp 327-52.
4. Gray, W. R.; Olivera, B. M.; Cruz, L. J. *Annu. Rev. Biochem.* **1988**, *57*, 665-700.
5. Kohn, A.J.; Saunders, P. R.; Wiener, S. *Annls. N.Y. Acad. Sci.* **1960**, *90*, 706-25.
6. Saunders, P. R.; Wolfson, F. *The Veliger* **1961**, *3*, 73-5.
7. Nybakken, J. *The Veliger* **1968**, *10*, 55-7.
8. Halstead, B. *Poisonous and Venomous Marine Animals of the World*; Darwin Press, Inc.: Princeton, N. J., 1978; p. 188.
9. Pali, E. S.; Tangco, O. M.; Cruz, L. J. *Bull. Phil. Biochem. Soc.* **1979**, *2*, 30-51.
10. Jiminez, E. C.; Olivera, B. M.; Cruz, L. J. *Toxicon* **1983**, Suppl. *3*, 199-202.
11. Kohn, A. J. *Ecological Monographs* **1959**, *29*, 47-90.
12. Lyman, F. *Shell Notes* **1948**, *2*, 78-82.
13. Alcala, A. C. *Toxicon* **1983**, Suppl. *3*, 1-3.
14. Endean, R.; Parish, G.; Gyr, P. *Toxicon*, **1974**, *12*, 131-38.
15. Whyte, J. M.; Endean, R. *Toxicon* **1962**, *1*, 25-31.
16. Olivera, B. M.; McIntosh, J. M.; Cruz, L. J.; Luque, F. A.; Gray, W. R. *Biochemistry* **1984**, *23*, 5087-90.
17. Cruz, L. J.; Johnson, D. S.; Imperial, J. S.; Griffin, D.; LeCheminant, G. W.; Miljanich, G. P.; Olivera, B. M. *Current Topics in Membranes and Transport* **1987**, *33*, 417-29.
18. McCleskey, E. W.; Fox, A. P.; Feldman, D. H.; Cruz, L. J.; Olivera, B. M.; Tsien, R. W.; Yoshikami, D. *Proc. Natl. Acad. Sci. USA* **1987**, *84*, 4327-31.

19. Cruz, L. J.; Imperial, J. S.; Johnson, D. S.; Olivera, B. M. *Abstr. Soc. Neurosci.* **1987**, *13*, 1011.
20. Spence, I.; Gillessen, D.; Gregson, R. P.; Quinn, R. J. *Life Sci.* **1978**, *21*, 1759-70.
21. Cruz, L. J.; Gray, W. R.; Olivera, B. M.; Zeikus, R. D.; Kerr, L.; Yoshikami, D.; Moczydlowski, E. *J. Biol. Chem.* **1985**, *260*, 9280-88.
22. Nakamura, H.; Kobayashi, J.; Ohizumi, Y.; Hirata, Y. *Experientia* **1983**, *39*, 590-91.
23. Yanagawa, Y.; Abe, T.; Satake, M. *Neurosci. Lett.* **1986**, *64*, 7-12.
24. Ohizumi, Y.; Nakamura, H.; Kobayashi, J.; Catterall, W. A. *J. Biol. Chem.* **1986**, *261*, 6149-52.
25. Cruz, L. J.; Kupryszewski, G.; LeCheminant, G. W.; Gray, W. R.; Olivera, B. M.; Rivier, J. *Biochemistry* **1989**, *28*, 3447-42.
26. Yanagawa, Y.; Abe, T.; Satake, M. *J. Neurosci.* **1987**, *1*, 1498-1502.
27. Cruz, L. J.; Gray, W. R.; Olivera, B. M. *Arch. Biochem. Biophys.* **1978**, *190*, 539-48.
28. Gray, W. R.; Luque, A.; Olivera, B. M.; Barrett, J.; Cruz, L. J. *J. Biol. Chem.* **1981**, *256*, 4734-40.
29. McManus, D. B.; Musick, J. R.; Gonzalez, C. *Neurosci. Lett.* **1981**, *24*, 57-62.
30. Zafaralla, G. C.; Ramilo, C.; Gray, W. R.; Karlstrom, R.; Olivera, B. M.; Cruz, L. J. *Biochemistry* **1988**, *27*, 7102-05.
31. Yanagawa, Y.; Abe, T.; Satake, M.; Odani, S.; Suzuki, J.; Ishikawa, K. *Biochemistry* **1988**, *27*, 6256-62.
32. Cruz, L. J.; de Santos, V.; Zafaralla, G. C.; Ramilo, C. A.; Zeikus, R.; Gray, W. R.; Olivera, B. M. *J. Biol. Chem.* **1987**, *262*, 15821-24.
33. McIntosh, J. M.; Olivera, B. M.; Cruz, L. J.; Gray, W. R. *J. Biol. Chem.* **1984**, *259*, 14343-46.
34. Rivier, J.; Galyean, R.; Simon, L.; Cruz, L. J.; Olivera, B. M.; Gray, W. R. *Biochemistry* **1987**, *26*, 8508-12.
35. Haack, J.; Rivier, J.; Parks, T.; Mena, E. E.; Cruz, L. J.; Olivera, B. M. *J. Biol. Chem.*, in press.
36. Hillyard, D. R.; Olivera, B. M.; Woodward, S.; Corpuz, G. P.; Gray, W. R.; Ramilo, C. A.; Cruz, L. J. *Biochemistry* **1989**, *28*, 358-61.

RECEIVED September 27, 1989

Chapter 21

Sea Anemone Polypeptide Toxins Affecting Sodium Channels

Initial Structure—Activity Investigations

William R. Kem[1], Michael W. Pennington[2], and Ben M. Dunn[2]

[1]Department of Pharmacology and Therapeutics, University of Florida
College of Medicine, Gainesville, FL 32610
[2]Department of Biochemistry and Molecular Biology, University of Florida
College of Medicine, Gainesville, FL 32610

Sea anemone polypeptides affecting action potentials are of considerable interest as probes of the sodium channel and as models for
designing new therapeutic and pesticidal compounds. Eleven 5,000
dalton homologous toxin variants have been sequenced. They are
classified into two groups because of differences in sequence,
antigenic determinants, and binding sites on the sodium channel.
Multivariate analysis and chemical modification approaches with the
natural toxins, plus solid phase synthesis of monosubstituted toxin
analogs, are being used to determine the receptor binding domains
of the two toxin types.

Over 40 different types of polypeptide toxins have been found in marine
animals (*1*). Many of these toxins are exquisitely selective in their actions,
affecting a single process or receptor at minute concentrations. So far the sea
anemone and gastropod (*Conus*) toxins have attracted the most attention as
molecular probes of ion channels. In this chapter, we discuss several
approaches which are being used to investigate, at the molecular level, the
interactions of the sea anemone neurotoxic polypeptides with sodium channels.
    All sea anemone neurotoxins which have been investigated electrophysiologi-
cally inhibit the process of sodium channel closing (inactivation), thereby pro-
longing the action potential. Although this has deleterious (sometimes lethal)
consequences for predators and prey of the sea anemone, low concentrations of
certain of these polypeptides have also been found to stimulate mammalian heart
contractility with a more favorable therapeutic index than the most commonly
used drugs for treating congestive heart failure (*2*). Catterall and Béress (*3*) first
showed that *Anemonia sulcata* toxin II (As II) binds in a voltage-dependent
manner to the same site on the sodium channel as scorpion α-toxins. Although
this observation has since been corroborated by several laboratories (*4*), some
uncertainty still exists regarding the equivalence of the binding site for As II
with the scorpion α-toxin binding site (*5*). More recently, it was found that a
group of scorpion polypeptide toxins homologous with the α-toxins, now known
as β-toxins, binds to a separate site on the sodium channel in a voltage-
independent fashion and selectively enhances channel opening (activation). The

0097–6156/90/0418–0279$06.00/0
© 1990 American Chemical Society

two sites at which the scorpion polypeptide toxins bind to the sodium channel protein have been designated as sites 3 (α-toxins) and 4 (β-toxins).

In the past five years, a new group of sea anemone toxins has been investigated by several laboratories (Figure 1). Although homologous with As II and other related toxins like the anthopleurins, these toxins differ structurally, immunologically, and pharmacologically from the toxins isolated from the family Actiniidae (6–8). The actiniid toxins have been designated as type 1 toxins and the stichodactylid sea anemone toxins as type 2 (7). Structurally, the type 2 toxins lack a single N-terminal residue and the two histidinyl residues found in the first type, possess three consecutive acidic residues at positions 6–8 and four consecutive basic residues at the C-terminus, and usually possess a single tryptophan at position 30. Rabbit polyclonal antibodies prepared with type 1 toxins failed to react with type 2 toxins and vice versa (6–8). These two types of sea anemone toxin also bind to separate sites on the sodium channel (6). Schweitz et al. (6) observed competition between *Heteractis paumotensis* toxins I and II and *Androctonus australis* toxin II (scorpion α-toxin) for binding to rat brain synaptosomal sodium channels; they concluded that the same site (Catterall's site 3) binds both the type 2 sea anemone toxins and the scorpion α-toxins, but that the type 1 sea anemone toxins bind to another site. However, we have observed no competition between another type 2 toxin, *Stichodactyla helianthus* toxin I (Sh I), and the same scorpion α-toxin (Pennington et al., submitted). Thus, future structure-activity investigations upon sea anemone neurotoxins must experimentally assess the interactions of these toxins with these different binding sites.

In addition to the multiplicity of binding sites for sea anemone neurotoxic polypeptides on a single sodium channel, there is also another pharmacological dimension which must be considered: the multiplicity of sodium channels within organisms. The rat brain has been shown to possess at least three genes for the major subunit of the sodium channel; cardiac and skeletal muscles probably possess additional gene variants (9). Catterall and Coppersmith (10) have shown that rat cardiac muscle sodium channels possess an exceptionally high affinity for As II relative to the scorpion α-toxins, whereas with rat brain sodium channels the affinity for scorpion α-toxin is highest. There are also major pharmacological differences between the neuronal sodium channels of arthropods, vertebrates, and molluscs, particularly with respect to their sensitivity to the polypeptide toxins. For instance, the squid giant axon has been found insensitive to at least two sea anemone toxins type 1 toxins (11,12), and displays a different response (less effect on inactivation and a significant decrease in peak sodium conductance) to many scorpion α-toxins (13–14). Thus, differences between sodium channels in various tissues and organisms must also be considered during structure-activity investigations. Bioassay data obtained for one sodium channel must be considered separately from data obtained from a different sodium channel.

Sea Anemone Polypeptide Structure

Amino acid sequences of eleven homologous sea anemone polypeptides have been elucidated. All possess three disulfide bonds. The six half-cystine residues always occur in the same positions (7,8). Initial studies concerning the toxin secondary and tertiary structures relied upon circular dichroism, laser Raman, and, to a lesser extent, fluorescence spectral measurements (15–18). The circular dichroism spectra of the four toxins so far examined are essentially superimposable and thus indicate a common secondary structure. The only peak observed, a negative ellipticity at 203 nm, largely results from a non-regular ("random")

```
                    1           10            20              30            40
                    A                    P          T     F        S    G A      K
              (NC)  A S       K         R T     S          L          EK K N      Y
Type 1:                                   N                               I
              (C)   G V P C L C D S D G P S V R G N T L S G I I W L A G C P S G W H N C K A H G P T I G W C C K Q
                                          M                               N

                    G A              Y        T     Y       N  W S   A      TA    I     S
              (NC)  A N C K C D D E G P D V R T A P L T G T V D L G Y C N E G W E K C A S Y Y T P I A E C C R K K K
Type 2:                             S         F        F              S V D
              (C)                 D     N I S
```

Figure 1. Most common (consensus) sequences of the two types of sea anemone toxins. Bold letters represent residues which both toxin types have in common. Letters above each sequence are nonconservative substitutions, while letters below each sequence are conservative substitutions. A nonconservative substitution was defined as one in which (a) electronic charge changed, (b) a hydrogen-bonding group was introduced or removed, (c) the molecular size of the sidechain was changed by at least 50%, or (d) the secondary structure propensity was changed drastically from b to h or vice versa (Ref. 28, Table VII). A total of seven type 1 toxins and four type 2 toxins were compared. Complete sequences of the toxins considered here are cited in reference 7.

structure and from β-turn residues (16). A small plateau at 215–225 nm probably results from β-pleated sheet residues. Analysis of the Raman amide I vibrations of As II, *Anthopleura xanthogrammica* I (Ax I, or anthopleurin A), and *Heteractis macrodactylus* III (Hm III) indicated the occurrence of small proportions of α-helix and β-sheet in these toxins. During the mid-1970's, when these studies were being done, computer methods for analyzing the CD and Raman spectra to obtain reliable quantitative estimates of secondary structure had not yet been developed. Raman analysis also permitted certain inferences about the geometry of the disulfide bonds and the environmental exposure of certain aromatic amino acid residues (15,16).

More recent studies on the folded toxin structure by Norton and colleagues have utilized ^{1}H- and ^{13}C-NMR techniques (19,20). By using 2D-FT-NMR, it was possible to localize a four stranded, antiparallel β-pleated sheet "backbone" structure in As II, Ax I, and Sh I (21,22). In addition, Wemmer et al. (23) have observed an identical β-pleated structure in Hp II. No α-helix was observed in these four variants. In the near future, calculated solution conformations of these toxins, utilizing distance measurements from extracted Nuclear Overhauser Enhancement (NOE) effects should greatly stimulate structure-activity investigations.

Sea Anemone Toxin Activity

Ultimately, comparison of the equilibrium dissociation constants (K_D's) for binding of the sea anemone toxins to the same sodium channel receptor site is the best means of assessing the structural requirements for toxin action. Some data of this type already exist (6,24, Pennington et al., submitted). These measurements are not easy and are complicated by voltage-dependent binding (i.e., sealed membrane vesicles must be prepared capable of maintaining a considerable resting potential). Consequently, whole animal toxicity (LD_{50}) data are mostly available at this time. Data for the actiniid type 1 toxins are presented in Figure 2. It should be kept in mind that LD_{50} values are not necessarily proportional to K_D estimates (25), particularly when considering toxins with low K_D's (high affinity). Also, it is likely that other factors such as susceptibility to endogenous proteases affect the observed LD_{50} estimates. Nevertheless, the LD_{50} data are in semi-quantitative agreement with the K_D estimates available (6,24, Pennington et al., submitted). Note that the mammalian toxicity varies more greatly than the crustacean toxicity in this group of toxins.

Comparison of Natural Variants

Toxin variants occurring in the same species often resemble each other rather closely in structure. Occasionally (the *Anemonia sulcata* toxins I and II are an example), a large toxicity difference nevertheless exists between two structurally similar variants. On the other hand, toxins from taxonomically distant species are more likely to display large differences in sequence. It seems that pharmacological comparisons between toxins varying only slightly in sequence are most illuminating, at least in terms of deciphering the influence of a particular sidechain upon toxicity. Since natural toxins differing in only a single amino acid residue of interest would be difficult to find, one must utilize an approach capable of using toxins with multiple differences to provide inferences regarding the importance of single sidechains.

A multivariate analytical approach has recently been applied to smaller molecules, including peptides (26,27). Recently, we have applied this approach to the sea anemone type 1 toxins. About 90% of the differences observed in

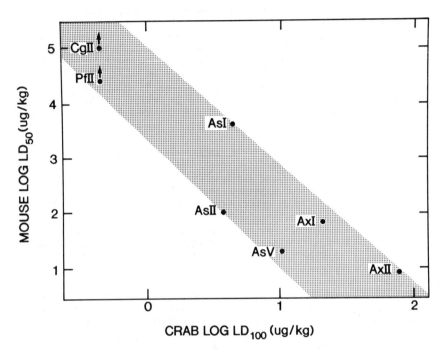

Figure 2.   Relative toxicity (LD_{50} and LD_{100}) estimates for actiniid sea anemone toxins upon crabs (*Carcinus maenas*) and mice. Values for *Anemonia sulcata* (As) and *Anthopleura xanthogrammica* (Ax) toxins are from ref. 24; data for *Condylactis gigantea* and *Phyllactis flosculifera* toxins are unpublished (Kem). The arrows indicate that the real mouse LD_{50} values for Cg II and Pf II must exceed the values indicated in the figure. Although insufficient data are presently available to quantitatively define a relationship between mammalian and crustacean toxicity, it seems that there is usually an inverse relationship, which may be approximately defined by the stipple zone.

toxicity between the different type 1 toxins could be accounted for (Hellberg and Kem, unpublished results). Certain amino acid residues seem important for explaining toxicity differences between the variant toxins. Since the method ignores contributions of sidechains which do not vary in the sample group, it cannot predict the contribution of non-variant residues towards toxicity. A large number of toxin variants would have to be considered in order for this method to confidently predict the importance of most residues. Since Nature is unlikely to provide a randomly substituted group of toxins, this method seems limited as long as it is only applied to naturally occurring toxin variants.

Comparisons between these toxins allow delineation of the variability of each position in the sequence. For instance, the residues which are extremely invariant (conservative) for both types of sea anemone toxin are the half-cystines, certain glycyl residues which are expected to be involved in β-turns, and only a few other residues — Asp 5 or 6, Arg 13 or 14, and Tryp 30 or 31 (the numbering depends upon the toxin type) — expected to be important for folding or receptor binding. Rather surprising is the variation in the residues which NMR studies (22,23) have shown are involved in formation of the four stranded β-pleated sheet.

Figure 1 summarizes the degree of variation found along the sequences of the type 1 and 2 toxins. It should be noted that since there are fewer type 2 toxin sequences currently available, the variability profile for this type is particularly likely to change as new sequences appear. Since HPLC resolves many sea anemone toxins into a much greater number of variants than was previously expected, many more natural toxin variants should become available for analysis in the future. Unfortunately, the naturally occurring polypeptide sequences are expected to be those with high toxicity; substitutions which are greatly deleterious for toxicity would be eliminated by natural selection. Thus, examination of the natural polypeptides is expected to only reveal what is successful.

Sidechain conservatism may be split up into at least two kinds: 1) substitutions which conserve sidechain bonding forces — providing similar electrostatic, hydrophilic, or hydrogen bonding interactions, and 2) substitutions conserving secondary structure propensity. For instance, substitution of glutamic acid with aspartic acid conserves charge, but this could have a considerable effect upon the secondary structure propensity of the peptide.

Chemical Modification

This approach is primarily limited by the group rather than residue selectivity of modifications and by the relatively small number of selective reagents available for use. Modification reactions carried out with large polypeptides or proteins generally lead to a complex mixture of products which cannot be completely resolved and individually analyzed. In this respect, the sea anemone toxins are relatively promising since in many cases only 1 or a few similar residues are present. Often the reaction can be directed towards the production of a very few products by manipulating pH, time, reagent concentration, or other conditions. Unfortunately, the importance of non-reactive residues (about half of the amino acids) cannot be investigated with this method.

Some chemical modification studies on the sea anemone toxins have unfortunately been less than rigorous in analyzing the reaction products. Consequently, results from many of these studies can only provide suggestions, rather than firm conclusions, regarding the importance of particular sidechains. Many such studies also have failed to determine if the secondary and tertiary structures of the toxin products were affected by chemical modification.

The actiniid toxins As II and Ax II both possess only three carboxyl moieties. One aspartyl carboxyl (either 7 or 9) has an abnormally low pK_a of 2 and is thought to participate in a buried salt-bridge with a cationic group. The other two carboxyls have normal pK_a's, the higher one (3.5) presumably being at the C-terminus. The only chemical modification reported for these carboxyls has been by carbodiimide-catalyzed reaction with glycine methyl ester. Modification of all three carboxyls destroyed toxicity (29,30). However, Barhanin et al. (30) have reported that their As II product behaved as a competitive antagonist. Several partially modified As II products were resolved, but they unfortunately were not chemically analyzed. Gruen and Norton (31) also modified Ax II in the same manner and found that when two glycine groups were added that this polypeptide lost its native conformation (measured by both ^{1}H-NMR and CD spectroscopy) as well as its inotropic activity on guinea pig heart.

The most rigorous chemical modification study done on a sea anemone toxin was that of Barhanin et al. (30). Besides the above-mentioned carboxyl modifications, they also studied the consequences of modifying the various basic amino acid sidechains. Reaction of Arg 14 in As II with cyclohexanedione abolished its toxicity for crabs as well as mice. This contrasts with an earlier report (29) that reaction of the same residue in Ax II with phenylglyoxal did not significantly affect its inotropic activity. Several possible explanations for this apparent discrepancy can be considered, including the possibility that the cardiac Na^+ channel receptor is less affected by altering the guanidinyl group. Since only one arginyl group resides on most type 1 toxins it should be possible to further assess the importance of this group by additional chemical modification investigations.

Carbethoxylation of the two histidinyl sidechains at positions 32 and 37 of As II resulted in at least a 20-fold loss in mouse toxicity and a 5-fold loss of crab toxicity (30). It is not yet clear if both residues equally contribute to this effect. Both residues are conserved in all except one of the reported actiniid toxin sequences, but are absent in the stichodactylid toxins (8). Newcomb et al. (29) also reacted Ax I with diethyl pyrocarbonate in essentially the same manner. When both histidinyl groups were modified, the resulting toxin derivative retained full inotropic activity upon the guinea pig heart. In apparent contradiction with the results of Barhanin et al. (30), Kolkenbrock et al. (32) found that modification of the two histidinyl residues of As II with this reagent actually increased inotropic activity on the guinea pig heart. Further experiments are needed to clarify the importance of the histidinyl residues for activity.

The contributions of individual primary amino groups are not yet known in much detail. Acetylation or reaction with fluorescamine of all three groups in As II reduced toxicity in crabs about 8- to 10-fold, but inotropic activity was reduced more than 20-fold by both of these modifications (32). Guanidation of the two ε-amino groups with o-methyl isourea and then acetylation of the α-amino group reduced both activities about 2-fold, suggesting that acetylation of the α-amino group by itself had little effect on toxicity. Stengelin et al. (31) utilized Schiff base formation of the amino groups with pyridoxal phosphate, followed by reduction of the imine bond with sodium borohydride. They isolated two monosubstituted derivations of As II. The Gly 1 derivative possessed only 2% of the native toxin activity, whereas the Lys 35 adduct displayed only 1% of the activity. Since this modification, besides introducing an aromatic group, also substitutes a negative charge for a positive charge, a greater loss of toxicity could be expected, compared to the acetyl derivatives. This investigation thus adds further support for the idea that several amino acid sidechains near the N-terminus are critical for toxicity.

Only a single paper has so far appeared in which a type 2 polypeptide toxin was chemically modified (*34*). Modification of *Heteractis macrodactylus* I (Hm I) at Tryp 30 (and possibly the tyrosyl residues) caused a 4.5-fold loss of mouse toxicity. Alkylation with Koshland's reagent (*35*), a more selective modification of Try 30, reduced toxicity only 2-fold. It was concluded that this sidechain is not essential for mouse toxicity. Modification of 2.7 lysyl residues with 2,4-pentanedione led to a 10-fold loss of mouse toxicity. The CD spectrum of the modified toxin was not significantly different from that of the natural toxin (*34*). Exposure of this modified toxin derivative with 0.1 M hydroxylamine at pH 6.0 fully restored the toxicity of Hm I; since any modified arginyl sidechains would not have been affected by this nucleophile, it was concluded that the original decrease in toxicity was entirely due to amino group modification. Reaction of Hm I with malonic dialdehyde at Arg 13 resulted in a 20-fold loss in toxicity. Hm III arginyl modification by phenylglyoxal or cyclohexanedione resulted in a 5-fold loss in mouse toxicity. The guanidyl group of Arg 45 failed to react with either reagent under the conditions employed. It was concluded that the Arg 13 sidechain is not essential for toxicity in these two (Hm I and Hm III) type 2 toxins (*34*).

Chemical Synthesis

In order to avoid some of the ambiguities mentioned above for extracting structure-activity relations from a series of natural or chemically modified polypeptide variants, we decided to use a synthetic approach to obtain monosubstituted *Stichodactyla* neurotoxin variants. We chose to investigate the influence of single substitutions near the N-terminus, since the acidic residues there had been previously implicated as essential for toxicity. We originally intended to use semi-synthesis, coupling a variety of N-terminus tridecapeptide analogs to the natural toxin 14–48 fragment, to be obtained by tryptic or clostripain cleavage. However, we were unsuccessful in obtaining a selective cleavage of the native toxin at Arg 13.

At this time, a solid phase automatic peptide synthesizer was acquired, so we attempted total synthesis of Sh I. A solid phase synthesis of Ax I (anthopleurin A) had previously been reported in an abstract (*36*). The synthetic Ax I possessed only 11% of the toxicity of the natural toxin.

First, we investigated whether Sh I could be reduced under denaturing conditions, exposed to the hydrogen fluoride cleavage procedure we intended to use, then reoxidized and refolded successfully. The toxicity of the resulting polypeptide (50% yield) was the same as that of the untreated natural toxin.

For the synthesis, a stable phenylacetamidomethyl (PAM) linkage was selected to anchor the peptide chain to the resin. A double-coupling protocol was also adopted for amino acids which were either β-branched or possessed large protecting groups (overall coupling efficiency was 95%). After deprotection and decoupling with the low-high hydrogen fluoride procedure (*37*), thiol and thioether scavenge compounds were removed from the synthetic peptide extraction with ether-ethyl acetate (1:1, w/w). Low molecular weight impurities were further eliminated by passage of the toxin through a G-50 Sephadex column equilibrated with 10% acetic acid. After dialysis at 4°C, the synthetic peptide was reoxidized by the glutathione method employed by Ahmed et al. (*38*) with scorpion toxins. The resulting sample was centrifuged to remove insoluble constituents and then subjected to phosphocellulose column chromatography. Elution with a linear gradient of ammonium formate at pH 4.0, as for purification of the natural toxin (*8*), yielded six sharp 280 nm absorbing peaks; the last peak, which eluted at the same conductivity as the native toxin, was the only peak displaying crab toxicity. This peak contained 24% of the initial mass placed on the column. After final purification with a C_{18} HPLC column, the synthetic toxin was found to be equivalent to the purified natural toxin by a wide variety of criteria. Chemically, its amino acid composition, N-terminal sequence (ten residues), and isoelectric point were identical to those of the natural toxin. The CD spectra of the two samples were superimposable, indicating the same secondary structure. The one-dimensional high resolution spectra of the two samples also coincided, except for some minor impurities. The fluorescence activation and emission spectral properties, attributed mainly to Tryptophan 30, were the same. When injected into crabs and mice, the two toxin samples were found to be of equal toxicity (*39*).

Having synthesized the natural Sh I sequence, we have now embarked upon a program of examining the consequences of substituting single amino acid residues. Our initial focus has been upon ionizable residues, particularly those with carboxylate-containing sidechains. Substitution of glutamic acid at position 8 by glutamine decreased Sh I toxicity over 10,000-fold. Substitution of asparagine at either position 6 or 7 similarly abolished crab toxicity. Replacement of aspartic acid at position 11 with asparagine reduced toxicity about 300-fold. Substitution of ε-N-acetyl lysine for lysine at position 4 reduced toxicity about 1,000-fold. Since the CD spectra of these analogs did not differ significantly from that of the natural toxin, it is unlikely that the substitutions affected polypeptide chain folding (*40*). The secondary and tertiary structures of these analogs will be further analyzed by ^{1}H-NMR when adequate samples become available.

Another means of investigating the receptor binding domain would be to determine if antibodies specific to single antigenic regions interfere with binding of the toxin to its receptor. Ayeb et al. (*41*) have recently utilized this approach to investigate the interaction of As II with rat brain sodium channels. They found that As II when bound to its receptor site remained fully accessible to a rabbit polyclonal antibody which binds to two of three acidic residues in this toxin (Asp 7, Asp 9, Gln 47). This result seems surprising in comparison with our toxicity data on synthetic Sh I analogs, which strongly implicate acidic residues at positions 6–8 in receptor binding. However, As II and Sh I do not bind to a common receptor site. Therefore, the structural requirements for binding to these two sites are unlikely to be the same.

Concluding Remarks

Combining the various approaches (isolation of a variety of polypeptide natural variants, chemical modification, chemical synthesis, immunochemical measurements) discussed above with NMR and crystallographic approaches for elucidating the native toxin structure should soon provide molecular models for the receptor binding domains of the two types of sea anemone toxins. Such models will guide future toxin analog syntheses aimed at delineating in sufficient detail these two receptor binding domains. Once the area and topography of each domain is recognized, it should then be possible to design simpler molecules mimicking the receptor binding surface of the sea anemone polypeptide. This could eventually lead to the design of safer, more selective drugs and pesticides acting upon sodium channels.

Acknowledgments

Our contributions to the research discussed in this paper were supported by NIH GM32848. We thank Ms. J. Adams for typing the manuscript.

Literature Cited

1. Kem, W. R. In *Biomedical Importance of Marine Organisms*; Dunn, D. F., Ed.; California Academy of Science: San Francisco, 1988; pp 69–84.
2. Scriabne, S.; Van Arman, C. G.; Morgan, G.; Morris, A. A.; Bennett, C. D.; Bohidar, N. R. *J. Cardiovasc. Pharmacol.* 1979, *1*, 571–583.
3. Catterall, W. A.; Béress, L. *J. Biol. Chem.* 1978, 253, 7393–7396.
4. Couraud, F.; Rochat, H.; Lissitzky, S. *Biochem. Biophys. Res. Comm.* 1978, *83*, 1525–1530.
5. Vincent, V. P.; Balerna, M.; Barhanin, J.; Fosset, M.; Lazdunski, M. *Proc. Natl. Acad. Sci., USA* 1980, 77, 1646–1650.
6. Schweitz, H.; Bidard, J-N.; Frelin, C.; Pauron, D.; Vijverberg, H. P. M.; Mahasneh, D. M.; Lazdunski, M. *Biochem.* 1985, *24*, 3554–3561.
7. Kem, W. R. In *The Biology of Nematocysts*; Hessinger, D.A.; Lenhoff, H. M., Eds.; Academic: New York, 1988; pp 375–406.
8. Kem, W. R.; Parten, B.; Pennington, M. W.; Dunn, B. M.; Price D. *Biochem.*, 1989, *28*, 3483–3489.
9. Noda, M.; Ikeda, T.; Kayano, T.; Suzuki, H.; Takeshima, H.; Kurasaki, M.; Takahashi, H.; Numa, S. *Nat.* 1985, *320*, 188–192.
10. Catterall, W. A.; Coppersmith, *J. Mol. Pharmacol.* 1981, 20, 533–542.
11. Narahashi, T.; Moore, J. W.; Shapiro, B. I. *Science* 1969, 163, 680–681.
12. Romey, G.; Abita, J-P.; Schweitz, H.; Wunderer, G.; Lazdunski, M. *Proc. Natl. Acad. Sci. USA* 1976, *73*, 4055–4059.
13. Romey, G.; Chicheportiche, R.; Lazdunski, M.; Rochat, H.; Miranda, F.; Lissitzky, S. *Biochem. Biophys. Res. Commun.* 1975, *64*, 115–121.
14. Gillespie, J. I.; Meves, H. *J. Physiol. (London)* 1980, 308, 479–499.
15. Prescott, B.; Thomas, G. J.; Béress, L.; Wunderer, G.; Tu, A. T. *FEBS Lett.* 1976, *64*, 144–147.
16. Ishizaki, H.; McKay, R. H.; Norton, T. R.; Yasunobu, K. T.; Lee, J.; Tu, A.T. *J. Biol. Chem.* 1979, *254*, 9651–9656.
17. Nabiullin, A. A.; Odinokov, S. E.; Kozlovskaya, E. P.; Elyakov, G. B. *FEBS Lett.* 1982, *141*, 124–127.
18. Nabiullin, A. A.; Odinokov, S. E.; Vozhova, E. I.; Kozlovskaya, E. P.; Elyakov, G. B. *Bioorg. Khim.* 1982, *8*, 1644–1648.
19. Norton, R. S.; Norton, T. R. *J. Biol. Chem.* 1979, *254*, 10220–10226.
20. Norton, R. S.; Zwick, J.; Béress, L. *Eur. J. Biochem.* 1980, *113*, 75–83.
21. Gooley, P. R.; Béress, L.; Norton, R. S. *Biochem.* 1984, *23*, 2144–2152.
22. Gooley, P. R.; Norton, R. S. *Biochem.* 1986, *25*, 2349–2356.
23. Wemmer, D. E.; Kumar, N. V.; Metrione, R. M.; Lazdunski, M.; Drobny, G.; Kallenback, N. R. *Biochem.* 1986, *25*, 6842–6849.
24. Schweitz, H.; Vincent, J. P.; Barhanin, J.; Frelin, C.; Linden, G.; Hughes, M.; Lazdunski, M. *Biochem.* 1981, *20*, 5245–5252.
25. Ishikawa, Y.; Menez, A.; Hori, H.; Yoshida, H.; Tamiya, N. *Toxicon* 1977, *15*, 477–488.
26. Hellberg, S.; Sjöström, M.; Skagerberg, B.; Wold, S. *J. Med. Chem.* 1987, *30*, 1126–1135.
27. Hellberg, S.; Sjöström, M.; Wold, S. *Acta Chem. Scand.* 1986, B40, 135–140.
28. Levitt, M. *Biochem.* 1979, *17*, 4277–4282.
29. Newcomb, R.; Yasunobu, K. T.; Seriguchi, D.; Norton, T. R. In *Frontiers in Protein Chemistry*; Liu; Mamiya; Yasunobu, K. T., Eds.; 1980, pp 539–550.

30. Barhanin, J.; Hughes, M.; Schweitz, H.; Vincent, J. P.; Lazdunski, M. *J. Biol. Chem.* **1981**, *256*, 5764–5769.
31. Gruen, L. C.; Norton, R. S. *Biochem. Intern.* **1985**, *11*, 69–76.
32. Kolkenbrock, H. J.; Alsen, C.; Asmus, R.; Béress, L.; Tschesche, H. *Proc. 5th Eur. Symp. Animal, Plant, and Microbial Toxins* **1983**, p 72.
33. Stengelin, S.; Rathmayer, W.; Wunderer, G.; Béress, L.; Hucho, F. *Anal. Biochem.* **1981**, *113*, 277–285.
34. Kozlovskaya, E.; Vozhova, H.; Elyakov, G. In *Chemistry of Peptides and Proteins*; Voelter, W.; Wünsch, E.; Ovchinnikov, J.; Ivanov, V., Eds.; Walter de Gruyter: New York, 1982; pp 379–387.
35. Hoare, D. G.; Koshland, D. E. *J. Biol. Chem.* **1967**, 242, 2447–2453.
36. Matsueda, G. R. Intern. *J. Pept. Prot. Res.* **1982**, *20*, 26.
37. Tam, J. P.; Heath, W. F.; Merrifield, R. B. *J. Am. Chem. Soc.* **1983**, *105*, 6445–6451.
38. Ahmed, A. K.; Schaffer, S. W.; Wetlaufer, D. B. *J. Biol. Chem.* **1975**, *250*, 8477–8482.
39. Pennington, M. W.; Kem, W.R.; Dunn, B. M. In *Macromolecular Sequencing and Synthesis. Selected Methods and Applications*; Schlesinger, D., Ed.; Alan B. Liss Publishers, Ch 19, pp 243–250.
40. Pennington, M. W.; Dunn, B. M.; Kem, W. R. In *Peptides: Chemistry and Biology*; Marshall, G. R., Ed.; ESCOM, Leiden, pp 264–266.
41. El Ayeb, M.; Bahraoui, E. M.; Granier, C.; Béress, L.; Rochat, H. *Biochem.* **1986**, *25*, 6755–6762.

RECEIVED June 23, 1989

Chapter 22

Solution Structure of Sea Anemone Toxins by NMR Spectroscopy

N. Vasant Kumar[1,3], Joseph H. B. Pease[1], Hugues Schweitz[2], and David E. Wemmer[1]

[1]Department of Chemistry, University of California at Berkeley, and Chemical Biodynamics Division, Lawrence Berkeley Laboratory, Berkeley, CA 94720
[2]Centre de Biochimie du Centre National de la Recherche Scientifique, Faculté des Sciences Université de Nice, Parc Valrose, 06034 Nice Cedex, France

Numerous organisms, both marine and terrestrial, produce protein toxins. These are typically relatively small, and rich in disulfide crosslinks. Since they are often difficult to crystallize, relatively few structures from this class of proteins are known. In the past five years two dimensional NMR methods have developed to the point where they can be used to determine the solution structures of small proteins and nucleic acids. We have analyzed the structures of toxins II and III of *Radianthus paumotensis* using this approach. We find that the dominant structure is β-sheet, with the strands connected by loops of irregular structure. Most of the residues which have been determined to be important for toxicity are contained in one of the loops. The general methods used for structure analysis will be described, and the structures of the toxins RpII and RpIII will be discussed and compared with homologous toxins from other anemone species.

Sea anemones use a variety of small proteins (ca. 5 kD) as part of their powerful defense and feeding systems (*1*). These proteins function as neurotoxins and/or cardiotoxins by binding to sodium channels and thereby altering ion-conducting characteristics of the channels. Binding of toxin slows down the inactivation process of the channels and hence prolongs the duration of action potential (*2*). Toxicity of these proteins depends on the type of organism against which they act. For example, some toxins are very potent against crustacean and mammalian channels whereas some are effective only on crustaceans. Tetrodotoxin (TTX)-resistant channels in general seem to have higher affinity than the TTX-sensitive channels for these toxins. In fact, such differences in affinities of these toxins for different channels have been exploited to find the subtypes of sodium channels.

[3]Current address: SmithKline & French, Physical and Structural Chemistry, Box 1539, King of Prussia, PA 19406-0939

Despite the vast amount of data on the pharmacological properties, very little about the conformation of the proteins has been known until recent NMR studies. 2D-NMR results have provided detailed information about the secondary structure of several related anemone toxins.   ATX I from *Anemonia sulcata* (*3,4*) and AP-A from *Anthopleura xanthogrammica* (*5,6*) have been studied by Gooley and Norton, and more recently Widmer et al.   have further purified the *A. sulcata* toxins and obtained complete sequence specific assignments for ATX Ia (*7*).   Our laboratory, on the other hand, has studied the structures of RpII and RpIII from *Radianthus paumotensis* (*8*).

Schweitz et al.   purified four related toxins (RpI, RpII, RpIII, and RpIV) from sea anemone *Radianthus paumotensis* (Rp) and studied their pharmacological properties (*9*).   During the course of initial NMR studies, the reported sequence of RpII was found to have errors, and was redetermined (*8*).   Subsequently Metrione et al. determined the sequence of RpIII as well (*10*).   The other two Rp toxin sequences are yet to be determined.   Sequences of the Rp, and several other sea anemone toxins, are shown in Table I.   We have used a two letter code to denote the species consistently and this notation differs from the earlier designations of Norton and Wüthrich groups.   In our notation, As Ia and Ax I correspond to ATX Ia and AP-A, respectively.   From alignment of the cystines in these sequences, it is clear that Rp toxins have three disulfide bonds, as do the other toxins.

The sequences of RpII and RpIII are very similar to one another, but differ more significantly from the other As and Ax toxins.   Schweitz et al.   showed that these toxins are functionally similar to other sea anemone toxins, that is, toxins slow down the inactivation process of sodium channels.   However, Rp toxins do not affect the binding of *A. sulcata* toxins to the sodium channels, and it is also interesting that the Rp toxins compete with the scorpion toxin AaII from *Androctonus australis* for binding sites on sodium channels, even though their primary sequences appear completely different.   They further demonstrated that Rp toxins are immunologically distinct from *A. sulcata* (As) or *A. xanthgrammica* (Ax) toxins.   Antibodies raised against RpIII recognize all other Rp toxins but not AsII, AsV, AxI, AxII.   Conversely, antibodies to AsII or AsV do not recognize any of the Rp toxins.   Without a knowledge of the structure of these proteins it is difficult to interpret these results.   We have undertaken the study of the structure of Rp toxins by 2D-NMR and distance geometry techniques to try to understand these functional relationships.   Here we review the underlying methods briefly, discuss the structures of these toxins, and compare them with what is known about the other related toxins.

2D-NMR Techniques

The first step in determination of a structure by NMR spectroscopy involves assignment of individual proton resonances.   Development of high-field spectrometers and the use of a second dimension (2D-NMR) along with isotopic substitution (*11*) and sophisticated pulse sequences (*12*) make it possible to almost completely assign the proton spectrum of proteins of about 15 kD molecular weight (*13–17*).   Some 2D-pulse sequences commonly used in the study of macromolecules are shown in Figure 1.

All these experiments involve at least three distinct time periods: preparation (tp), evolution (t1), and detection (t2); these periods are usually separated by rf pulses.   Some experiments (e.g., NOESY, RELAY) further contain an additional "Mixing" period, tm, between the evolution and detection periods.

The spin system is perturbed by rf pulses during the preparation period to create the desired coherences which are then allowed to evolve during the time t1,

Table I

Sequences of Sea Anemone Toxins

Species	Sequence
Radianthus paumotensis II (RP II)	-ASCKCDDDGPDVRSATFTGTVDFWN--CNEGWEKCTAVYTPVASCCRKKK
Radianthus paumotensis III (RP III)	-GNCKCDDEGPNVRTAPLTGYVDLGY--CNEGWEKCASYYSPIAECCRKKK
Radianthus macrodactylus III (RM III)	-GNCKCDDEGPYVRTAPLTGYVDLGY--CNEGWEKCASYYSPIAECCRKKK
Anemonia sulcata Ia (AS Ia)	GAACLCKSDGPNTRGNSMSGTIWVF--GCPSGWNNCEGRA-IIGYCCKQ
Anemonia sulcata Ib (AS Ib)	GAPCLCKSDGPNTRGNSMSGTIWVF--GCPSGWNNCEGRA-IIGYCCKQ
Anemonia sulcata II (AS II)	GVPCLCDSDGPSVRGNTLSGIIWLA--GCPSGWHNCKKHGPTIGWCCKQ
Anemonia sulcata V (AS V)	GVPCLCDSDGPSVRGNTLSGILWLA--GCPSGWHNCKKHKPTIGWCCK
Anthopleura xanthogrammica (AX I)	GVSCLCDSDGPSVRGNTLSGTLWLYPSGCPSGWHNCKAHGPTIGWCCKQ
Anthopleura xanthogrammica (AX II)	GVPCLCDSDGPRPRGNTLSGILWFYPSGCPSGWHNCKAHGPNIGWCCKK

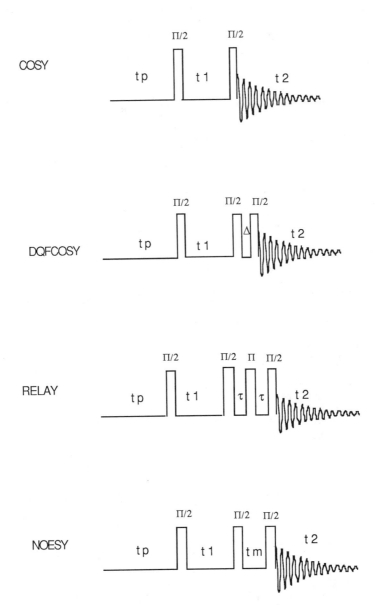

Figure 1. Pulse sequences of some typical 2D-NMR experiments. COSY = COrrelation SpectroscopY, DQFCOSY = Double Quantum Filtered COSY, RELAY = RELAYed Magnetization Spectroscopy, and NOESY = Nuclear Overhauser Effect SpectroscopY.

frequency labeling them. The signal (or free induction decay, FID) is detected during the time t2 and it carries information about behavior of spin system during the time t1, and also about how the information was exchanged by spins during the mixing time.

The evolution period t1 is systematically incremented in a 2D-experiment and the signals are recorded in the form of a time domain data matrix S(t1,t2). Typically, this matrix in our experiments has the dimensions of 512 points in t1 and 1024 in t2. The frequency domain spectrum $F(\omega 1, \omega 2)$ is derived from this data by successive Fourier transformation with respect to t2 and t1.

2D-spectra are usually represented by contour plots (Figures 2–4). Peaks with the same frequency in both the dimensions ($\omega 1 = \omega 2$) are called the diagonal peaks and correspond to the resonances in the 1D-spectrum. In cross peaks, for which $\omega 1 \neq \omega 2$, the intensity and structure contain important information about the topology and dynamics of the spin system. For example, a cross peak at ($\omega 1, \omega 2$) in a COSY or a multiple quantum filtered COSY spectrum implies that the spin A with resonance frequency $\omega 1$ and another spin B with frequency $\omega 2$ are coupled through ≤ 3 covalent bonds, and the fine structure in DQFCOSY obtained under high digital resolution contains information about the coupling constants involved.

Similarly, cross peak positions in RELAY spectra give valuable information about remote connectivities (spins coupled to common partner but not to each other) and are very useful in resolving chemical shift degeneracies. An AMX spin system with no coupling between A and X shows a cross peak at (ω_A, ω_X) in a RELAY spectrum with appropriate delay 2τ in contrast to the COSY cross peaks at (ω_A, ω_M) and (ω_M, ω_X). The intensity of this peak depends on the magnitude of coupling constants J_{AM} and J_{MX} in a nonlinear manner. As a result, it is difficult to obtain information about all the remote connectivities from a single RELAY experiment.

A related experiment TOCSY (Total Correlation Spectroscopy) gives similar information and is relatively more sensitive than the RELAY. On the other hand, intensity of cross peak in a NOESY spectrum with a short mixing time is a measure of internuclear distance (less than 4Å). It depends on the correlation time τ_c and varies as $<r^{-6}>$. It is positive for small molecules with short correlation time ($\omega\tau_c < < 1$) and is negative for macromolecules with long correlation time ($\omega\tau_c > > 1$) and goes through zero for molecules with $\omega\tau_c \simeq 1$. Relaxation effects should be taken into consideration for quantitative interpretation of NOE intensities, however.

Sequence-Specific Assignments

Systematic analysis of interproton distances in proteins led Wüthrich and colleagues to formulate the sequential assignment algorithm (18, 19). In this method sets of scalar coupled spins corresponding to the individual amino acids are identified using various correlation experiments (COSY, RELAY, TOCSY, etc.), and then sequence specific assignments are obtained by analyzing NOE cross peaks which correspond to neighbors in the primary sequence. We have assigned RpII and RpIII by basically following this procedure.

The DQFCOSY spectrum of RpII in D_2O is shown in Figure 2. Each cross peak in this spectrum identifies a pair of coupled spins of the amino acid side chains. Since couplings are not propagated efficiently across amide bonds, all groups of coupled spins occur within individual amino acids. The chemical structure of an amino acid side chain is reflected in the characteristic coupling network and chemical shifts (13). Valine spin system ($CH-CH-(CH_3)_2$) is explicitly shown in Figure 2 as an example.

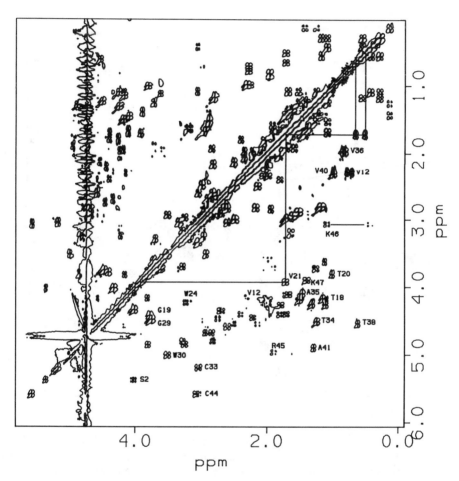

Figure 2.   DQFCOSY spectrum of RpII in D_2O.   Cross peaks corresponding to α-β protons of various spin systems are labeled.   The valine spin systems are shown explicitly.

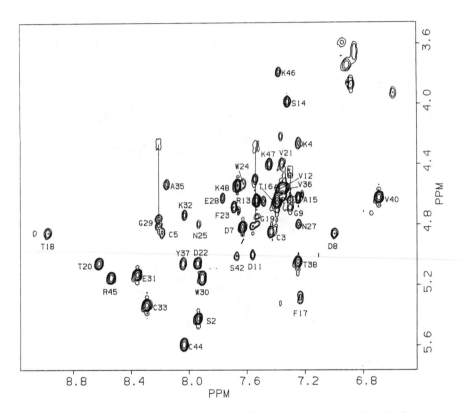

Figure 3.   Fingerprint region of the COSY spectrum of RpII in H₂O
showing the amide to α proton connectivities.   Cross peaks are labeled
with the identity of the amino acid in the sequence from which they
arise.

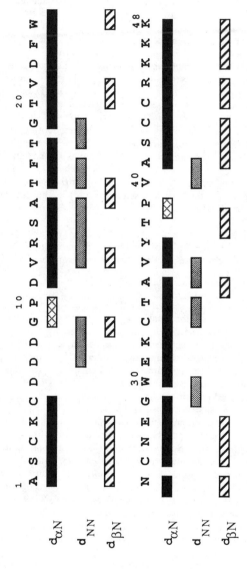

Figure 4. Summary of the sequential NOEs observed in RpII. A bar connecting the residues means that the indicated type of connectivity was observed between those residues.

It was straightforward to identify the spin systems of four valines, five threonines, four alanines, three glycines, two glutamates, and AMX type residues (Asp, Cys, Phe, Tyr, Trp, and Ser) of this protein from the COSY, RELAY, and NOESY spectra in D_2O solution. The RELAY spectrum was particularly useful in identification of Val, Ala, Thr, and Glu spin systems.

To identify the resonances of amide protons (which can exchange with solvent) additional experiments were carried out in H_2O solution. The fingerprint region, corresponding to cross peaks between amide and alpha protons, is shown in Figure 3. Degeneracies in alpha proton chemical shifts were partly resolved by finding the β, β' peaks corresponding to the given amide chemical shift in the RELAY spectrum, and also by varying the experimental conditions such as temperature.

Sequence-specific assignments were obtained by analysis of the NOESY and COSY spectra obtained under the similar conditions in H_2O. The three important classes of NOE cross peaks for this purpose are designated $d_{\alpha N}$, $d_{\beta N}$, and d_{NN}. In this notation, a d_{XY} connectivity indicates an NOE crosspeak in a short mixing time NOESY between X and Y protons belonging to neighboring residues i and i+1, and hence corresponding to a distance of ≤ 3.2 Å between them. The type of NOE which is observable in any segment of the protein is dependent on the local conformation of the backbone (19). Wüthrich and colleagues showed that for essentially any allowed conformation at least one of these NOEs is observable between nearest neighbors in the primary sequence (18). The sequential NOEs which were observed in RpII, and used to establish assignments, are shown schematically in Figure 4.

All the residues except Asp6 are assigned in RpII. Typically more than one type of connectivity was observed for most residues. Such multiple NOEs between the same pair of neighboring residues further verifies the assignments.

Secondary Structure

As mentioned above, which of the sequential NOEs $d_{\alpha N}$, $d_{\beta N}$, and d_{NN} is observed depends on the conformation of the backbone for the residues involved. Repetition of a particular type of connectivity for a sequence of amino acids often occurs in regions of regular secondary structure (19). For example a stretch of $d_{\alpha N}$-type NOEs is a signature of extended conformation, whereas a sequence of d_{NN}-type NOEs is characteristic of helical conformation. Turns, on the other hand, are characterized by short, distinct patterns of d_{NN} and $d_{\alpha N}$ connectivities.

Figure 4 shows that predominant NOEs observed in RpII are of $d_{\alpha N}$ type with relatively few d_{NN} type. Collectively these NOEs indicate that RpII primarily consists of extended chains connected by tight turns, and no regular helical structure. Furthermore, several NOE crosspeaks between the α-protons belonging to non-neighbors were observed in RpII. Spatial proximity of α-protons, indicated by the presence of these cross peaks, indicates an antiparallel β-sheet conformation. Analysis of long range α-α and amide-α NOEs resulted in identification of a four strand β-sheet in RpII. Residues W30–C33 form the outer edge of the sheet, pairing with residues C43–K46. A third strand involves residues G19–F23 pairing to the A41–C44 segment, with the fourth strand made up of residues A1–C3 paired to G19–V21. This structure is summarized schematically in Figure 5.

We also determined which amide protons are protected against solvent exchange by carrying out COSY experiments shortly after dissolving a sample of protonated protein in D_2O. Only those amides which are hydrogen bonded in the structure remain long enough to be observed. These measurements give further support for the β-sheet structure described. The amide protons of residues 5, 19–22, 30, 31, 33, 44, and 45 are observed, most are involved in this sheet

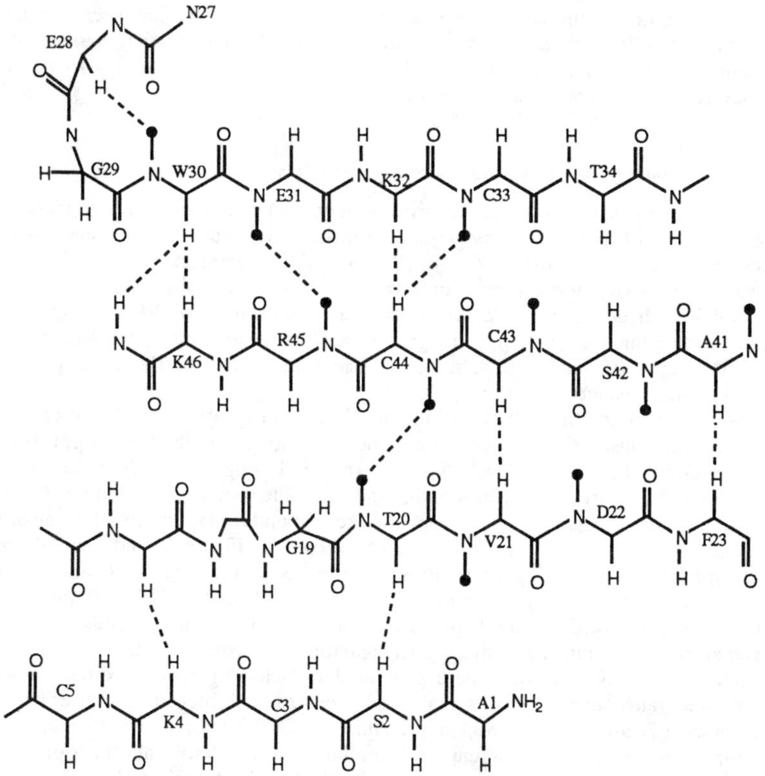

Figure 5.   Schematic representation of four-stranded β-sheet in RpII. Observed NOEs which were used to identify this secondary structure are shown by dashed lines.   Dots indicate slowly exchanging amide hydrogens.

structure. The protection of the amide of W30, and the NOEs in this region indicate that it is most probably involved in a type II β-turn, with G29 in the corner of the turn.

Although the β-sheet structure as shown in Figure 5 gives the correct local interactions, it is not a complete picture. The cystine pairing of C3, C5, and C26 with C43, C33, and C44, respectively, requires that the sheet be highly twisted. The relative intensities of cross strand α-α NOEs indicates that there is significant distortion between the central two strands. The NOE patterns in the remainder of the molecule do not indicate any other regions of regular secondary structure.

The assignment and secondary structure analysis procedure has been carried out for RpIII as well, with overall very similar results. It is clear that basic structural elements of the four strand sheet, and the β-turn, are also present. However, we see several extra cross peaks which indicate that there are differences in the segment containing residues 17-19, which will be discussed further below.

Distance Geometry Algorithm and Tertiary Structure

While the analysis described above gives a good indication of the secondary structure, it does not provide a complete picture of the structure. We have used distance geometry algorithm (20-22) (the DSPACE implementation from Hare Research) to derive tertiary structures consistent with the observed NOE distances (23). Wüthrich et al. have demonstrated that classification of NOE intensities into "short", "medium" and "long" range classes, corresponding to distance limits 2.0–2.5Å, 2.0–3.0Å, and 2.0–4.0Å, is sufficient to reproduce the tertiary structure of BPTI using distance geometry (24).

We have used this classification of NOEs, with appropriate corrections as described previously (25), to describe upper and lower limits for interproton distances. Other information including known bond lengths, bond angles, van der Waals radii, chirality, hydrogen bonds, and disulfide bonds, in addition to NOE-derived distances, were used to define the "bounds matrix" in this algorithm. Hydrogen bonds in the β-sheet of RpII are shown in Figure 5, and disulfide bonds were assumed to be analogous to those in the As II toxin (25). Conservation of Cys residues in **all** the sea anemone toxins sequenced so far (Table I) justifies this assumption for disulfide bond pattern. Information in the bounds matrix was extended by smoothing the matrix by triangulation procedures (21).

Several starting structures were generated by selecting random distances within the allowed range in the bounds matrix and "embedding" them into Cartesian space. The choice of random distances, in principle, avoids bias for any particular starting structure, and the local minimum problem inherent in almost all the optimization techniques. These embedded structures were then refined against the constraints in the bounds matrix by conjugate gradient and pseudo molecular dynamics methods. Residual violations of interproton distances from their set bounds after refinement reflect the quality of the final structure to some extent. A family of related structures are obtained in this process and five typical structures of RpII are shown in Figure 6. Most of the violations in these structures are within 0.2Å.

These structures show that RpII essentially consists of the small core of four-stranded β-sheet and three relatively large loops. Residues 6–16, 23–30, and 35–40 form loops 1, 2, and 3, respectively, and the chain reversals are accomplished by tight turns involving residues D8–D11, E28–E31, and V36–P39. Segments involved in β-sheet strands and loops alternate in the primary sequence of Rp toxins. As mentioned above, these structures indicate that the β-sheet is highly twisted in order to form the correct disulfide bonds. Comparison of different refined structures of RpII indicate that the structure of loop 1 in RpII is less well defined than

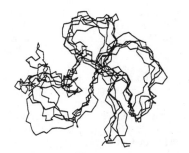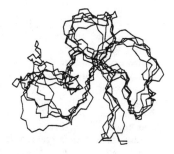

Figure 6.   Stereo pairs of distance geometry structures of RpII.   Only backbone atoms are shown.

other regions. The nature of the NOEs in this segment indicate that it could be flexible, and undergoing conformational averaging, but do not clearly establish that it is.

Barhanin et al. (26) chemically modified the side chains of several residues to correlate structure and function in As II. Their results established the importance of charged residues for the function of the toxin. They showed that Arg-13 is essential for binding to the sodium channels as well as for toxicity, while the aspartate, glutamate, and lysine residues in the N-terminal segment of the protein are crucial for toxicity only. The structures of Rp toxins show that these residues are located in easily accessible regions of the protein. In particular, Asp-6, Asp-8, and Arg-13 in RpII are part of the looped structures. Other work by Kem and coworkers (these proceedings) have also demonstrated the importance of residues in this region of the protein. X-ray crystallographic studies on α-cobratoxin also revealed that functionally important residues are part of the loops (27).

Comparision with Other Anemone Toxins

We limit the discussion to the level of secondary structures of sea anemone toxins since the tertiary structure of no other anemone toxins is known at present. Widmer et al. (7), on the basis of high resolution 2D-NMR results, have provided the most detailed information about the secondary structure of the A. sulcata toxin ATX 1a. Gooley and Norton (4-6) have partially assigned the toxins ATX I and AP-A, and compared their secondary structures. The NMR results are consistent and show that the anemone toxins all consist of a four strand antiparallel β-sheet core and have no significant helical structure.

When sequences are aligned using the cystine positions, all of the toxins have the same residues involved in the core of the sheet, including the pairing of amino acids on opposite strands. The only differences observed are at the N-terminus, and involve residues 17 and 18. The data of Widmer et al. indicate that there is a β-bulge around residue 18 in ATX 1a. Interestingly, RpIII does have the β-bulge around residue 18, and thus seems similar to ATX 1a in this region.

The Rp toxins also have an additional residue at the N-terminus which extends the sheet one residue in that direction. This may be an indication that the structural difference arises from the substitution of F17 for L17 in RpII, because there is also leucine at position 17 in the A. xanthogrammica toxins. The structure of the core sheet is remarkably similar among the different toxins, considering that there is essentially **no** conservation of the residues involved in forming it, aside from the cystines.

It is also interesting to note that the only other highly conserved residues in addition to the cystines are the Gly–Pro at residues 9–10, Arg-13, Gly-19, and Gly–Trp at 29–30. Of these only Arg-13 is implicated directly in toxicity. The Gly–Trp forms a tight turn, and it is likely that these residues have been conserved for structural reasons. The role of the Gly–Pro, and the single Gly at 19 are less clear. It is possible that they are required for correct folding of loop 1, either in the sense of folding into an active structure, or in being capable of forming the remainder of the secondary structure. Perhaps structural studies on proteins with these amino acids replaced can be carried out to address these questions.

Conclusions

We have sequenced RpII and studied the structures of RpII and RpIII in solution by 2D-NMR and distance geometry methods. The resonances are almost completely assigned, and secondary and tertiary structures have been determined. Our results indicate that Rp toxins have a four strand anti-parallel β-sheet and no α-helix. Functionally important residues are found to be located in looped regions of the toxins. These features are consistent with the structures of the sulcata and xanthogrammica toxins.

Acknowledgments

This work was supported by the Office of Energy Research, Office of Health and Environmental Research, Health Effects Research Division of the U.S. Department of Energy under Contract NO. DE-AC03-76SF00098. We are grateful to Prof. Neville Kallenbach and Michel Lazdunski for collaboration in the early work, and to Professors Robert Metrione and Ken Walsh for their interest in the sequencing, and to Dr. Dennis Hare for providing the DSPACE distance geometry program, and helping in its implementation.

Literature Cited

1. Beress, L. *Pure and Appl. Chem.* **1982**, *54*, 1981–1994.
2. Romey, G.; Abita, J. P.; Schweitz, H.; Wunderer, G.; Lazdunski, M. *Proc. Natl. Acad. Sci. U. S. A.* **1976**, *73*, 4055–59.
3. Gooley, P. R.; Beress, L.; Norton, R. S. *Biochemistry* **1984**, *23*, 2144–2152.
4. Gooley, P. R.; Norton, R. S. *Biopolymers* **1986**, *25*, 489–506.
5. Gooley, P. R.; Norton, R. S. *Eur. J. Biochem.* **1985**, *153*, 529–539.
6. Gooley, P. R.; Norton, R. S. *Biochemistry* **1986**, *25*, 2349–2356.
7. Widmer, H.; Wagner, G.; Schweitz, H.; Lazdunski, M.; Wüthrich, K. *Eur. J. Biochem.* **1988**, *171*, 177–192.
8. Wemmer, D. E.; Kumar, N. V.; Metrione, R. M.; Lazdunski, M.; Drobny, G.; Kallenbach, N. R. *Biochemistry* **1986**, *25*, 6842–6849.
9. Schweitz, H.; Bidard, J.-N.; Frelin, C.; Pauron, D.; Vijverberg, H. P. M.; Mahasneh, D. M.; Lazdunski, M.; Vilbois, F.; Tsugita, A. *Biochemistry* **1985**, *24*, 3554–61.
10. Metrione, R. M.; Schweitz, H.; Walsh, K. A. *FEBS Lett.* **1987**, *218*, 59–62.
11. LeMaster, D. M.; Richards, F. M. *Biochemistry* **1988**, *27*, 142–150.
12. Ernst, R. R.; Bodenhausen, G.; Wokaun, A. *Principles of Magnetic Resonance in One and Two Dimensions*, Clarendon Press, New York, 1987.
13. Wüthrich, K. *NMR of Proteins and Nucleic Acids*, John Wiley, New York, **1986**.
14. Di Stefeno, D. L.; Wand, A. J. *Biochemistry* **1987**, *26*, 7272–7281.
15. Weber, P. L.; Brown, S. C.; Mueller, L. *Biochemistry* **1987**, *26*, 7282–7290.
16. Redfield, C.; Dobson, C. M. *Biochemistry* **1988**, *27*, 122–136.
17. Wemmer, D. E.; Reid, B. R. *Annu. Rev. Phys. Chem.* **1985**, *36*, 105–137.
18. Billiter, M.; Braun, W.; Wüthrich, K. *J. Mol. Biol.* **1982**, *155*, 321–346.
19. Wüthrich, K.; Billiter, M.; Braun, W. *J. Mol. Biol.* **1984**, *180*, 715–740.
20. Crippen, G. M. *Distance Geometry and Conformational Calculations*, Research Studies Press, Wiley, New York, **1981**.
21. Havel, T. F.; Kuntz, I. D.; Crippen, G. M. *Bull. Math. Biol.* **1984**, *45*, 665–720.
22. Havel, T. F.; Wüthrich, K. *Bull. Math. Biol.* **1984**, *46*, 673–698.
23. Havel, T. F.; Wüthrich, K. *J. Mol. Biol.* **1985**, *182*, 281–294.
24. Patel, D. J.; Shapiro, L.; Hare, D. *Annu. Rev. Biophys. Biophys. Chem.* **1987**, *16*, 423–454.
25. Wunderer, G. *Hoppe-seyler's Z. Physiol. Chem.* **1978**, *359*, 1193–1201.
26. Barhanin, J.; Hughues, M.; Schweitz, H.; Vincent, J. P.; Lazdunski, M. *J. Biol. Chem.* **1981**, *256*, 5764–5769.
27. Martin, B. M.; Chibber, B. A.; Maelicke, A. *J. Biol. Chem.* **1983**, *258*, 8714–8722.

RECEIVED August 14, 1989

Chapter 23

Cytolytic Peptides of Sea Anemones

Alan W. Bernheimer

Department of Microbiology, New York University School of Medicine, New York, NY 10016

The characteristics and possible mechanisms of action of cytolytic peptides isolated from sea anemones during the past 20 years are described. These agents fall into three categories: *(1)* sphingomyelin-inhibitable basic peptides of molecular weights between 15,000 and 21,000, present in at least 16 species; (2) metridiolysin, a cholesterol-inhibitable, thiol-activated peptide of molecular weight ca. 80,000, known so far to be present only in *Metridium senile*; (3) Aiptasiolysin, a hemolytic system involving the cooperative action of phospholipase A and two other proteins, known so far to be present only in *Aiptasia pallida*.

At a conservative estimate about 65 cytolytic peptides have been isolated from natural sources and characterized (*1*). They are found (a) as products of a variety of bacterial species, frequently although not always pathogens, and most commonly extracellularly, but sometimes intracellularly, (b) as constituents of the basidiocarps of various kinds of higher fungi, (c) as components of snake venoms, (d) in the venoms of hymenopteran insects, (e) in a variety of marine invertebrates, particularly cnidarians, and (f) in at least one vertebrate, the flatfish *Pardachirus marmoratus*. Among the most thoroughly studied cytolytic peptides are melittin from the honey bee, staphylococcal alpha-toxin, and the thiol-activated cytolysins of streptococci and other bacteria.

The present survey is concerned with cytolytic peptides occurring only in sea anemones. However, the existence of cytolytic peptides or proteins of importance to public health, in the Atlantic Portugese Man-of-war (*Physalia physalix*) and in the North Australian coastal cubomedusan, *Chironex fleckeri*, is worthy of mention in passing. Much attention has been devoted to anemone peptides having molecular weights in the range of 2700 to 7000, especially because of their inotropic effect on heart muscle and because of their potential usefulness in the design of agents for the treatment of cardiopathies. In our experience these peptides are not cytolytic, but at least one has been reported to be (*2*). It is evident that these peptides constitute a class of agents distinct from those considered in this paper.

Among several thousand species of sea anemones the toxins of only a few dozen have been investigated. Knowledge of the sea anemone peptides permits them to be categorized as (a) sphingomyelin-inhibitable cytolysins, (b) metridiolysin, and (c) *Aiptasia* lysin.

Sphingomyelin-Inhibitable Group

Cytolytic peptides derived from 16 species of anemones currently comprise this group (Table I). It is likely that as additional species are examined in the future

0097–6156/90/0418–0304$06.00/0

Table I.   Sphingomyelin-Inhibitable Peptides

Proposed Name	Source	Detection and/or Characterization Reference No.
Gigantolysin	*Condylactis gigantea*	(3,4)
Kentolysin	*Stoichactis kenti*	(5,6)
Equinolysin	*Actinia equina*	(7,8)
Heliantholysin	*Stoichactis helianthus*	(9–11)
Lofotensolysin	*Tealia lofotensis*	(12)
Epiactolysin A	*Epiactis prolifera*	(12,13)
Epiactolysin B	*Epiactis prolifera*	(12,13)
Xantholysin	*Anthopleura xanthogrammica*	(12)
Carilysin	*Actinia cari*	(14)
Koseirolysin	*Radianthus koseirensis*	(15)
Gyrolysin	*Gyrostoma helianthus*	(15)
Variolysin	*Pseudactinia varia*	(16)
Flagolysin	*Pseudactinia flagellifera*	(16)
Michaelsenolysin	*Anthopleura michaelseni*	(16)
Parasolysin	*Parasicyonis actinostoloides*	(17)
Japonicolysin	*Anthopleura japonica*	(17)
Felinolysin	*Tealia felina*	(18)

SOURCE: Adapted and reproduced with permission from Ref.   26.
Copyright 1986 Pergamon Journals, Ltd.

other toxins will be added. Individual members of this group have been characterized to various degrees. The first to be described was the lysin of *Condylactis gigantea* (3) and the most thoroughly studied is heliantholysin from *Stoichactis (Stichodactyla) helianthus*. All are single, basic polypeptide chains having molecular weights between 15,000 and 21,000. They are broadly similar in amino acid composition but exhibit individual differences. Because of convenience and rapidity, capacity to lyse mammalian erythrocytes in vitro is commonly used for assaying activity, but it is clear that these peptides are also toxic for other cells, such as cultured fibroblasts and blood platelets. Some or all are lethal for whole animals such as mice, rats, and crustaceans. Their biological effects, or more accurately those that have been examined, are specifically inhibited by low concentrations of sphingomyelin (Table II).

Heliantholysin. The major form of heliantholysin is a basic polypeptide chain (pI in the region of 9.8) having a molecular weight of 16,600. Its amino acid sequence has been determined (11). It is powerfully hemolytic for washed erythrocytes derived from a variety of animals, those of the cat being the most sensitive, and those of the guinea pig the most resistant (10). As is true of most hemolytic systems, the biochemical basis for the very large differences in sensitivity of erythrocytes from different animal species is unknown.

The capacity of sphingomyelin to inhibit the hemolytic activity of heliantholysin suggests that this lipid may be the receptor for the toxin. This idea is supported by the fact that prior treatment of erythrocytes with sphingomyelinase renders them resistant to lysis. Washed erythrocyte membranes also inhibit the toxin from lysing red cells, and prior treatment of membrane suspensions with enzymes known to destroy sphingomyelin abolishes the inhibition (10). Lysis appears to depend upon the formation of transmembrane channels that are formed by aggregation of toxin molecules in the bilayer (19-20). There are observations suggesting that lysis may also occur by way of a detergent-like action (21).

Kentolysin Compared to Heliantholysin. *Stoichactis helianthus* occurs in the Caribbean region whereas another species, *Stoichactis kenti* is distributed in the Indo-Pacific area. The latter produces a toxin, kentolysin, that is similar to, but not identical with heliantholysin (6). The amino acid compositions of the two polypeptides show a distinct resemblance but appear to differ significantly in number of residues of lysine, methionine, tyrosine and histidine. IgG from a rabbit immunized against heliantholysin neutralizes both heliantholysin and kentolysin, but neutralization of the homologous toxin is more efficient (Table III). It can be seen that in the concentrations used, the IgG failed to neutralize the related lytic peptides of *Condylactis gigantea* and *Epiactis prolifera*.

Multiple Toxins from a Single Anemone Species

Stoichactis (Stichodactyla) helianthus. It has recently been shown by CM-cellulose chromatography that heliantholysin consists of four isotoxins having different N-terminal amino acid sequences (Kem and Dunn, in press). Designated I to IV in order of increasing isoelectric pH, toxins I and II had one additional amino acid at the amino terminus than did toxin III. Toxin IV had a seven residue extension at the amino terminus relative to toxin III. Toxin III and toxin II contributed 83% and 14% of the total hemolytic activity, respectively, and toxins I and IV together about 3%.

Epiactis prolifera. Extracts of this relatively common species contain two toxins, epiactolysins A and B (13). They have the same molecular weight, namely about 19,500, but differ in pI values as determined by isoelectric focusing. Both lack

Table II. Inhibition of Heliantholysin by Sphingomyelin

Inhibitor	Amount required to inhibit 2/3 the test amount[a] of toxin (μg)
Phosphatidylcholine	>500
Phosphatidylethanolamine	>500
Phosphatidylserine	>500
Phosphatidylinositol	>500
Phosphatidylglycerol	>500
Diphosphatidylglycerol	>500
Sphingomyelin (prep. A)	1.5
Sphingomyelin (prep. B)	2.0
Ceramide (prep. A)	70
Ceramide (prep. B)	35
Sphingosine	>500
Cholesterol	>500
Cerebrosides	>500
Partially purified brain gangliosides[b]	>500
Chloroform–methanol-extracted brain lipids	2.5
Human serum	0.008[c]
α-Globulin fraction of human serum	5
β-Lipoprotein fraction of human serum	10
Human serum albumin	>500
Concavanalin A	>500

SOURCE: Reproduced with permission from Ref. 10. Copyright 1976 National Academy of Science.
[a] Three hemolytic units.
[b] Two additional specimens from commercial sources gave the same result.
[c] Measured in μL.

Table III. Neutralizing Capacity of Anti-helianthin IgG for Helianthin, Kentin and Lysins from Other Sea Anemones

Source of cytolysin	Volume of IgG solution required to neutralize 2 hemolytic units of test lysin (mL)	Reciprocal of initial dilution of IgG solution required to neutralize 2 hemolytic units of test lysin
S. helianthus	0.006	167
S. kenti	0.029	34
C. gigantea	>0.1	<10
E. prolifera (epiactin A)	>0.1	<10
E. prolifera (epiactin B)	>0.1	<10

SOURCE: Reproduced with permission from Ref. 6. Copyright 1985 Pergamon Journals, Ltd.

methionine and proline, and the amino acid compositions of the two toxins are generally similar and do not explain the difference in pI values. Epiactin B is the major toxin. It is inactivated by dialysis against distilled water, by freeze-drying, and by exposure to 50° C for 30 minutes, whereas epiactolysin A is stable to these treatments. A third toxin, epiactolysin C, seen in electrofocusing, appears to be a variant form of epiactolysin B—perhaps a dimer or oligomer.

Actinia cari. When an extract of the tentacles of this species was subjected to gel filtration, a single protein peak having hemolytic and lethal activities was obtained. Further fractionation by CM-cellulose ion-exchange chromatography yielded three well separated peaks (CTI, CTII, CTIII) which appeared to be similar in biological activity (*14*).

Metridiolysin

Homogenates of *Metridium senile*, possibly the world's most common large sea anemone, yield extracts that are powerfully hemolytic for washed mammalian erythrocytes (*22*). The active substance, metridiolysin, is a protein of molecular weight approximately 80,000. In contrast to the sphingomyelin-inhibitable toxins, metridiolysin is an acidic protein having a pI of about 5. It is thermolabile and is inactivated by proteolytic enzymes. The optimal pH for hemolysis is between 5 and 6, and at pH 8 the lysin is inactive. It can be dissociated into two subunits of unequal size. Besides being cytolytic in vitro, metridiolysin is lethal when injected intravenously into mice. As shown in Table IV erythrocytes from the horse or dog are about a hundred times as sensitive to lysis as those from the mouse, and erythrocytes from other animals tested are intermediate in sensitivity.

Unlike heliantholysin and congeners, the toxicity of metridiolysin is not prevented by sphingomyelin, but is inhibited by cholesterol in low concentration, as well as by certain related sterols (*23*). In addition, metridiolysin is activated by thiols such as dithiothreitol, and is reversibly inactivated by compounds having an affinity for SH-groups, such as p-hydroxy-mercuribenzoate. A third notable feature is that the action of metridiolysin on membranes involves, or is associated with, the formation of 33 nm rings demonstrable by electron microscopy of negatively stained preparations.

In the three properties mentioned (thiol activation, cholesterol inhibition, and ring formation) metridiolysin resembles the members of a large group of cytolytic toxins produced by streptococci, pneumococci, bacilli and clostridia. These much studied cytolysins are immunologically similar to each other as shown by cross-reactions in neutralization and precipitin tests. In contrast, metridiolysin, despite its biochemical and biological similarity to them, is antigenically distinct from this group of toxins.

Regarding the thiol-activated bacterial lysins, it is generally agreed that the initial step in their action involves combination of the lytic protein with membrane cholesterol. It is also established that the ring structures are each composed of 18 to 20 subunits, and that each subunit is a single toxin molecule. Two different models have been proposed to explain the mechanism whereby erythrocytes, and presumably other kinds of cells, are lysed by thiol-activated lysins. According to one model, sequestration of cholesterol from certain regions of the membrane causes a lipid phase transition to occur, so that phospholipid domains develop with reduced molecular cohesion and altered permeability properties. This in turn is thought to result in disruption and fragmentation of the membrane. In this model the ring structures are incidental polymeric by-products that do not participate in the lytic process itself. An alternative model postulates that the rings themselves demarcate large pores through which water and salts can pass, eventuating in the

Table IV.   Relative Sensitivity to Metridiolysin of
Erythrocytes from Various Animal Species

Animal	Hemolytic Activity of metridiolysin (hemolytic units per mL)
Horse	335
Dog	335
Sheep	33
Guinea pig	15
Ox	12
Human	11
Rat	10
Rabbit	7.5
Goat	6.7
Mouse	3.3

SOURCE: Reproduced with permission from Ref. 22.
   Copyright 1978 Elsevier Science Publishers, B. V.
NOTE: The numbers are the dilutions of a solution of
   metridiolysin which will liberate half the hemoglobin
   contained in a ~ 0.35% (v/v) suspension of
   erythrocytes under standard conditions.

classical Wilbrandt type of osmotic hemolysis. It seems likely that the model that ultimately proves correct for the bacterial lysins will also explain the toxicity of metridiolysin.

Aiptasia Lysin

Venom obtained from acontial nematocysts of *Aiptasia pallida* was found to be lytic for washed rat erythrocytes, and hemolysis required the presence of Ca^{2+} (24). Phospholipase A activity was subsequently demonstrated to be present in venom, and it too required Ca^{2+} (25). DEAE-cellulose fractionation yielded four proteins, two of which were phospholipase A and hemolytic, and two of which had neither phospholipase A nor hemolytic activities. Either of the latter two proteins enhanced to various degrees the hemolytic activity of either of the two phospholipases. The findings suggest considerable analogy with synergistic mechanisms underlying the hemolytic action of the venoms of a number of snakes.

Summary

During the past 20 years much progress has been made in delineating the broad biochemical features of sea anemone cytolysins. Future advances involving amino acid sequencing and determination of tertiary structures should lead to a better understanding of the precise manner in which these toxins interact with membranes.

Acknowledgment

I am greatly indebted to Dr. W. R. Kem for making available prior to publication, his results on heliantholysin.

Literature Cited

1. Bernheimer, A. W.; Rudy, B. *Biochim. Biophys. Acta* **1986**, 864, 123–141.
2. Krebs, H. C.; Habermehl, G. G. *Naturwissenschaften* **1987**, 74, 395–396.
3. Shapiro, B. L. *Toxicon* **1968**, 5, 253–259.
4. Bernheimer, A. W.; Avigad, L. S.; Lai, C. Y. *Archs. Bichem. Biophys.* **1982**, 214, 840–845.
5. Norton, T. R.; Kashiwagi, M. *J. Pharm. Sci.* **1972**, 61 1814–1817.
6. Bernheimer, A. W.; Lai, C. Y. *Toxicon* **1985**, 23, 791–799.
7. Ferlan, I.; Lebez, D. *Toxicon* **1974**, 12, 57–61.
8. Ferlan, I.; Lebez, D. *Bull. Inst. Pasteur* **1976**, 74, 121–124.
9. Devlin, J. P. *J. Pharm. Sci.* **1974**, 63, 1478–1480.
10. Bernheimer, A. W.; Avigad, L. S. *Proc. Nat. Acad. Sci. U.S.A.* **1976**, 73, 467–471.
11. Blumenthal, K. M.; Kem, W. R. *J. Biol. Chem.* **1983**, 258, 5574–5581.
12. Bernheimer, A. W.; Avigad, L. S. *Toxicon* **1981**, 19, 529–534.
13. Bernheimer, A. W.; Avigad, L. S. *Archs. Biochem. Biophys.* **1982**, 217, 174–180.
14. Macek, P.; Sencic, L.; Lebez, D. *Toxicon* **1982**, 20, 181–185.
15. Mebs, D.; Liebrich, M.; Reul, A.; Samejima, Y. *Toxicon* **1983**, 21, 257–264.
16. Bernheimer, A. W.; Avigad, L. S.; Branch, G.; Dowdle, E.; Lai, C. Y. *Toxicon* **1984**, 22, 183–191.
17. Shioma, S.; Tanaka, E.; Yamanaka, H.; Kikuchi, T. *Toxicon* **1985**, 23, 865–874.
18. Elliot, R. C.; Konya, R. S.; Vickneshwara, K. *Toxicon* **1986**, 24, 117–122.
19. Michaels, D. W. *Biochim. Biophys. Acta* **1979**, 555, 67–78.
20. Shin, M. L.; Michaels, D. W.; Mayer, M. M. *Biochim. Biophys. Acta* **1979**, 555, 79–88.
21. Varanda, W.; Finkelstein, A. *J. Membrane Biol.* **1980**, 55, 203–211.

22. Bernheimer, A. W.; Avigad, L. S. *Biochim. Biophys. Acta* **1978**, *541*, 96–106.
23. Bernheimer, A. W.; Avigad, L. S.; Kim, K. S. *Toxicon* **1979**, 17 69–75.
24. Hessinger, D. A.; Lenhoff, H. M. *Archs. Biochem. Biophys.* **1973**, *159*, 629–638.
25. Hessinger, D. A.; Lenhoff, H. M. *Archs. Biochem. Biophys.* **1976**, *173*, 603–613.
26. Bernheimer, A. W. *Toxicon* **1986**, *24*, 1031–1032.

RECEIVED May 25, 1989

Chapter 24

Toxins from Marine Invertebrates

M. J. A. Walker and V. L. Masuda

Department of Pharmacology and Therapeutics, Faculty of Medicine,
The University of British Columbia, 2176 Health Sciences Mall, Vancouver,
British Columbia V6T 1W5, Canada

It is increasingly realized that many toxins are very site-specific in their actions and hence are of value as biological tools. As a result, there is an increasing exploration of toxins from marine sources, especially those from invertebrates. Plant toxins were of paramount importance in the elucidation of the mechanisms by which tissues communicate (e.g., neurotransmitters and autacoids). More recently, animal toxins (e.g., tetrodotoxin and alpha bungarotoxin) have been of equal value in analyzing communication across the cell membrane, i.e., through receptors and ion channels. Against such a background, the objectives of this chapter include: 1) Outlining the extent of knowledge concerning marine invertebrate toxins; 2) Ensuring that most marine invertebrate toxins receive some consideration; 3) Generalizing upon molecular mechanisms of actions; 4) Outlining search strategies for identifying new toxins and their possible mechanisms of action.

As a result, this chapter will review, although not comprehensively, marine invertebrate toxins not considered in detail elsewhere. In addition, it will speculate on relationships between toxins from the phylla which make up the marine invertebrates. The relationships to be considered are those concerned with the use toxins are put to in defense, offense, and digestion. Their use as defensive or offensive weapons cannot be discussed without considering the target at which they are directed since, to be effective, a toxin has to utilize the underlying physiology of the target organism. For example, a toxin which specifically blocks sodium channels would be expected to be effective only against a species in which such channels are vital. Thus, it is useful to consider the nature and actions of toxins within the context of the physiology of target species.

Over the last few years, there has been widespread recognition that toxins contain very site-specific molecules and as a result there has been a surge of interest in using them for analyzing cellular mechanisms. However, the upsurge in screening toxic animals for the presence of site-specific toxins has not been correspondingly large. Thus this chapter will specifically consider the need to screen biologically all marine invertebrates for toxins. Appropriate consideration of the pharmacological basis of various screens, and of the relationships between a toxin and its biological function, should make it possible to identify putatively the cellular mechanism which a toxin targets.

In other chapters of this volume considerable attention is given to marine toxins whose cellular sites of action have been identified. For example, saxitoxin, brevetoxin, and sea anemone toxins are prototypes of toxic molecules whose chemical structure is known, and whose actions on ionic channels in the cell membrane have been elucidated. Recent additions to such toxins are the piscivorus cone

0097–6156/90/0418–0312$06.25/0
© 1990 American Chemical Society

toxins; these startlingly exemplify the principle that toxins which are used as offensive weapons can be targeted to fundamental physiological mechanisms responsible for locomotion in the prey species.

In contradistinction to site-specific molecules there are other toxins which, in a sense, create their own specificity. Thus ionophore toxins, by opening up the cell membrane to sodium and calcium ions, create their own specificity. Prime examples of this are found in cytolytic toxins in general, and jellyfish toxins and palytoxin in particular.

Before considering marine invertebrate toxins in detail, it is useful to define some of the terms to be used. Many of these terms are teleological, nevertheless they have a usefulness. A **toxin** is any single (or multiple) chemical entity produced in one species capable of producing pathological changes in a second species. No clear distinction can be made between toxins and venoms although the latter are often considered to be actively injected by the secreting species. Toxins that have evolved function in **offense, defense,** and **digestion.** Bakus et al. (*1*) have given an overview of marine chemical ecology in which the emphasis was placed on the value toxins have in preventing predation, ensuring a clear and free environment, giving species dominance, etc.

While a particular toxin may only fulfill a single role of offense, defense, or digestion, there is no a priori reason against a toxin having multiple roles. Two examples illustrate this point, the previously mentioned conotoxins have a main role of offense. They rapidly immobilize a victim so that it may be engulfed and digested but appear to have no direct role in digestion; any defensive role is probably fortuitous. Olivera et al. (*2*) have commented in detail upon this topic. On the other hand, jellyfish toxins play all three roles with equal facility. Thus, jellyfish (nematocyst) toxins are highly effective in subduing prey species and ensuring that prey become entangled in tentacles for subsequent transportation into the oral cavity. Such cytotoxic (and cytolytic) nematocyst toxins also begin the process of digestion and, in addition, deter other animals from preying upon jellyfish.

At least two types of defense can be recognized: **active** defense where toxin is actively injected into the attacking or intruding organism, and **passive** defense, which includes environmental defense, whereby an organism dissuades competing organisms from living in its territory. Antibiotic-type toxins are probably examples of passive defense.

Targets for toxins can be considered to exist at various levels. Toxins may evolve which subdue prey by either blocking the systems responsible for locomotion, circulation, or for central coordination, in the potential victim. In order to disable these systems, advantage is often taken of the fact that their physiology depends upon specific transmembrane channels such as those for sodium, potassium, and calcium ions.

The **chemical nature** of marine toxins is extremely varied. Molecules such as saxitoxin and palytoxin are varyingly complex organic molecules while sea anemone toxins are polypeptides. One may speculate as to why there is such variety. There are advantages associated with polypeptide and protein toxins. For example, their structure readily responds to genetic reshuffling with resulting improved specificity; in order to achieve the same sort of change with non-peptide organic molecules, a whole panoply of synthesizing enzymes has to be changed in order to change the parent molecule. However, polypeptides and proteins are readily broken down (as in the gut) and are absorbed with difficulty. Small organic molecules are readily absorbed and passed easily from one species to another, as is the case with the paralytic shellfish toxins.

Toxic Marine Invertebrates

In the animal kingdom, many toxic species are to be found among marine invertebrates. The marine invertebrate kingdom consists of the following phyla, Protozoa, Porifera, Ectoprocta, Coelenterata, Echinodermata, Platyhelminthes, Rhynchocoela, Annelida, Sipunculida, Mollusca, and Anthropoda. These phylla range from the simplest organisms to animals such as the cephalopods which have complex circulation and nervous systems. Each phyllum consists of a number of classes and extensive orders and families. For example, the class Crustacea, in the phyllum Anthropoda (25,000 species), contains over 18 orders. The phyllum Mollusca has over 10,000 species divided into 7 classes. By their very nature, most of these invertebrates are potential victims of the faster, more mobile, and more intelligent, vertebrates. In partial compensation, the organisms that have prevailed through evolution are in many cases equipped with an impressive array of toxins for defense and offense.

Some idea of the spectrum of toxins to be found in marine invertebrates is shown in Table I while the review of Halstead and Vinci (3) gives a comprehensive overview of which phylla and species are toxic. Even within a single order, a host of toxins differing in their actions and structure may be found. Prominent toxin-containing phylla include Coelenterata, Echinodermata, Annelida, and Mollusca and within each phylla, all classes and orders contain species which are toxic to vertebrates or invertebrates. However, no simple rule allows us to predict which are likely to be toxic. An additional complication is that some species seem to acquire toxicity from their environment and thus contain exogenous toxins. A classical example are the molluscs which accumulate toxins, such as the saxitoxins, from dinoflagellates living in their environment. Although this example is perhaps the most well-known, it is tempting to speculate as to how common is such acquired toxicity.

The following section outlines the range of toxins found in the various phylla that constitute the marine invertebrates. The toxins and toxicity of some of the these phylla have been reviewed recently (4-7).

Protozoa and Porifera. The Protista, which lie somewhere between plants and animals, include protozoans, algae, bacteria, yeasts, and fungi. Some 80 marine species have been found to be toxic to humans and other animals; most are dinoflagellates, free-living motile single cell organisms. Dinoflagellates are a source of toxins that accumulate in many fish and shellfish (see Chapters 3, 8, and 11 in this volume). Aside from dinoflagellates, toxic protists include the marine alga, *Spirulina subsala*, which kills shrimps. Other blue green marine algae are capable of producing a contact dermatitis due, in some cases, to production of the inflammatory toxin, debromoaplysiatoxin (8). However, most of the toxic blue green algae are found in fresh water. A full list of protista found to contain or release toxin or to be responsible in someway for deaths of marine species are to be found in Table I in Russell's monograph (9).

The sponges (Porifera) are simple colonial animals (over 8,000 species) with a silica calcinate skeleton. The phyllum Porifera consists of the subphylla Gelatinosa and Nuda. The Gelatinosa has three classes and over 10 orders. Table III in Russell's monograph (9) lists all of the known toxic sponges which are most common in the family Haliconidae. Some sponges kill fish when placed in close proximity. More than 70% of coral sponges appear to be toxic in this manner. Only 30% of cryptic sponges are similarly toxic (10-12) suggesting that the extrusion of toxin is part of a sponge's defense system. A few teleost fish can safely eat such toxic sponges and some nudibranchs even acquire toxicity by feeding upon them (13,14). LD_{50} values can be determined for alcoholic extracts of sponges by

Table I. Some Marine Toxins

Toxin	Source	M.W.	LD_{50}[a] ug/kg	Amino Acid Composition
Saxitoxin	Dinoflagellate	309	3.0	
Brevetoxin	Dinoflagellate	900	95	
Maitotoxin	Dinoflagellate	3,300	0.13	
Ciguatoxin	Dinoflagellate	1,111	0.45	
Okadaic acid	Coelenterate	786	192	
Palytoxin	Coelenterate	3,300	0.15	
Conotoxins				
α	Conidae (Mollusca)			13-15
μ	Conidae (Mollusca)			22
ω	Conidae (Mollusca)			25-29
K	Conidae (Mollusca)			25

[a]LD_{50} estimates were obtained in mice after ip injection.

observing the behavior of fish in an aquarium with the end points being behavioral changes and ultimately death. Bakus and Thun (*10*) found that of 54 Caribbean sponges examined in this manner, 31 were toxic.

Ectoprocta and Bryozoa. These phylla contain species which exist as sessile tufted organisms or branched cell colonies. They are often found attached to bethonic animals, kelp, shells, or rocks. Clinically, the most important is curty-weed or sea chervil (*A. gelatinosum* or *A. hirsutum*) which are responsible for the Dogger Bank syndrome found in North Sea fishermen. This allergic condition involves sensitization to the hapten 2-OH ethyl dimethylsulphoxonium.

Coelenterates and Echinoderms. In the phylla Coelenterata and Echinodermata approximately 90 species have been investigated for toxicity (*see* Tables II and III). Only 20 or so have been extensively studied (e.g., sea anemones, sea cucumber, and jellyfish). Even so, while relatively complete studies have been made on isolation, characterization, and elucidation of mechanisms of action, in no one species have all of the toxins present been identified. Thousands of species have not been subjected to even the most cursory examination.

The coelenterates, the first organisms to have signs of a mesoderm, divide into two sub-phylla, Medusozoa and Anthozoa. The former has the classes Hydrozoa and Scyphozoa (true jellyfish). The latter divides into Alcyonaria (Octocorallia), Zoantharia (Hexacorallia) and Ceriantipatharia. Schyphozoan toxins are mainly contained within stinging organelles (nematocysts) from which they are expressed via a hollow stinging thread. Typical coelenterate toxins are considered in this volume by Frelin and Burnett (*see* Chapters 13 and 25 in this volume). About 70 of the 10,000 coelenterates are either venomous or toxic to humans (*15*).

The sub-phyllum Anthozoa (6,500 species) contains the classes Alcyonaria (soft corals, sea fans, sea pens, sea pansies), Zoantharia (sea anemones and true corals), and Ceriantipatharia. The Scleroactinia (Madreporaria-true or stony corals) build the massive coral reefs and atolls which occur in tropical waters. According to Hashimoto (*15*), toxicity to humans is mainly found in the fire or stinging corals (*Millepora* sp.) and, to a lesser extent, in the stony corals (*Goniopora* sp.).

The echinoderms (5,900 species) are spiny skinned animals generally found on the sea floor. They have a 5-rayed, symmetrical body varying from 5 mm to 1.0 m in size; some species are covered with stalk-like appendages (pedicellariae) which have pincers at their tips. These pedicellariae serve cleaning, offensive, and defensive roles. The sub-phylum Crinozoa (sea-lillies, feather stars) contains about 600 species while the Asterozoa (3,600 species) is composed of sub-classes, the Asteroides (starfish) and Ophiuroidea (brittlestars). The sub-phylum Echinozoa contains two classes, the Echinoidea (sea urchins, heart urchins, and sand dollars) and Holothurioidea (sea cucumbers). In sea urchins, toxins are found on spine-tips and in pedicellariae. Spine-tips of some urchins (e.g., *Diadema* sp.) are covered by a sac lined with toxin-containing cells. Hashimoto (*15*) recorded that spines of sea urchins such as *Diademata*, *Phormosoma*, *Acanthaster*, and *Echinothrix* are capable of producing injuries in humans while the pedicellariae of Temnopleuridae and Toxopneustidae can cause envenomation.

Sea cucumbers (Holothurioidea) contain highly toxic saponins which presumably play a role in protection since the relatively unprotected bodies of these species make them particularly vulnerable to predation. When "attacked", a sea cucumber expels Cuverian glands whose projections elongate and enmesh an assailant while, at the same time, exuding ichthyotoxic saponins to discourage further attacks. Some 30 species of sea cucumbers contain saponins which are highly toxic to other marine creatures as well as fish.

Table II. Partial List of Coelenterates from which Toxic Substances Have Been Isolated

Subphylum	Class	Order	Family	Species
		hydroids		
Medusozoa	Hydrozoa	Siphonophora	Physallidae	*Physalia physalis*
		fire coral		
Medusozoa	Hydrozoa	Athecata	Milleporidae	*Millepora* sp.
		jellyfish		
Medusozoa	Scyphozoa	Semaostomeae	Pelagidae	*Pelagia noctiluca*
Medusozoa	Scyphozoa	Semaostomeae	Pelagidae	*Chrysaora quinquecirrha*
Medusozoa	Scyphozoa	Semaostomeae	Cyanidae	*Cyanea* sp.
Medusozoa	Cubozoa	Cubomedusae	Carybdeidae	*Carybdea rastoni*
Medusozoa	Cubozoa	Cubomedusae	Chirodropidae	(*Chironex fleckeri*) *Chiropsalmus quadrigatus*)
		sea anemones		
Anthozoa	Zoanthraria	Actiniaria	Actinidae	*Actinia equina* (*Anemonia sulcata, Anthopleura xanthogrammica, Condylactis* sp., *Tealia felina*)
Anthozoa	Zoanthraria	Actiniaria	Stoichactiidae	*Stoichactis* sp. (*Radianthus kosierensis, Gyrostoma helianthus*)
Anthozoa	Zoanthraria	Actiniaria	Metridiidae	*Metridium* sp.
Anthozoa	Zoanthraria	Actiniaria	Aiptasiidae	*Aiptasia* sp.
Anthozoa	Zoanthraria	Actiniaria	Aiptasiidae	*Parasicyonis* sp.
Anthozoa	Zoanthraria	Zonanthiniaria	Zoanthidae	*Palythoa* sp.
		soft corals		
Anthozoa	Alcyonaria	Alcyonacea	Alcyonidae	*Sinularia* sp.
Anthozoa	Alcyonaria	Alcyonacea	Alcyonidae	*Sarcophyton glaucum*
		sea whips		
Anthozoa	Alcyonaria	Gorgonacea	Gorgoniidae	*Lophogorgia* sp.
		stone or true corals		
Anthozoa	Zoanthraria	Scleractinia	Acroporidae	*Acvropora palmata* (Astrocoeniina)
Anthozoa	Zoanthraria	Scleractinia	Poritidae	*Goniopora* sp. (Fungiina)

SOURCE: Reproduced with permission from Ref. 5. Copyright 1988 Marcel Dekker.

Table III. Partial List of Echinoderms from which Toxic Substances Have Been Isolated

Subphylum	Class	Order	Family	Species
		starfish		
Asterozoa	Stelleroides	Forcipulatida	Asteriidae	*Asterias forbesi* (Asteroidea)
		sea urchins		
Echinozoa	Echinoidea	Diadematacea	Diadematidae	*Diadema antillarum* (Euechinoidea)
Echinozoa	Echinoidea	Echinaces	Toxopneustidae	*(Toxopneustes pileolus Tripneustes gratilla)*
		sea cucumbers		
Echinozoa	Holothurioidea	Aspidochirotida	Holothurioidae	*Holothuria* sp (Aspidochirotacea)

Marine Worms. (**Platyhelminthes, Rynchocoela, Annelida, Sipunaelida.**) A variety of species from worm phylla have been found to contain toxins. There are approximately 56,000 species of worms (14,000 annelids, 25,000 platyhelminthes, 15,000 nematodes, and 800 nemertines), and of these, most of the toxic species are found in the nemertines. The most well-known toxin is nereisotoxin which has been modified to form a very useful insecticide.

The annelids include the bristle worms and blood worms in which toxicity is associated with bristle-like setae and/or biting jaws. In the order Polychaetae, toxicity is usually found in three genera (Chloeia, Eurythoe, Hemodice). The platyhelminthes are not associated with many cases of human toxicity. The only class of platyhelminthes in which toxicity can readily be found is in the Turbellaria. In the Rhynchocaela (ribbon worms), toxic species include *Lineus* sp. Some platyhelminthes (e.g., *Planocera multitenta*) have been found to contain tetrodotoxin (*16*).

Mollusca and Athropoda. Of the 80,000 species in the 5 classes of molluscs, only 85 are known to be toxic to humans. The phyllum Mollusca contains two notable classes with toxic species; these are Gastropoda and Cephalopoda. Their relevant toxicology and zoology have been reviewed by Fange (*17*). In the class Gastropoda, there are a variety of orders and families in which toxic genera utilize toxic substances to subdue their prey. As originally discovered in the 1860's, many tonnacean gastropods have the ability to secrete free sulfuric acid from their salivary glands (*18,19*). In addition to secreting acid there is evidence of the presence of neurotoxins in their secretions. There has been a report of tetrodotoxin being present (*20*) as well as other neurotoxins (*21*).

In the order Muricidae these carnivorous snails attack molluscs or sessile crustaceans by boring holes in their shells using softening secretions. Hypobranchial and other glands contain small molecular weight substances with pharmacological activity (e.g., murexine, 5HT, etc.).

The order Mesogastropoda contains the family Conidae (400 species) whose venom apparatus consists of a barbed hollow tooth through which venom is expressed from a venom duct using a bulb as the source of pressure (*see* Chapter 20 in this volume). The cone shells, as discussed previously, are divided on the basis of their prey species be it fish, worms, or other shellfish.

In the order Buccinidae (whelks) the salivary glands contain large amounts of tetramine and other small molecular weight substances such as choline esters. However, there is little evidence for their use as toxins according to zoological observations (*22*).

In the order Opisthobranchia (sea slugs, sea horse) the nudibranchs (3,000 species) are shell-less and appear in desperate need of protection. They achieve some protection by excreting from their skins various toxins. These are sometimes obtained by grazing on toxic sponges, for example, as occurs with *Phyllidia varicosa* which grazes on *Hymeniacidon* sp. However, toxins have also been isolated from the digestive glands of nudibranchs (*23*). Among the 33,000 other gastropods, toxic species are found among *Aplysia, Haliotis, Murex, Thais,* and *Neptunae.* The ingestion of *Haliotis* sp. (abalone) has been associated with a photo-sensitizing dermatitis (*15*).

The cephalopods often secrete venom from their posterior salivary glands and hence secretion of venom probably forms part of the process of digestion. Cephalopods include species, such as octopi, capable of injecting various venoms including simple amines such as tetramine. On the other hand, maculotoxin isolated from the octopus *H. maculosa* was eventually determined to be tetrodotoxin (*24*).

The multiple-jointed animals in the phyllum arthropoda do not usually use toxins as offensive weapons. However, a variety of crabs have occasionally been shown to be the cause of food poisoning (*25,26*).

Chemical Nature of Toxins

In this section the range of chemical structure found in marine invertebrate toxins is considered. The different phyla are considered as in the previous section.

Protista and Porifera. A wide variety of substances which are toxic or potentially toxic have been isolated from Porifera. Hashimoto (15) lists over 60 individual or classes of compounds many of which (e.g, amino acids, nucleosides, etc.) are found in most biological tissue. A variety of antibiotic substances such as terpenoids, furanoterpenes, brominated phenols, and pyrroles have all been isolated from sponges (15). Polyethers are increasingly found in various species. Many are similar to ionophoric polyethers and crown ethers. This type of compound has the ability to form membrane spanning rings through which ions can shuttle backwards and forwards across the cell membrane. An ionophoric polyether monocarboxylic acid, okadaic acid, has been isolated from *Halichondria okadai* (27-29). Bryostatin is a macrolide ester discovered in bryozoans (30). Small molecular weight amines and amino acid derivatives, such as N-acyl-2-methylene beta alanine from *Fasciospongia cavernosa* (31), have been obtained from porifera. The sponge *Hymeniacidon* sp. produces a sesquiterpene (9-isocyanopupukaenane) and this compound is concentrated in nudibranchs which graze upon them (28).

A saponin-like substance, agelasine, was isolated from *Agelas dispar* (10). Suberitine from *Suberites domcunculus* was originally isolated by Richet (32) and was shown to be heat labile and have a MW of 28,000 by Cariello et al. (33). A cyclic peptide, dolastin, was isolated from *Dolabella auricularia* (34).

Coelenterates and Echinoderms. Coelenterate and echinoderm toxins range from small molecular weight amines, to sterols, to large complex carbohydrate chains, to proteins of over 100,000 daltons. Molecular size sometimes reflects taxonomy, e.g., sea anemones (Actiniaria) all possess toxic polypeptides varying in size from 3,000 to 10,000 daltons while jellyfish contain toxic proteins (ca. 100,000 daltons). Carotenoids have been isolated from *Asterias* species (starfish), Echinoidea (sea urchins), and Anthozoans such as Actiniaria (sea anemones) and the corals. These are sometimes complexed with sterols (35).

According to Tursch et al. (36) small molecular weight terpenoids have only been isolated from the subclass Alcyonaria (soft corals, gorgonians) of coelenterates. Examples include sarcophine from *Sarcophyton glaucum* (37) and stylatulide from *Stylatula* species (38). Some of these terpenoids are toxic to mice and fish (e.g., sarcophine) but others are not. Antineoplastic and antifungal activity have been reported for coelenterate terpenoids. Lophotoxin from sea whips (*Lophogorgia* sp.) is a neuromuscular blocker that belongs to the cembrene class of diterpenoids (39).

Goad (40) and others have extensively reviewed coelenterate and echinoderm sterols including the saponins found in starfish and sea cucumbers. Cholesterol is a common sterol in most families, except for gorgonians and zoanthids; some soft corals contain polyhydroxylated sterols. The amount of variation associated with phylogeny is illustrated in the echinoderms by the fact that crinoids, ophuiroids, and echinoids contain Δ 5 sterols while holothuriodeans and asteroids contain Δ 7 sterols. Some classes contain uniquely structured sterols.

Saponins are complexes of sugars and steroids or triterpenoids (as aglycones) which occur widely in plants but are rarer in animals. The saponins from sea cucumbers are triterpenoids, whereas those from starfish are steroidal saponins (41). A variety of aglycones have been isolated from Holothurioidea (42,43), which vary little in their structure. The ubiquitous nature of the distribution of the aglycones and holothurins in Holothurioidae and Actinopyga is exemplified in Tables IV-XII of Hashimoto's review (15). The sugar moieties, D-xylose, D-glucose, D-quinovose, and 3-O-methyl-D-glucose attached to the aglycone, confer solubility.

Of the five classes of echinoderms, both the Asteroidea (starfish) and Holothurioidea (sea cucumbers) invariably contain saponins, with their content varying with species. The active compounds from the Cuverian glands are given the general name of holuthurins. Holothurinogen was recognized to be the prototypical aglycone by Russell (*44*), while Roller et al. (*45*) gave a rigorous proof of the structure of holothurinogens. A variety of similar substances has been isolated from many sea cucumbers [e.g., Holotoxin A and B from *Stichopus* sp. (*46*)]. The asterosaponins which can be isolated from starfish were first recognized by Hashimoto and Yasumoto (*47*) on the basis of extracts which repelled molluscs and fishes.

Large molecular weight toxins have been isolated from sea anemones. Equinatoxin, a lethal cytolytic toxin from *A. equina* is one such toxin (*48,49*). Molecular weight estimates for parasitoxin, a lethal hemolysin, were 19,000 by polyacrylamide gel electrophoresis and 17,000 by sedimentation (*49*). Three lethal and hemolytic toxins from *Actinia cari* (caritoxin I, II and III) are basic proteins of 20,000-35,000 daltons (*51*). Similar toxins are found in related species (*52*). A basic toxic protein of 18,000 daltons has been isolated from *Stoichactis kenti* (*53*) while *S. helianthus* has a 17,000 dalton basic protein cytolysin, and two 5,000 molecular weight neurotoxins of unusual structure (*54*). Haemolysin from *S. helianthus* (cytolysin III) has 153 amino acid residues and both alpha helix and beta chains (*55*). Variolysin from the anemone *Pseudactinia varia* has a molecular weight of 19,500 (*56*). Mebs et al. (*57*) have isolated ichthyotoxic hemolysins of approximately 10,000 daltons (48 amino acids) from *Gyrostoma helianthus*, *Radianthus koseirensis*, and *Rhodactis rhodostoma*. A 31,800 toxic protein occurs in the nematocysts of *Aiptasia* and *Pachycerianthus* sp. (*58*). Aldeen et al. (*59*) described 12,000-14,000 dalton extracts from *Tealia felina* that had hemolytic actions. Further purification yielded a protein of 7,800 daltons (*60*).

Both sea anemones (order Actinaria) and stony corals (order Scleractinia) belong to the class Zoantharia and sub-phylum Anthozoa and therefore it is not surprising that toxic polypeptides of 12,000 and 30,000 daltons have also been isolated from Goniopora corals.

Mollusca and Arthropoda. The basic polypeptides utilized by Conidae are discussed elsewhere by Olivera et al. (*2*, Chapter 20 in this volume). Tetramine and tetramine-like quaternary ammonium compounds have been found in Tonnacean gastropods; in the Muricidae, such compounds include small molecular weight substances such as murexine and dihydromurexine, which are similar to other pharmacologically active cholines such as senecioylcholine, acrylylcholine, and seneciloylcholine. These substances can be extracted from various glands (salivary, proboscis, hypobranchial) of genera such as Murex, Thais, and Neptunea (*61*).

Toxicological and Pharmacological Actions

Protozoa and Porifera. The pharmacology and toxicology of the dinoflagellate toxins which act upon the voltage- and time-dependent sodium channel found in nerves of vertebrates and invertebrates, and the skeletal muscle of vertebrates, are discussed in other chapters in this volume.

Suberitine, a small protein from the sponge *Suberites domcuncula*, has a variety of actions. It is not very toxic but causes hemolysis in human erythrocytes, flaccid paralysis in crabs and depolarization of squid axon and abdominal nerve of crayfish. A variety of extracts from Porifera have been shown to be toxic to fish and generally have cytotoxic and hemolytic actions (*62,63*). As discussed previously, a variety of sponges exude substances that are toxic to fish.

Cytotoxic and hemolytic activity has been obtained from the sponge *Pachymatisma johnstonii* (63). Debromoaplysiatoxin, which causes contact dermatitis and a pustular folliculitis in humans and severe inflammation in rabbits, is one of the most potent skin irritants known (64).

Coelenterates and Echinoderms. Toxic proteins have also been found in echinoderms. Alender et al. (65) found a 67,000 dalton basic protein in pedicellarean toxin from *Tripneustes*. Fleming and Howden (66) suggested that toxin from *Tripneustes* was an acidic protein. The partial purfication of a toxic protein from *Toxopneustes pileolus* has been described by Nakagawa and Kimura (67) while Mebs (68) isolated one of 25,000 daltons from *T. gratilla*. Three toxic proteins (hemoagglutinins and hemolysins) have been isolated from coelemic fluid obtained from *Holothuria polii* (69).

Venom from the globiferous pedicellariae of sea urchins is lethal to mice, rabbits, crabs, lobsters, and worms (70). Seasonal changes in toxicity of such toxins (71) have been observed. The LD_{50} estimate (mice) for toxic fractions from the urchin *Tripneustes gratilla* ranged from 0.05-0.5 mg/kg (70).

The pharmacology of echinoderm toxins centers on the pharmacological actions of saponins from sea cucumbers and starfish. Such pharmacological actions have been reviewed many times. In general, many of the toxicological actions of saponins are consistent with their ability to lyse cells via a detergent action. For example, Ruggieri and Nigrelli (72) discussed the variety of actions such saponins produce, including those on growth and differentiation (especially on sea urchin eggs). Holothurins also immobilize sperm, arrest growth, and lyse a variety of tissues. Similar actions may account for the antifungal and antiparasitic actions of holothurins. Holothurins can cause contraction of skeletal muscle and irreversibly block neuromuscular transmission, as well as disturb electrogenesis in cardiac tissue. Asterosaponins isolated from starfish share such actions (73). The toxicological actions of the various saponins isolated from starfish and sea anemones have been extensively discussed by various authors. The major findings are that saponins have toxic actions on mammals, fish, and invertebrates. The hemolytic actions of the saponins can contribute to the lethal and toxicological actions of these substances if substantial intravascular hemolysis occurs. In addition to saponins, the sea cucumbers and starfish contain other pharmacologically active substances. Both clotting and irreversible smooth muscle contracting activities can be isolated from the coelomocytes of *Asterias forbesi* (74).

Sea urchin toxins extracted from spines or pedicellariae have a variety of pharmacological actions, including electrophysiological ones (75). Dialyzable toxins from *Diadema* caused a dose-dependent increase in the miniature end-plate potential frequency of frog sartorius muscle without influencing membrane potential (76). A toxin from the sea urchin *Toxopneustes pileolus* causes a dose-dependent release of histamine (67). Toxic proteins from the same species also cause smooth muscle contracture in guinea pig ileum and uterus, and are cardiotoxic (77).

Lophotoxin, a triterpenoid toxin isolated from sea whips (*Lophogorgia* sp.) appears to have specific actions on the neuromuscular junction (78). It abolishes both miniature and end-plate potentials in rat diaphragm and frog cutaneous pectoris muscles, as well as the depolarization induced by direct application of acetylcholine (79). Lophotoxin is a nicotinic receptor blocker which possibly acts at a site remote from the acetylcholine recognition site (80). However, its binding to nicotinic receptors can be prevented by both nicotinic agonists and antagonists (81).

Marine Worms (Platyhelminthes, Rynchocoela, Annelida, and Sipunaelida).
Nereistoxin ($C_5H_{11}NS_2$) kills fish and insects but is relatively non-toxic to homeotherms. The LD_{50} value in mice varies from 30-1000 mg/kg depending on

the route of injection. However, the toxicity to insects approaches that seen with DDT and pyrethrin (*15*). Large groups of proteins from the genus *Glycera* are proteolytic and cause paralysis of insect heart. Similarly, proteolytic enzymes which cause sustained contraction of whelk muscle have been obtained from Annelids (*82*).

Mollusca and Arthropoda. A variety of pharmacological actions are induced by the toxins found in molluscs (*17*). For example, surugatoxin is a potent mydriatic (*83*), ganglion blocker (*84*), and a potent hypotensive agent in cats.

Aplysin has marked effects in mammalian tissue and produces hypotension, bradycardia, neuromuscular paralysis, and contracture of intestinal smooth muscle (*85*). Aplysiatoxin causes an elevation in blood pressure resistant to adrenoceptor blockade (*86*).

Murexine and related compounds have marked actions on the nicotine receptor as expected from choline esters (*87-89*). Toxins from the digestive glands of nudibranchs have marked effects on the cardiovascular system of the rat (*23*). Antiviral and antibacterial substances have been obtained from molluscs (*90,91*).

The pharmacology of the amines such as octopamine, 5HT, and noradrenaline isolated from cephalopods has been well documented and the mechanisms of action are well understood. The more complex toxins such as cephalotoxin have more complex actions in that they are often species specific (*15*). Eledoisin has a multiplicity of actions. It is a hypotensive secretagogue capable of affecting vascular smooth muscle. Erspamer (*92*) investigated the actions of eledoisin and found that it caused vasodilation but stimulated smooth muscle and salivary glands. Kem and Scott (*93*) found a similarly acting protein in the squid (*Loligo peali*). The octopus, *Octopus vulagaris*, produces two paralyzing proteins from its posterior salivary glands (*94*).

Sites of Action

In order to ensure that a toxin used for offense incapacitates a prey species so as to render it available for digestion, it is essential that the toxin "attacks" mechanisms that are critical for locomotion in the prey species. For vertebrate prey such critical mechanisms can be readily identified. The activation of skeletal muscle locomotion depends critically upon axonal transmission, neurotransmitter release, neurotransmitter reception at the end-plate, excitation-contraction coupling, and finally contraction. Contraction mechanisms are hidden beneath the cell membrane and therefore are not such a suitable site for toxin action. However, okadaic acid may be an exception to this in that it appears to have direct actions on the contraction mechanism (*95*). At each of the other sites noted above, there are a number of fundamental molecular sites at which a toxin might act. For axons these are the voltage- and time-dependent sodium and potassium channels. At nerve endings there is, in addition, a calcium channel whose activation ensures transmitter release. The action of the transmitter, acetylcholine, at the nicotinic skeletal muscle receptor is an obvious site of toxin attack, a site notably attacked by the snake toxin, alpha bungarotoxin. Excitation-contraction coupling depends critically upon the maintenance of a normal resting membrane potential and thus on a variety of ionic channels and membrane pump systems.

Invertebrate prey species contain analogous, but not identical, sites to those considered above. In many phylla, calcium channels play the role normally ascribed to sodium channels in vertebrates. In addition, the peripheral locomotor neurotransmitter is not acetylcholine but amino acids such as gamma amino butyric acid (GABA). In other phylla, the channels which underly locomotion remain poorly understood.

In all higher species, locomotion is controlled by a central nervous system and, therefore, it might be argued that this system would provide an "ideal" target for toxins. However, when the nervous system is centrally located there is often in-built protection from blood-borne toxins and this "blood-brain" barrier offers protection, especially against large molecular weight toxins.

With respect to toxins which target specific sites, insight can be obtained from the anomolies that are observed. For example, both puffer fish and tetrodotoxin-containing crabs (96) are insensitive to tetrodotoxin. The investigation of such insensitivities can provide information about membrane channels and their toxin binding sites.

In the light of such considerations, it is possible to discuss toxins which have already been analyzed in terms of their sites of action. Such a discussion is best conducted by categorizing the various possible cellular sites at which a toxin might act. The most obvious sites are the membrane channels for ions, receptors for neurotransmitters, membrane pumps, and the membrane itself. Invertebrate toxins acting on membrane channels include the conotoxins (10) and several of the sea anemone toxins (97).

Another possible target for toxins are the receptors for neurotransmitters since such receptors are vital, especially for locomotion. In vertebrates the most strategic receptor is that for acetylcholine, the nicotinic receptor. In view of the breadth of action of the various conotoxins it is perhaps not surprising that alpha-conotoxin binds selectively to the nicotinic receptor. It is entirely possible that similar blockers exist for the receptors which are vital to locomotion in lower species. As mentioned previously, lophotoxin effects vertebrate neuromuscular junctions. It appears to act on the end plate region of skeletal muscle (79,80), to block the nicotinic receptor at a site different from the binding sites for other blockers (81).

The fact that toxins target a specific membrane system has been commented upon many times. Where a toxin does not have such a specific site of action a general cytotoxic strategy appears to have been selected. Cytotoxicity can be regarded as occurring through one of two actions. The first involves an attack on the cell membrane so as to destroy its function as a barrier, while the second is to attack the fundamental chemistry of the cell. In many cases the fundamental chemistry that is attacked is that involved in nucleic acid chemistry, i.e., growth and differentiation. With regard to toxins which act on the membrane in general, and not on a particular structure, they can either disrupt the membrane completely or alter the membrane in such a manner as to disrupt its function. Saponins, and similar cytolytic toxins, can completely disrupt cellular integrity; however, ionophoric toxins create selective or non-selective channels which, as a result of "ion-loading" in the cell, may cause cell lysis.

A variety of toxins are capable of forming their own channels in cell membranes. For example, okadaic acid has been shown to be an ionophoric polyether (27,29).

The actions of proteins isolated from sea anemones, or other coelenterates, involve mechanisms different from those described for saponins. Thus, hemolysins from sea anemone *R. macrodactylus* are capable of forming ion channels directly in membranes (98). The basic protein from *S. helianthus* also forms channels in black-lipid membranes. These channels are permeable to cations and show rectification (99). This ability of *S. helianthus* toxin III to form channels depends upon the nature of the host lipid membrane (100). Cytolysin *S. helianthus* binds to sphingomyelin and this substance may well serve as the binding site in cell membranes (101-106).

Toxin from the jellyfish, *Chrysaora quinquecirrha*, creates large cation-selective channels with 760 pS conductance and no rectification. The channels are equally permeable to sodium, lithium, potassium, and cesium (*104*). Cobbs et al. (*105*) have also reported that *C. quinquecirrha toxins* forms pores (31 pS conductance) for monovalent cations in black-lipid membranes.

There has been considerable discussion regarding the mode of action of the sea cucumber and starfish saponins. Both the triterpene and steroidal glycosides inhibit both Na/K ATPase and Ca/Mg ATPase (*06*) possibly as a result of their aglycone structures. However, their detergent properties cause membrane disruption which will influence the activity of membrane-bound enzymes such as the ATPases. In investigating the actions of saponins on multilamellar liposomes, it was found that cholesterol serves as the binding site for such saponins and that cholesterol-free lipsomes are not lysed by saponins (*107*).

Many of the toxins obtained from coelenterates and echinoderms, because of their hemolytic or cytotoxic actions, are assumed to have a general disruptive action on cell membranes. However, since many of these toxins are capable of forming pores or channels in the plasma membrane of cells, their cytolytic actions may be a result of this highly selective action. On the other hand, the saponins from starfish and sea cucumbers have a direct lytic action as a result of their detergent action on the integrity of cells.

Strategies for Discovering and Identifying Marine Toxins and for Elucidating Possible Mechanisms of Action

In the absence of armies of toxicologists, pharmacologists and zoologists, it will be very difficult to systematically screen all marine invertebrates for toxins. There are, after all, hundreds of thousands of species that can be investigated and even the most cursory of examinations consumes months of research activity. For example, it took many years to elucidate the mechanisms of action of tetrodotoxin. In the past, tracing the route from lethality to underlying biochemical or cellular mechanisms was both tortuous and prolonged. The route involved first determining lethality and then actions on organ systems. After identifying target organs, the next step was to identify the target cells and finally the underlying mechanism. While it is intellectually satisfying to trace such a route, in many recent studies analysis of cellular mechanisms has preceded analysis of actions in whole tissue, isolated organs or the intact animal. However, it should not be assumed that techniques for analysis of biochemical or cellular mechanisms have completely superceded the stepwise analysis of actions in the whole animal, component organs, tissues, and cell types. Despite the availablity of more and more techniques, there are no universal shortcuts for elucidating the nature and actions of toxins. What are needed are strategies and procedures for both identifying and analysing mechanisms of actions of toxins. Just as obvious is the need to identify toxic species.

Approaches are required for the following stages: identification of toxic species, screening techniques for detecting toxicity, techniques for purifying toxins and provisionally identifying chemical nature, and techniques for tentatively identifying mechanisms of action.

Once these four stages are passed, it becomes a relatively easy process to dissect out the molecular site and mechanism by which a toxin may act. Consideration of the four stages listed above, indicates that sets of reasonable guidelines can be used to direct a study through the various stages.

Identification of Toxic Species. In trying to identify toxic species suitable for further studies, careful consideration should be given to zoology and, where applicable, clinical toxicology. Zoological knowledge allows the identification of the species which use toxins for purposes of offence and/or defense. For example, observations of carnivorous fish showed that many appear able to identify toxic species and avoid eating them. Similarly, zoological observation showed that cone shells could be divided into classes on the basis of prey species. Variation in prey species suggested that different toxins were evolved by different cone shells in order to confer specificity for particular prey species. Thus, zoological knowledge allows identification of toxic species on the basis of interaction between species. This knowledge leads to identification of prey species, identification of avoidance behavior on the part of species and effects on species which live in the same environment. In the case of humans being the interacting species, toxicity manifests itself by a variety of responses to oral ingestion or contact.

An interesting example of both acquired defensive toxicity and the importance of biological observation is the acquisition of toxicity from the sponge (*Hymeniacidon* sp.) by the nudibranch *Phylida varicosa*. The original observation was that a specimen of the nudibranch placed in an aquarium was sufficiently toxic to kill most of the other occupants. It was subsequently discovered that the material could be milked from the surface of the nudibranch to the extent that it could be literally milked dry of the toxic material, 9-isocyanopupukeanane (Scheuer, personal communication). Later field observations showed that the nudibranch aquired its toxicity by grazing on sponges.

Screening Techniques for Detecting Toxicity. Simple toxicity screening techniques are necessary to identify toxic species and to monitor the efficacy of isolation and purification procedures used to purify toxins. Atterwill and Steele (*108*) have recently comprehensively reviewed in vitro methods for toxicology and so much of the following is in the nature of a general overview.

In attempting to quickly elucidate the possible mechanisms of action of a toxin, there are a number of obvious routes to take. In the case of a toxin which has a rapid and acute lethal action, the route is especially obvious, but with toxins having effects on growth and differentiation, the best approach is more obscure.

In order to establish the presence of a toxin in a particular species, or in an extract from that species, a variety of procedures can be followed. Where a toxin is specific for a particular prey species, it is best to use that prey species for assessing toxicity. Unfortunately, marine prey species are often not readily available and routinely available laboratory animals have to be used as substitutes. There are advantages in using whole animals to test for toxicity since careful observation, plus monitoring of vital functions, may give useful clues as to possible mechanims of action. Different routes of administration can give clues as to the chemical identity of toxins. For example, proteins and polypeptides are usually inactived when given orally. Charged and/or large molecular weight toxins often appear less toxic when injected into a site (i.e., subcutaneous) with poor absorption. Use of whole animals also avoids complications which may arise from the presence of contaminating, but not really toxic, substances present in crude extracts. Such contaminants can badly obscure analysis in many in vitro systems, such as cultured cells. When more than one toxic molecule is present in an extract, complications are multiplied although careful observations should give clues as to the possible presence of multiple toxins.

In attempting to analyze the actions of toxins by means of observing behavioral changes, care has to be exercised to avoid misinterpreting behavioral observations. The literature is replete with false deductions drawn from such observations. One example is the sometimes erroneous interpretation of convulsions in rodents as being due to central nervous system actions. A lethal dose of a cardiotoxin will produce convulsions in both mice and rats. Such convulsions are secondary to cerebral hypoxia which occurs as a result of ventricular fibrillation. To avoid such errors, the investigator should have experience in monitoring behavioral responses, in different test species, to lethal injections of a variety of toxins with different but unambiguous mechanisms of action.

When lethal doses of a toxin are determined by a number of different routes of injection, further clues as to how a toxin might be working are obtained. In addition to the time required to produce death, the injection site should be examined for local signs of irritation and tissue damage. In any determination of lethal doses it is obviously important to perform a thorough autopsy. A full histological examination may not be necessary, but the autopsy should include a meticulous examination for hemorrhage, lung edema, exudates, cardiac state, hypoxia, etc. For example, where death is of a non-cardiac origin, the heart, or at least the atria, can often be observed to be slowly beating if examined immediately after death. Where a toxin kills by cardiotoxic actions, ventricular fibrillation and/or cardiac contracture is often observed.

It is generally true that if a toxin kills a particular species by one mechanism it will kill other species in the same phyllum by the same mechanism. However mechanisms of lethality cannot necessarily be expected to cross phylla. Such considerations of zoology should play a role in assessing toxicity especially where the species used to test for toxicity is the prey-species. In such a situation one would expect rapid lethality and a toxin targeted to a fundamental physiological function of that species.

Marine toxins are not always acutely toxic. This may be particularly so for toxins which are used to deter competing occupants for living space since they often have comparatively slow actions on growth. With such toxins, the procedures for the evaluation of acutely lethal toxins cannot apply. However, interesting discoveries may be made by using the simplest of screen of alcoholic extracts for cytolytic actions as exemplified in Table I of Shier (*109*).

The above considerations also apply to the next step in the analysis of a toxin's action, which is analysis in a anesthetized animal in which variables such as blood pressure, EKG, EMG, and respiration. are monitored. Lee (*110*) has considered in some detail the manner in which toxins can be investigated in such preparations. He has demonstrated, with various classes of toxins, that the spectrum of toxicological action in anesthetized rats and mice is characteristic for different classes of toxins. Thus cardiac toxins produce markedly different responses than those which kill through actions on nerve conduction. The preparations used by Lee are relatively simple in that anesthetized animals are cannulated for blood pressure recording and iv injection together with EKG and phrenic nerve recordings. Where appropriate, the animal can be ventilated artificially. Only a moderate degree of skill is required to record from a phrenic nerve within the chest of animal while still allowing the animal to breathe spontaneously. If these and similar procedures were to be standardized, it would be relatively simple to ascertain how a toxin kills acutely.

Techniques for Purifying Toxins and Provisionally Identifying Chemical Nature. Chemical techniques for the isolation, purification and elucidation of the structure of toxins have evolved to the extent that it is frequently a routine procedure to identify the chemical nature of a newly discovered toxin once it has been purified, although difficulties arise when the toxin is a very large polypeptide, protein, or a very complex organic molecule. However, it is sometimes found that a toxin becomes progressively more labile and stabilizing contaminants are removed by the purification processes. An example of this is *Cyanea* toxic material which becomes increasingly labile with each purification step (*111*).

Techniques for Tentatively Identifying Mechanisms of Action. Once the mechanism by which a toxin kills has been assessed, and toxin reasonably purified, it becomes relevant to try and ascertain as efficiently as possible the cellular mechanisms "targetted" by the toxin. This is a necessary step before final analysis of action using pure toxin and site-specific procedures such as the patch-clamp technique.

In trying to isolate the cellular mechanism by which a toxin might work, it is useful to consider what mechanisms are known to be influenced by toxins. Prime examples are ionic channels, membrane pumps, membrane carrier systems and the integrity of the cell membrane. Actions at these sites generally produce marked effects on cellular electrogenesis and subsequent intracellular potentials; thus, an electrophysiological analysis may often provide rapid insights into the actions of a toxin. Such an analysis should not be confined to one type of tissue in view of the varying participation of different ion channels in different cell types. It is practical to record intracellular potentials from cardiac, skeletal muscle and certain types of neuronal tissue. The neuromuscular junctions of different phyla gives insights into mechanisms responsible for neuronal transmission as well as transmitter release and reception. With a panoply of tissues it should be relatively easy to make an educated guess as to which membrane system(s) a toxin might be working on.

While electrophysiological preparations have great usefulness, conventional in vitro pharmacological preparations should not be ignored. However, some of these preparations should be approached with caution since their apparent simplicity often hides a wealth of complexity. One example is the guinea pig ileum. This tissue not only consists of complex smooth muscle but also contains so many complex neuronal elements that it is inherently difficult to analyze even a simple contraction. Brevetoxin causes contraction of smooth muscle but contraction is mediated via the release of transmitters as a result of the toxin's action on nerves (112). It is best to use tissue whose physiology has fewer component parts, and which is better understood. Thus the classic neuro-skeletal muscle preparations are relatively simple and the sites at which toxins might work are better understood. Similarly, red blood cells are relatively simple systems which can be damaged in relatively well-understood ways. Cardiac tissue is also useful in that its physiology is comparatively well understood and secondary effects mediated by nerves, etc., are generally not a complicating factor. The inotropic, chronotropic, and electrical effects of toxins are easily recorded in such tissue and are related to the underlying physiology in a fairly well-understood manner. For example, the positive inotropic effects of sea anemone polypeptide toxins are not unexpected in view of their effects on sodium channel activation (113).

The above considerations were based on the premise that toxins act primarily on the cell membrane in an acute manner. However, we know that certain toxins do not have actions at the cell membrane but rather act directly upon the processes responsible for cell metabolism and growth. Such toxins may act directly on nuclear systems (at the level of DNA or RNA) while others may disrupt the process of miosis and mitosis, for example, as occurs with saponins and plant toxins such as colchicine. When dealing with such toxicity, reliance has to be placed on the use of cell cultures. A full array of cell cultures ranging from well-differentiated primary to poorly differentiated immortal cultures are now available. There are many cell lines derived from neoplasms that are readily used for screening for effects on growth. One note of caution is that a huge variety of acutely acting toxins will cause the death of cultured cells. Any toxin which has a major action on cell membranes has the potential for killing cells by virtue of disturbing the intracellular milieu. Such an action should not be confused with more specific actions on cell growth and differentiation.

Conclusions

Studies into the toxic substances that can be extracted from marine invertebrates are growing at an increasing rate. Despite this, we are a long way from having complete explanations of the zoological importance of known toxins in terms of their usefulness as defensive, offensive, or digestive chemicals. Similarly, we have not identified all of the cellular mechanisms by which toxins may produce their actions and the toxins which influence them. Some of the small molecular weight substances that have been isolated from marine invertebrates are well understood. For example, the chemistry and biology of amines from molluscs and the saponins isolated from echinoderms have been extensively studied. The list of pharmacologically interesting substances that have been isolated continues to grow. Good examples include the polypeptides isolated from sea anemones, conotoxins, etc. Some of these afford both specific labels for ion channels and tools for specifically altering channel activity.

The chief impetus for research into marine invertebrate toxins appears to arise from the realization that such toxins are often exquisitively site-selective in their actions and thus are useful in helping to solve problems in physiology and biochemisty. In addition there are still outstanding problems concerning the zoological usefulness of toxins as well as the treatment of human cases of poisoning and envenomation. In addition marine toxins may well provide paradigms and prototypes for the development of new drugs. In the past, toxins and poisons have provided the main source of new drugs and they continue to be a useful source although in a less direct manner. One of the most recent notable examples has been the development of the newer antihypertensive drug, captopril. This drug was developed as a result of investigations into the bradykinin potentiating actions of a polypeptide fraction from snake venom. In particular, the marine invertebrate toxins may provide the impetus for developing new channel blocking drugs. Such toxins as omega-conotoxin may lead to potent and selective drugs acting on the N-type calcium channel.

Literature Cited

1. Bakus, C. J., Targett, N. M.; Schulte, B. *J. Chem. Ecol.* **1986**, *12*, 951-988.
2. Olivera, B. M., Gray, W. R., Zeikus, R., McIntosh, J. M., Varga, J., Rivier, J., de Santos, V.; Cruz, L. J. *Science* **1985**, *230*, 1338-1343.
3. Halstead, B. W.; Vinci J. M. In *Marine Toxins and Venoms. Handbook of Natural Toxins*; Tu, A. T., Ed.; Marcel Dekker: New York, 1988; Vol 3., pp 1-30.
4. Stonik, V. A.; Elyakov, G. B. In *Marine Toxins and Venoms. Handbook of Natural Toxins*; Tu, A. T., Ed.; Marcel Dekker: New York, 1988; Vol 3., pp 107-120.
5. Walker, M. J. A. In *Marine Toxins and Venoms. Handbook of Natural Toxins*; Tu, A. T., Ed.; Marcel Dekker: New York, 1988; Vol 3, pp 279-325.
6. Olivera, B. M.; Gray, W. R.; Cruz, L. J. In *Marine Toxins and Venoms. Handbook of Natural Toxins*; Tu, A. T., Ed.; Marcel Dekker: New York, 1988; Vol 3., pp 327-352.
7. Kem, W. R. In *Marine Toxins and Venoms. Handbook of Natural Toxins*; Tu, A. T., Ed.; Marcel Dekker: New York, 1988; Vol 3., pp 353-378.
8. Mynderse, J. S., Moore, R. E., Kashiwagi, M.; Norton, T. R. *Science* **1977**, *196*, 538-540.
9. Russell, F. E. *Adv. Marine. Biol.* **1984**, *2*, 60-217.
10. Bakus, G. J.; Thun, M. *Biol. Spongiaries* **1979**, *291*, 417-422.
11. Bakus, G. J. *Science* **1981**, *211*, 497-499.
12. Green, G. *Mar. Biol.* **1977**, *40*, 207-215.
13. Graham, A. *Proc. Malac. Soc. (Lond.)* **1955**, *31*, 144-158.
14. Randall, J. E.; Hartman, W. D. *Mar. Biol.* **1968**, *1*, 216-225.
15. Hashimoto, Y. *Marine Toxins and Other Bioactive Marine Metabolites*; Japan Scientific Societies Press: Tokyo, 1979; pp. 1-151.

16. Miyazawa, K.; Jeon, J. K.; Noguchi, T.; Ito, K.; Hashimoto, K. *Toxicon* **1987**, *25*, 975-980.

17. Fange, R. In *Toxins, Drugs and Pollutants in Marine Animals*; Bolis, L.; Zadwaisky, I.; Gilles, R., Eds.; Springer Verlag: New York, 1984; pp 47-62.

18. Houbrick, J. R.; Fretter, V. *Proc. Malac. Soc. (Lond)* **1969**, *38*, 415-429.

19. Fange, R.; Lidman, U. *Comp. Biochem. Physiol.* **1976**, *53*A, 101-103.

20. Narita, H.; Noguchi, T.; Maruyama, J.; Ueda, Y.; Hashimoto, K.; Watanabe, Y.; Hida, K. *Bull. Jpn. Soc. Fish.* **1981**, *47*, 935-942.

21. Endean, R. In *Chemical Zoology*; Academic Press: New York, 1972; Vol. 7, pp 421-466.

22. Pearce, J. B.; Thorson, G. *Ophelia* **1967**, *4*, 277-314.

23. Fuhrman, F. A.; Fuhrman, G. J.; DeRoemer, K. *Biol. Bull.* **1979**, *156*, 289-299.

24. Sheumack, D. D.; Howden, M. E.; Spence, I.; Quinn, R. J. *Science* **1978**, *199*, 188.

25. Bagnis, R. *Clin. Tox.* **1970**, *3*, 585-588.

26. Banner, A. H.; Stephens, B. J. *Nat. Hist. Bull. Siam. Soc.* **1966**, *21*, 197-203.

27. Tachibana, K., Scheuer, P. J., Tsukitani, Y., Kikuchi, H., Van Engen, D., Clardy J., Gopichand, Y.; Schmitz, F. J. *J. Amer. Chem. Soc.* **1981**, *103*, 2469-2471.

28. Burreson, B. J.; Scheuer, P. J.; Finer P. J.; Clardy, J. *J. Am. Chem. Soc.* **1975**, *97*, 4763-4764.

29. Shibata, S. *Trends in Autonomic Pharm.* **1982**, *3*, 301-317.

30. Pettit, G. R., Herald, C. L., Doubek, D. L.; Herald, D. L. *J. Am. Chem. Soc.* **1982**, *104*, 6846-6848.

31. Das, N. P., Lim, H. S.; Teh, Y. F. *Comp. Gen. Pharm.* **1971**, *2*, 473-475.

32. Richet, C. *C. R. Soc. Biol.* **1906**, *61*, 686.

33. Cariello, L., Zanetti, L.; Rathmayer, W. In *Natural Toxins*; C. Eaker; T. Wadetrom, Eds.; Pergamon Press: Oxford, U. K., 1980; pp. 631-636.

34. Pettit, G. R., Kamano, Y., Brown, P., Gust, D., Irone, M.; Herald, C. L. *J. Am. Chem. Soc.* **1982**, *104*, 905-957.

35. Liaaen-Jensen, S. In *Marine Natural Products: Chemical and Biological Perspectives VII*; Scheuer, P. J., Ed.; Academic Press: New York, 1978; pp. 1-73.

36. Tursch, B., Braekman, J. C., Daloze, D.; Kaisin, M. In *Marine Natural Products: Chemical and Biological Perspectives VII*; Scheuer, P. J., Ed.; Academic Press: New York; 1978; pp 247-296.

37. Neeman, I., Fishelson, L.; Kashman, Y. *Toxicon* **1974**, *12*, 593-598.

38. Wratten, S. J., Faulkner, D. J., Hirotsu, K.; Clardy, J. *J. Am. Chem. Soc.* **1977**, *99*, 2824-2825.

39. Fenical, W., Okuda, R. K., Bandurraga, M. M., Culver, P.; Jacobs, R. S. *Science* **1981**, *212*, 1512-1518.

40. Goad, L. J. In *Marine Natural Products: Chemical and Biological Perspectives*; Scheuer, P. J., Ed.; Academic Press:New York; 1978, pp. 75-172.

41. Scheuer, P. J. *Fortschr. Chem. Org. Naturst.* **1969**, *27*, 322-339.

42. Habermehl, G.; Volkwein, G. *Justus Liebigs. Ann. Chem.* **1970**, *731*, 53-57.

43. Habermehl, G.; Volkwein, G. *Toxicon* **1971**, *9*, 319-326.

44. Russell, F. E. *Clin. Pharmacol. Ther.* **1967**, *8*, 849-873.

45. Roller, P., Djerassi, C., Cloetens, R.; Tursch, B. *J. Am. Chem. Soc.* **1969**, *91*, 4918-4920.

46. Kitagawa, I., Sugawara, T., Yosioka, I.; Kuriyama, K. *Chem. Pharm. Bull.* (Tokyo) **1976**, *24*, 266-274.

47. Hashimoto, Y.; Yasumoto, T. *Bull. Jpn. Soc. Scient. Fish.* **1960**, *26*, 1132-1138.

48. Ferlan, I.; Lebez, D. *Toxicon* **1974**, *12*, 57-61.

49. Sket, D., Dreslar, K., Ferlan, I.; Lebez, D. *Toxicon* **1974**, *12, 63-68.*
50. Shiomi, K., Tanaka, E., Yamanaka, H.; Kikuchi, T. *Toxicon* **1985**, *23*, 865-874.
51. Macek, P., Sencic, L.; Lebez, D. *Toxicon* **1982**, *20*, 181-185.
52. Macek, P.; Lebez, D. *Toxicon* **1988**, *26*, 441-451.
53. Bernheimer, A. W.; Lai, C. Y. *Toxicon* **1985**, *23*, 791-799.
54. Kem, W. R.; Dunn, B. M.; Parten, B.; Price, D. *Toxicon* **1985**, *23*, 580.
55. Blumenthal, K. M.; Kem, W. R. *J. Biol. Chem.* **1983**, *258*, 5574-5581.
56. Bernheimer, A. W., Avigad, L. S., Branch, G., Dowdle, E.; Lai, C. Y. *Toxicon* **1984**, *22*, 183-191.
57. Mebs, D., Liebrich, M., Reul, A.; Samejima, Y. *Toxicon* **1983**, *21*, 257-264.
58. Phelan, M. A.; Blanquet, R. S. *Comp. Biochem. Physiol. [B]* **1985**, *81*, 661-666.
59. Aldeen, S. I., Elliott, R. C.; Sheardown, M. Br. *J. Pharmacol.* **1981**, *72*, 211-220.
60. Elliott, R. C., Konya, R. S.; Vickneshuwara, K. *Toxicon* **1986**, *24*, 117-122.
61. Blankenship, J. E., Langlais, P. J.; Kittredge, J. S. *Comp. Biochem. Physiol.* **1975**, *51*C, 129-137.
62. Mebs, D.; Weiler, I.; Heinke, H. F. *Toxicon* **1988**, *23*, 955-962.
63. Bourget, G., More, M. T., More, P., Guimbretiere, L., Le Boterff, J.; Verbist, J. F. *Toxicon* **1988**, *26*, 324-327.
64. Solomon, A. E.; Stoughton, R. B. *Arch. Dermatol.* **1978**, *114*, 1333.
65. Alender, C. B., Feigen, G. A.; Tomita, J. T. *Toxicon* **1965**, *3*, 9-17.
66. Fleming, W. J.; Howden, M. E. H. *Toxicon* **1974**, *12*, 447-448.
67. Nakagawa, H.; Kimura, A. *Jpn. J. Pharmacol.* **1982**, *32*, 966-968.
68. Mebs, D. A. *Toxicon* **1984**, *22*, 306-307.
69. Canicatti, C.; Parrinello, N. *Experientia* **1983**, *39*, 764-766.
70. Feigen G. A.; Hadji, L. In *Bioactive Compounds from the Sea*, Humm, H. J.; Lane, C. E., Eds.; Marcel Dekker: New York; 1974; pp. 37-97.
71. Kimura, A.; Nakagawa, H.; Hayashi, H.; Endo, K. *Toxicon* **1984**, *22*, 353-358.
72. Ruggieri, G. D.; Nigrelli, R. F. In *Bioactive Compounds from the Sea*, Humm, H. J.; Lane, C. E.; Eds.; Marcel Dekker: New York, 1974; pp. 183-195.
73. Goldsmith, L. A.; Carlson, G. P. In *Food-Drugs from the Sea Proceedings*; Webber, H. H.; Ruggieri, G. D.; Eds.; Marine Technology Society, 1974; pp. 354.
74. Marcum, J. A., Levin, J.; Prendergast, R. A. *Thromb. Haemost.* **1984**, *52*, 1-3.
75. Biedebach, M. C., Jacobst, G. P.; Langjahr, S. W. *Comp. Biochem. Physiol. [C]* **1978**, *59*, 11-12.
76. Anraku, M., Kihara, H.; Hashimura, S. *Jpn. J. Physiol.* **1984**, *34, 839-847.*
77. Kimura, A., Hayashi, H.; Kuramoto, M. *Jpn. J. Pharmacol.* **1975**, *25, 109-120.*
78. Culver, P.; Jacobs, R. S. *Toxicon* **1981**, *19*, 825-830.
79. Atchinson, W. D., Narahashi, T.; Vogel, S. M. Br. *J. Pharmacol.* **1984**, *82*, 667-672.
80. Langdon, R. B.; Jacobs, R. S. *Life Sci.* **1983**, *32*, 1223-1228.
81. Culver, P., Fenical, W.; Taylor, P. *J. Biol. Chem.* **1984**, *259*, 3763-3770.
82. Nilsson, A.; Fange, B. *Comp. Biochem. Physiol.* **1967**, *22*, 927-931.
83. Hayashi, E.; Yamada, S. *Br. J. Pharmacol.* **1975**, *53*, 207-215.
84. Hirayama, H., Gohgi, K., Urakawa, N.; Ikeda, M. *Jpn. J. Pharmacol.* **1970**, *20*, 311-312.
85. Winkler, L. R. *Pac. Sci.* **1961**, *15*, 211-214.
86. Watson, M.; Raygne, M. D. *Toxicon* **1973**, *11*, 269-276.
87. Roseghini, M., Erspamer, V., Ramorino, L.; Gutierrez, J. E. *Eur. Biochem.* **1970**, *12*, 468-473.
88. Bender, J. A., DeRiemer, K., Roberts, T. E., Rushton, R., Mosher, H. S. and Fuhrman, F. A. *Comp. Gen. Phamacol.* **1974**, *5*, 191-198.

89. Erspamer, V.; Glasser, A. *Br. J. Pharmacol.* **1957**, *12*, 176-184.
90. Pettit, G. R., Day, R. H., Hartwell, J. H.; Wood, H. B. J. W. *Nature* **1970**, *227*, 962-963.
91. Ruggieri, G. D. *Science* **1976**, *194*, 491-497.
92. Erspamer, V. *The tachykinin peptide family. TINS* **1981**, *4*, 267-269.
93. Kem, W. R.; Scott, J. D. *Biol. Bull. (Woods Hole)* **1981**, *159*, 475.
94. Cariello, L.; Zanetti, L. *Comp. Biochem. Physiol.* **1977**, *57C*, 169-173.
95. Ozaki, H.; Ishihara, H.; Kohama, K.; Nonomura, Y.; Shibata. S.; Karaki, H. *J. Pharmacol. Exp. Ther.* **1987**, *243*, 1167-1172.
96. Daigo, K., Noguchi, T., Miwa, A., Kawai, N.; Hashimoto, K. *Toxicon* **1988**, *26*, 485-490.
97. Kem, W. R.; Pennington, M. W.; Dunn, B. M. Chapter 21 in this volume.
98. Kozlovskaia, E. P., Ivanov, A. S., Mol'nar, A. A., Grigor'ev, P. A.; Monastyr-naia, M. M. *Dokl. Akad. Nauk. SSSR.* **1984**, *277*, 1491-1493.
99. Michaels, D. W. *Biochim. Biophys. Acta* **1979**, *555*, 67-78.
100. Shin, M. L., Michaels, D. W.; Mayer, M. M. *Biochim. Biophys. Acta* **1979**, *555*, 79-88.
101. Linder, R., Bernheimer, A. W.; Kim, K. S. *Biochem. Biophys. Acta* **1977**, *467*, 290-300.
102. Bernheimer, A. W.; Avigad, L. S. *Proc. Natl. Acad. Sci. USA* **1976**, *73*, 467-471.
103. Bernheimer, A. W. Chapter 23 in this volume.
104. Dubois, J. M., Tanguy, J.; Burnett, J. W. *Biophys. J.* **1983**, *42*, 199-202.
105. Cobbs, C. S., Gaur, P, K., Russo, A. J., Warnick, J. E., Calton, G. J. and Burnett, J. W. *Toxicon* **1983**, *21*, 385-391.
106. Gorshkov, B. A., Gorshkova, I. A., Stonik, V. A.; Elyakov, G. B. *Toxicon* **1982**, *20*, 655-658.
107. Yu, B. S.; Jo, I. H. *Chem. Biol. Interact.* **1984**, *52*, 185-202.
108. Atterwill, C. K.; Steele, C. E. In *In Vitro Methods in Toxicology*; Cambridge University Press: Cambridge, U. K., 1987.
109. Shier, W. T. In *Marine Toxins and Venoms. Handbook of Natural Toxins*; Tu, A. T., Ed.; Marcel Dekker: New York, 1988; Vol 3., pp. 477-492.
110. Lee, C. Y. *Toxicon* **1988**, *26*, 29.
111. Walker, M. J. A. *Toxicon* **1977**, *15*, 3-14.
112. Ishida, Y.; Shibata, S. *Pharmacology* **1985**, *31*, 237-240.
113. Shibata, S.; Izumi, T.; Seriguchi, D. G.; Norton, T. R. *J. Pharmacol. Exp. Ther.* **1978**, *205*, 683-692.

RECEIVED August 23, 1989

Chapter 25

Some Natural Jellyfish Toxins

Joseph W. Burnett

Division of Dermatology, Department of Medicine, University of Maryland School of Medicine, Baltimore, MD 21201

Recent investigations on natural jellyfish toxins have centered upon four species. The four animals studied most are: *Chrysaora quinquecirrha* (sea nettle), whose distribution is worldwide, appears seasonally in the Chesapeake Bay; *Physalia physalis* (Portuguese man-o'war) has a worldwide distribution with an annual occurrence in the southeast American, Atlantic, and Gulf Coasts; *Chironex fleckeri* (box jellyfish) is found in the northeast Australian and Indo Pacific areas; and *Pelagia noctiluca* (mauve baubler), whose distribution is also worldwide, can be abundant in the Mediterranean Sea. All of these animals deliver their venom by means of an intracellular organelle, the nematocyst, which everts a toxin-coated thread forcefully to penetrate into the human dermis.

The human reactions to the natural jellyfish toxins can be classified as fatal, systemic, chronic, or local (*1*) (Table I). Death can be produced by toxicity in a dose dependent manner through cardiotoxic, central respiratory, or renal effects. Anaphylaxis has been reported in a limited number of cases. The systemic reactions include a spontaneously remitting syndrome produced by *Carukia barnesi* called "Irukandji", which is characterized by muscle cramps, nausea, vomiting, tachycardia, and hypertension lasting for two days. Local reactions to envenomation include persistent eruptions, recurrent eruptions, eruptions distant to the site of envenomation, exaggerated local angioedema, papular urticaria, and respiratory acidosis. Chronic lesions such as keloids, pigmentation, fat atrophy, contractions, gangrene with vascular spasm, mononeuritis, autonomic nerve paralysis, urticaria, and bladder paralysis with ataxia occur.

Preparation of Venoms and Toxins

The methods of preparing the purified venom extracts vary according to the species. Presently *Chrysaora quinquecirrha* nematocysts are isolated by grinding, passage through a mesh, and centrifugation before being ruptured by sonic treatment. Organelles from *Chironex fleckeri* are prepared by homogenization and centrifugation or the venom is taken from a beaker which has been covered by a monolayer of amnion cells on top of which tentacles have been placed, moved, and electrically stimulated (*2*). Recently *Physalia* nematocysts have been purified by fluorescent automated cell sorters (FACS) into two sizes (*3*). Because of their stability, *Physalia* are the only species whose nematocysts have been purified in this manner. A problem with this technique is that discharged nematocysts with their long threads still attached to the capsule can disturb the size sorting of the organelles.

The best method of preparing individual toxins from the crude venom is by affinity immunochromatography utilizing monoclonal antibody or in one instance, polyclonal antibody (*4–6*). Monoclonal antibodies to both the fishing and mesentery

0097–6156/90/0418–0333$06.00/0

Table I. Important Disorders Produced by Jellyfish Envenomation

Acute local reactions	Chronic reactions
Eruption and pain due to toxin	Keloids
Exaggerated local angioedema	Pigmentation
Delayed reactions up to several hours	Fat atrophy
Recurrent eruptions up to 4 episodes	Contractions
Distant site reactions	Gangrene, vascular spasm[a]
Papular urticaria	Mononeuritis
Respiratory acidosis	Autonomic nerve paralysis
	Urticaria
Post episode dermatoses	**Fatal reactions**
Herpes simplex	1. Toxin-induced
Granuloma annulare	Immediate cardiac arrest
	Rapid respiratory arrest
	Later "shocked" kidney
	2. Anaphylaxis

[a] Two of these patients had operative repair of their vasospasm and both developed postoperative fever of 102°F.

tentacles of *Chrysaora quinquecirrha* have been prepared and yield toxins of 100,000 and 190,000 or 108,000, 160,000, and 175,000, respectively, for the fishing and mesentery tentacles. Earlier experiments utilizing ion exchange columns and molecular sieving techniques yielded a toxin of 240,000 molecular weight for *Physalia physalis* (7). However, monoclonal antibody immunoaffinity columns inoculated with the contents of FACS purified *Physalia* nematocysts yielded a toxin of 69,000 molecular weight (3).

Past experiments with *Chironex* using ion exchange and gel filtration chromatography have yielded toxins with molecular weights of 10,000 to 30,000, 75,000, 150,000 and 500,000 (2,8). Toxins in the 50,000 and 150,000 molecular weight range were obtained by immunoaffinity chromatography with polyclonal antisera (6). Monoclonal antibody columns separated *Chironex* toxins near 20,000 as well as other sizes (9). These results indicate that the venom within *Chironex* nematocysts contains a mixture of individual toxins with varying potency and activities.

Human Reactions to the Toxins

All the jellyfish venoms are toxic but also stimulate the cell mediated and humoral immunological systems of man. After injection of large doses of jellyfish venom into human skin, a perivascular mononuclear cell infiltration appears within the dermis. This infiltration is composed predominantly of helper inducer cells which produce suppressor activity. It appears that the NK enhancement of human leukocytes in patients envenomated by *Chrysaora quinquecirrha* is depressed when the clinical lesion is inflammatory (10). Recovery from this suppression follows the amelioration of the acute cutaneous reaction. In other instances, envenomated patients have abnormal macrophage migration tests (11).

Pain production is the most common injury inflicted on man. This noxious stimulus is perceived almost instantly after skin - tentacle contact. A subpopulation (30 - 40%) of visceral sensory C fibers denoting noscioception have been shown to be selectively excited experimentally in nerve ganglia preparations by a component of

Physalia nematocyst toxin (*12*). This phenomenon appears to be important in understanding the pathogenesis of toxin-induced cutaneous pain. Additionally, the importance of kinin-like mediators in pain production has also been hypothesized (*2*).

The toxic mechanism of action of these various jellyfish venoms is complex. The cardiotoxic reaction seems to focus on calcium transport and is blocked by the prior or post administration of therapeutic doses of verapamil (*13*). In neuronal tissue, *Chrysaora* venom induces large cationic selective channels which open and close spontaneously. These channels are permeable to Na^+, Li^+, K^+, and Cs^+, but not Ca^{2+}, and the channels are present in spite of the treatment with sodium and potassium inhibitors such as tetrodotoxin and tetraethylammonium (*14*).

Summary

Research on the chemistry, toxicology, immunology, and treatment of jellyfish venoms has been progressing steadily for the past 20 years. In spite of this fact, knowledge of these venoms and their activity is not as complete as that of other species. The reasons for this statement are: (1) some animals are not abundant and are found in isolated marine areas; (2) the venoms are thermolabile and appear to adhere to support media and to have a tendency to aggregate or dissociate; (3) interesting or unusual clinical cases are poorly reported, difficult to find, and occur in sparsely populated areas of the world. In addition it appears that each species has a venom with its own particular characteristics. Although there are some general common relationships between these venoms, it is apparent that both the first aid and definitive therapies of envenomation by jellyfish will be species specific to some degree. It is hoped that additional interest will result in better case reporting and further research both by biologists and physicians.

Literature Cited

1. Burnett, J.W.; Calton, G.J.; Burnett, H.W. *J. Amer. Acad. Dermatol.* **1986**, *14*, 100–106.
2. Burnett J.W.; Calton G.J. *Toxicon* **1987**, *25*, 581–602.
3. Burnett, J.W.; Ordonez, J.V.; Calton, G.J. *Toxicon* **1986**, *24*, 514–518.
4. Cobbs, C.S.; Gaur, P.K.; Russo, A.J.; Warnick, J.E.; Calton, G.J.; Burnett, J.W. *Toxicon* **1983**, *21*. 385–391.
5. Kelman, S.N.; Calton, G.J.; Burnett, J.W. *Toxicon* **1984**, *22*, 139–144.
6. Calton, G.J.; Burnett, J.W. *Toxicon* **1986**, *24*, 416–420.
7. Tamkun, M.M.; Hessinger, D.; Hessinger, D. *Biochim. Biophys. Acta.* **1981**, *667*, 87–92.
8. Olson, C.E.; Pockl, E.E.; Calton, G.J.; Burnett, J.W. *Toxicon* **1984**, *22*, 733–742.
9. Naguib, A.M.F.; Bonsal, J.; Calton, G.J.; Burnett, J.W. *Toxicon* **1988**, *26*, 387–394.
10. Burnett, J.W.; Hepper, K.P.; Aurelian, L.; Calton, G.J.; Gardepe, S.F. *J. Amer. Acad. Dermatol.* **1987**, *17*, 86–92.
11. Burnett. J.W.; Hepper, K.P.; Aurelian, L. *Toxicon* **1986**, *24*, 104–107.
12. Weinrich D.; Burnett, J.W.; unpublished data.
13. Burnett, J.W.; Calton, G.J. *Med. J. Aust.* **1983**, *2*, 192–194.
14. Dubois, J.M.; Tanguy, J.; Burnett, J.W. *Biophysics J.* **1983**, *42*, 199–202.

RECEIVED May 9, 1989

Chapter 26

Neurotoxins from Sea Snake and Other Vertebrate Venoms

Anthony T. Tu

Department of Biochemistry, Colorado State University,
Fort Collins, CO 80523

Neurotoxins present in sea snake venoms are summarized. All sea snake venoms are extremely toxic, with low LD_{50} values. Most sea snake neurotoxins consist of only 60-62 amino acid residues with 4 disulfide bonds, while some consist of 70 amino acids with 5 disulfide bonds. The origin of toxicity is due to the attachment of 2 neurotoxin molecules to 2 α subunits of an acetylcholine receptor that is composed of $\alpha_2\beta\gamma\delta$ subunits. The complete structure of several of the sea snake neurotoxins have been worked out. Through chemical modification studies the invariant tryptophan and tyrosine residues of postsynaptic neurotoxins were shown to be of a critical nature to the toxicity function of the molecule. Lysine and arginine are also believed to be important. Other marine vertebrate venoms are not well known. All evidence indicates that the fish venoms are composed of proteins.

There are many venomous marine vertebrates in the seas, notably sea snakes and fishes. Venoms of sea snakes have been studied much more thoroughly than fish venoms. In this chapter, sea snake venom is described in greater detail than fish venoms simply because there is much more scientific information available.

The sea snake is a marine-adapted serpent belonging to the family of Hydrophiidae. There are many varieties of sea snakes with different colors, shapes, and sizes. They are well adapted for the marine environment and have a flat tail and a salt gland. Sea snakes are widely distributed in tropical and subtropical waters along the coasts of the Indian and Pacific Oceans. They are not found in the Atlantic Ocean.

All sea snakes are poisonous and their venoms are extremely toxic. The LD_{50} for crude sea snake venom can be as low as 0.10 $\mu g/g$ mouse body weight (1). For purified toxin the LD_{50} is even lower, suggesting the high toxicity of sea snake toxins and venoms. This toxicity is derived from the presence of potent neurotoxins. Compared to snake venoms of terrestrial origin, sea snake venoms have been studied less. Different enzymes reported to be present or absent are summarized in Table I.

Primary Structure of Sea Snake Neurotoxins

Before discussing the structure of the neurotoxins, it is necessary to define the types of neurotoxins. Three types of neurotoxins have been found so far in snake venoms. The first one is a postsynaptic neurotoxin, the second is a presynaptic neurotoxin, and the last is a cholinesterase inhibiting neurotoxin. Most sea snake venoms seem to contain only the postsynaptic neurotoxin. Only in *Enhydrina*

0097–6156/90/0418–0336$06.00/0
© 1990 American Chemical Society

Table I.   Presence and Absence of Enzymes in Sea Snake Venoms

Enzyme	Venom	Comment	Reference
Acetylcholinesterase	*Enhydrina schistosa*	Activity detected	6, 33
	Hydrophis cyanocinctus	Activity detected	34
Hyaluronidase	*E. schistosa*	Activity detected	6
Leucine amino-peptidase	*E. schistosa*	Activity detected	6
5'-Nucleotidase	*E. schistosa*	Activity detected	33
	H. cyanocinctus	Activity detected	34
	Laticauda semifasciata	Activity detected	35
Phosphodiesterase	*E. schistosa*	Activity detected	6,35
	H. cyanocinctus	Activity detected	34
Phosphomonoesterase	*E. schistosa*	Activity detected	34
	H. cyanocinctus	Activity detected	34
	L. semifasciata	Activity detected	35
		Isolated	36
Phospholipase A	*E. schistosa*	Activity detected	6,37,38
		Isolated	39,40
	H. cyanocinctus	Activity detected	34,41
	L. semifasciata	Isolated	42–46
	Pelamis platurus	Activity detected	41
L-amino acid oxidase	*E. schistosa*	Activity not detected	33
	H. cyanocinctus	Activity not detected	34
Arginine esterase	*E. schistosa*	Activity not detected	33
	H. cyanocinctus	Activity not detected	34
Protease	*L. semifasciata*	Activity not detected	47

schistosa venom, which also possesses a postsynaptic toxin, was a presynaptic type found and identified as phospholipase A.

Toxins from the venoms of the subfamily Hydrophiinae are extremely similar in amino acid sequence (Table II). Similarly, the sequences in toxins from the subfamily of Laticaudinae are also very similar among themselves (Table II). Although comparison of the sequences of toxins from two subfamilies shows there are considerable similarities, differences also become noticeable.

There are four disulfide bonds in short-chain (Type I) neurotoxins. This means that there are eight half-cystines. However, all Hydrophiinae toxins have nine half-cystines with one cysteine residue. An extra cysteine residue can be readily detected from the Raman spectrum as the sulfhydryl group shows a distinct S-H stretching vibration at 2578 cm^{-1}. Some Laticaudinae toxins do not have a free cysteine residue as in the cases of *L. laticaudata* and *L. semifasciata* toxins. In long toxins (Type II) there are five disulfide bonds (Table III).

Another type of neurotoxin found in sea snake venoms is a hybrid type structurally situated between the short-chain and long-chain types. As can be seen in Table IV, two toxins shown here have a long stretch of segment 4, yet there is no disulfide bond in this portion.

Secondary Structure

In order to understand the exact mechanism of the neurotoxic action, it is important to know the secondary structure of the neurotoxins as well. It is now known that postsynaptic neurotoxins attach to the α-subunits of acetylcholine receptor (AChR).

It is a supposition that the β-sheet structure of neurotoxin is an essential structural element for binding to the receptor. The presence of β-sheet structure was found by Raman spectroscopic analysis of a sea snake neurotoxin (2). The amide I band and III band for *Enhydrina schistosa* toxin were at 1672 cm^{-1} and 1242 cm^{-1}, respectively. These wave numbers are characteristic for anti-parallel β-sheet structure. The presence of β-sheet structure found by Raman spectroscopic study was later confirmed by X-ray diffraction study on *Laticauda semifasciata* toxin b.

Sea snake short-chain toxins have a molecular weight of only 6,800. The small size with four disulfide bonds makes these toxins very compact and stable molecules. Therefore, when the *Pelamis platurus* toxin is subjected to heat treatment at 100° C and subsequent cooling, it does not change its conformation substantially. Amide I and III bands and S-S stretching vibration did not change by heat treatment.

Four disulfide bonds are clustered in one area and there is a protruding loop. It is suspected that this loop is the one that plays an important role in binding to the AChR.

The four disulfide bonds are believed to be important for maintaining the specific conformation and have been studied extensively. The conformation of the disulfide bond in C-C-S-S-C-C network is gauche-gauche-gauche conformation at the S-S stretching vibration appearing at 510-512 cm^{-1}.

Structure-Function Relationship

The amino acid residues in neurotoxins which are important for neurotoxic action are still not entirely clarified. Some neurotoxins contain one free SH group, while others do not. From this fact, it would be logical to assume the sulfhydryl group is not essential. This was actually proven to be the case.

When *N,N*-1,4-phenylenedimaleimide was used for modifying the sulfhydryl group in *Pelamis* toxin, 2 mol of toxins combined with 1 mol of the reagent. With the sulfhydryl group modified, the S-H stretching vibrational band at 2578 cm^{-1} disappeared. The modification of the single sulfhydryl group did not alter the binding ability to AChR or toxicity (3).

Disulfide bonds, however, are important in maintaining the particular toxin structure and have been shown to be essential for toxicity. When all four disulfide bonds are reduced and alkylated, the neurotoxin loses its toxicity (4).

The one residue most extensively studied is tryptophan. It is very easily modified, indicating that tryptophan residue is exposed (5-8). Raman spectroscopic analysis of a sea snake neurotoxin indicated that a single tryptophan residue is indeed exposed (2). The tryptophan residue lies in the important loop consisting of segment 4. Modification of the tryptophan residue induces the loss of AChR binding ability as well as the loss of toxicity (5-8).

There is only one tyrosine residue in some sea snake neurotoxins. This residue is usually quite difficult to modify, but once it is modified, the toxicity is lost (9). Histidine seems not to be essential as the chemical modification of this residue does not affect the toxicity (10).

Argine and lysine are believed to be important, but results are not clear because sea snake neurotoxins contain several residues of these amino acids (7).

A cloned complementary DNA to a neurotoxin precursor RNA extracted from the venom glands of *Laticauda semifasciata* was isolated and its nucleotide sequence was identified (11). The cloning of neurotoxin should aid the understanding of structure–function relationship eventually.

Comparison to Other Snake Toxins

The similarity of the primary structure of different sea snake venoms has already been discussed. Postsynaptic neurotoxins from Elapidae venom have been extensively studied. Elapidae include well-known snakes such as cobra, krait, mambas, coral snakes, and all Australian snakes. Like sea snake toxins, Elapidae toxins can also be grouped into short-chain (Type I) and long-chain (Type II) toxins. Moreover, two types of neurotoxins are also similar to cardiotoxins, especially in the positions of disulfide bonds. However, amino acid sequences between cardiotoxins and sea snake and Elapidae neurotoxins are quite different. In comparing the sequence of sea snake and Elapidae neurotoxins, there is a considerable conservation in amino acid sequence, but the difference is greater than among the various sea snake toxins.

Similarity of venoms among different sea snakes and Elapidae can also be detected immunologically. For instance, the antibody for *Enhydrina schistosa* showed cross reactivity with the venoms of *Hydrophis cyanocinctus*, *Lapemis hardwickii*, and *Pelamis platurus* (12). The sea snake antivenin not only neutralizes the toxicity of various sea snake venoms, but also *Naja naja atra* (Taiwan cobra) venom (13-16). The reverse is also true; namely, some Elapidae antivenins are also effective for neutralizing sea snake venom lethality (17-19).

Pharmacological and Biological Activities

When a nerve-muscle preparation is stimulated in the presence of a sea snake neurotoxin, there is no twitch. However, when the muscle itself is stimulated directly in the presence of a neurotoxin, the muscle contracts. This means that neurotoxin does not inhibit the muscle itself. Moreover, postsynaptic neurotoxin does not inhibit the release of acetylcholine from the nerve ending. Therefore, the site of snake toxin inhibition must be in the postsynaptic site (20). Later it was shown that a neurotoxin strongly binds to the acetylcholine receptor (AChR).

Table II. Amino Acid Sequence of Sea Snake Neurotoxins (Type I or Short Chain)

Snake	Toxin	Segment 1	Segment 2		Segment 3	Segment 4
Hydrophiinae						
Acalyptophis peronii	major	M T C	C N Q Q S S Q P K T T T N	C	A G N S C	Y K K T W S D H R G T I I E R G
	minor	M T C	C N Q Q S S Q P K T T T N	C	A G N S C	Y K K T W S D H R G T I I E R G
Astrotia stokesii	toxin a	M T C	C N Q Q S S Q P K T T T N	C	A G N S C	Y K K T W S D H R G T I I E R G
Enhydrina schistosa	toxin 4	M T C	C N Q Q S S Q P K T T T N	C	A E S S C	Y K K T W S D H R G T R I E R G
	toxin 5	M T C	C N Q Q S S Q P K T T T N	C	A E S S C	Y K K T W S D H R G T R I E R G
Hydrophis cyanocinctus	hydrophitoxin a	M T C	C N Q Q S S Q P K T T T N	C	A E S S C	Y K K T W S D H R G T R I E R G
	hydrophitoxin b	M T C	C N Q Q S S Q P K T T T N	C	A E S S C	Y K K T W S D H R G T R I E R G
Lapemis hardwickii	lapemis toxin	M T C	C N Q Q S S Q P K T T T N	C	A E S S C	Y K K T W S D H R G T R I E R G
Pelamis plaurus	pelamitoxin a	M T C	C N Q Q S S Q P K T T T N	C	A E S S C	Y K K T W S D H R G T R I E R G
	pelamis toxin b	M T C	C N Q Q S S E P K T T T N	C	A E S S C	Y K K T W S D H R G T R I E R G
Laticaudinae						
Aipysurus laevis	toxin a	L T C	C N Q Q S S S Q P K T T T D	C	A D N S C	Y K K T W Q D H R G T R I E R G
	toxin b	L T C	C N Q Q S S S Q P K T T T D	C	A D N S C	Y K M T W R D H R G T R I E R G
	toxin c	L T C	C N Q Q S S S Q P K T T T D	C	A D N S C	Y K K T W K D H R G T H I E R G
L. laticaudata	laticotoxin a	R R C	F N H P S S S Q P Q T N K S	C P P	G E N S C	Y N K Q W R D F R G T I I E R G
L. semifasciata	erabutoxin a	R I C	F N H S S S Q P Q T T K T	C P S	G Q S S C	Y N K Q W S D F R G T I I E R G
	erabutoxin b	R I C	F N Q H S S S Q P Q T T K T	C P S	G Q S S C	Y H K Q W S D F R G T I I E R G
	erabutoxin c	R I C	F N Q H S S S Q P Q T T K T	C P S	G Q S S C	Y H K Q W S D F R G T I I E R G
	toxin b	R I C	F N Q H S S S Q P Q T T K T	C P S	G Q S S C	Y H K Q W S D F R G T I I E R G

Continued on next page

Table II. *Continued*

Snake	Toxin	Segment 5	Segment 6	Segment 7	Segment 8	Ref.
Hydrophiinae						
Acalyptophis peronii	major	C G	C P Q V K S G I K L E C	C H T N E	C N N	48
	minor	C G	C P V K S G I K L E C	C H T N E	C N N	49
Astrotia stokesii	toxin a	C G	C P Q V K S G I K L L E C	C H T N E	C N N	50
Enhydrina schistosa	toxin 4	C G	C P Q V K P G I H K L L E C	C H T N E	C N N	51
	toxin 5	C G	C P Q V K S G I K L L E C	C H T N E	C N N	51
Hydrophis cyanocinctus	hydrophitoxin a	C G	C P Q V K K K G I K L L E C	C H T N E	C N N	52
	hydrophitoxin b	C G	C P Q V K S G I K L E C	C H T N E	C N N	52
Lapemis hardwickii	lapemis toxin	C G	C P Q V K K P G I K K L E C	C H T N E	C N N	53
Pelamis platurus	pelamitoxin a	C G	C P Q V K S G I H K L E C	C H T N E	C N N	54
	pelamis toxin b	C G	C P Q V K S G I K L E C	C H T N E	C N	55
Laticaudinae						
Aipysurus laevis	toxin a	C G	C P Q V K K P G I K K L E C	C K T N E	C N N	56
	toxin b	C G	C P Q V K K P G I H K L L E C	C K T N E	C N N	56
	toxin c	C G	C P Q V K K P G I H K L L T	C Q S E D C	C N N	56
L. laticaudata	laticotoxin a	C G	C P T V K P G I K K L S S C	C E S E V	C N N	57
L. semifasciata	erabutoxin a	C G	C P T V K P G I K K L S S C	C E S E V	C N N	50
	erabutoxin b	C G	C P T V K P G I N N L S S C	C E S E V	C N N	50
	erabutoxin c	C G	C P T V K P G I K L S S C	C E S E V	C N N	50
	toxin b	C G	C P T V K P G I K L S S C	C E S E R	C N N	58

Table III. Amino Acid Sequences of Sea Snake Neurotoxin (Type II, Long Chain)

Snake	Toxin	Segment 1	Segment 2	Segment 3	Segment 4a	Segment 4b	Segment 4c
Hydrophiinae							
Astroitia stokesii							
	toxin b	L S C	Y L G Y K H S Q T	C P P G E N V	C F V K T W	C D G F	C B T R G E R I I M G
	toxin c	L S C	Y L G Y K H S Q T	C P P G E N V	C F V K T W	C D A F	C S T R G E R I V G M
Laticaudinae							
Laticauda semifasciata	LS III	R E C	Y L N P H D T Q T	C P S G Q E I	C Y V K S W	C N A W	C S S R G K V L E F G

Table III. Continued

Snake	Toxin	Segment 5	Segment 6	Segment 7	Segment 8	Ref.
Hydrophiinae						
Astroitia stokesii						
	toxin b	C A A T	C P T A K S G V H I A	C C S T D N C	N I Y A K W G Ser-NH$_2$	59
	toxin c	C A A T	C P T A K S G V H I A	C C S T D N C	N I Y T K W G S G R-NH$_2$	59
Laticaudinae						
Laticauda semifasciata	LS III	C A A T	C P S V N T G T E I K	C C S A D K C	N T Y P	57

Table IV. Amino Acid Sequences of Sea Snake Neurotoxins (Hybrid Type)

Snake	Toxin	Segment 1	Segment 2	Segment 3	Segment 4
Laticauda colubrina	Lc_a	R I C	Y L A P R D T Q I C	A P G Q E I C	Y L K S W D D G T G F L K G N R L E F G
	Lc_b	R I C	Y L A P R D T Q I C	A P G Q E I C	Y L K S W D D C T G S I R G N R L E F G

Table IV. *Continued*

Snake	Toxin	Segment 5	Segment 6	Segment 7	Segment 8	Ref.
Laticauda colubrina	Lc_a	C A A T C	P T V K P G I D I K C	C S T D K C	N P H P K L A	60
	Lc_b	C A A T C	P T V K R G I H I K C	C S T D K C	N P H P K L A	60

The AChR is composed of five subunits, $\alpha_2\beta\gamma\delta$. A neurotoxin attaches to the α subunit. Since there are 2 mol of the α subunits, 2 mol of neurotoxins attach to 1 mol of AChR. A neurotransmitter, acetylcholine (ACh), also attaches to the α subunit. When the ACh attaches to the AChR, the AChR changes conformation, opening up the transmembrane pore so that cations (Na^+, K^+) can pass through. By this mechanism the depolarization wave from a nerve is now conveyed to a muscle. The difference between neurotoxin and ACh is that the former's attachment does not open the transmembrane pore. As a consequence, the nerve impulse from a nerve cannot be transmitted through the postsynaptic site (21).

At the moment, it is not known whether each toxin attaches to the same site in AChR.

Laticauda semifasciata venom added to the outside bathing solution of frog skin causes an increase in transmural potential difference and short-circuit current, indicating the change in the Na^+ transport system. The venom-induced stimulatory effects can be explained as being either due to an increase in Na^+ permeability of the outer membrane or by an increase in the activity of the Na^+-pump (22).

While most investigations show that sea snake neurotoxins are postsynaptic type, Gawade and Gaitonde (23) stated that *Enhydrina schistosa* major toxin has dual actions or postsynaptic as well as presynaptic toxicity. *E. schistosa* venom phospholipase A is both neurotoxic and myotoxic. Neurotoxic action of the enzyme is weak so that there is sufficient time for myonecrotic action to take place (24). Sea snake, *L. semifasciata*, toxin also inhibits transmission in autonomic ganglia, but has no effect on transmission in choroid neurons.

Other Vertebrate Venoms

Since sea snake venoms are discussed here, it is appropriate to review other vertebrate venoms also. Unfortunately, very few investigations have been done on the venoms of other marine vertebrates. It is known that some fish secrete venoms from their spines. The fishes known to have venoms are the scorpion fish (family: Scorpaenidae), weever fish (family: Trachinidae), catfish (order: Siluriformes; there are 31 families), stargazers (family: Uranoscopidae), toad fish (family: Batrachoididae), and stingrays (suborder: Myliobatoidea).

Venom is secreted from the dorsal, pelvic and anal spines. A review of original papers indicates that many papers have failed to specify from which spine the venom was obtained. Therefore, some publications are meaningless scientifically. Not a single component of fish venoms has been characterized for the amino acid sequence yet. Even the molecular weight of fish toxins is not clear. Deakins and Saunders (25) concluded that the molecular weight of *Scorpaena* toxin was 150,000, while Schaeffer et al. (26) concluded that it had a molecular weight range of 50,000 to 800,000.

There is 5-hydroxytryptamine in weever fish venom besides protein. It is believed that local pain is attributed to the presence of 5-hydroxytryptamine (27). Other small compounds such as histamine, adrenaline, and noradrenaline are also present in the weever fish (28).

As for the stingray venom, not much is known. There was a report on the presence of 5'-nucleotidase and phosphodiesterase in stingray, *Urolophus halleri*, venom (29).

Spine venom of catfish, *Ictalurus catus*, contains toxins with a molecular weight of 10,000 and isoelectric points of 3.8 and 7.8 (30). Pectoral venom of *Arius thallasinus* contains alkaline phosphatase (31). The venom is a mixture of at least 30 proteins.

Summary

From this brief review of marine vertebrate venoms, it is obvious that very few biochemical investigations have been done. The technology to study marine vertebrate venom components is available. There are simply not enough scientists interested enough to enter the field. The first task is to isolate the toxic principles and identify the amino acid sequences. Pharmacological investigation should be done on the purified toxic principle and not on the crude venom, which is a mixture of many proteins and nonproteins.

Acknowledgment

This work was supported by U. S. Army Medical Research and Development Contract DAMD17-81-C-6063.

Literature Cited

1. Tu, A. T. *Handbook of Natural Toxins*, Vol. 3, *Marine Toxins and Venoms*; **1988**, 379–444.
2. Yu, N. T.; Lin, T. S.; Tu, A. T. *J. Biol. Chem.* **1975**, *250*, 1782.
3. Ishizaki, H.; Allen, M.; Tu, A. T. *J. Pharm. Pharmacol.* **1984**, *36*, 36.
4. Tu, A. T.; Lin, T. S.; Bieber, A. L. *Biochemistry* **1975**, *14*, 3408.
5. Seto, A.; Sato, S.; Tamiya, N. *Biochim. Biophys. Acta* **1970**, *214*, 483.
6. Tu, A. T.; Toom, P. M. *J. Biol. Chem.* **1971**, *246*, 1012.
7. Tu, A. T.; Hong, B. S. *J. Biol. Chem.* **1971**, *246*, 2772.
8. Tu, A. T.; Hong, B. S.; Solie, T. N. *Biochemistry* **1971**, *10*, 1295.
9. Raymond, M. L.; Tu, A. T. *Biochim. Biophys. Acta* **1972**, *285*, 498.
10. Sato, S.; Tamiya, N. *J. Biochem.* **1970**, *68*, 867.
11. Tamiya, T.; Lamouroux, A.; Julien, J. F.; Grima, B.; Mallet, J.; Fromageot, P.; Menez, A. *Biochimie* **1985**, *67*, 185.
12. Tu, A. T.; Ganthavorn, S. *Am. J. Trop. Med. Hyg.* **1969**, *18*, 151.
13. Tu, A. T.; Salafranca, E. S. *Am. J. Trop. Med. Hyg.* **1974**, *23*, 135.
14. Kaire, G. H. *Med. J. Aust.* **1964**, *2*, 729.
15. Gawade, S. P.; Budak, D. P.; Gaitonde, B. B. *Ind. J. Med. Res.* **1980**, *72*, 747.
16. Okonogi, T.; Hattori, Z.; Watanabe, M.; Amagai, E. *Snake*, **1970**, *2*, 18.
17. Coulter, A. R.; Harris, R. D.; Sutherland, S. K. *Proc. Melbourne Herp. Symp.* **1981**, 39.
18. Baxter, E. H.; Gallichio, H. A. *Toxicon* **1976**, *14*, 347.
19. Madsen, T.; Lundstroem, H. *Toxicon* **1979**, *17*, 326.
20. Tu, A. T. *Venoms: Chemistry and Molecular Biology*, John Wiley, New York, **1977** .
21. Cash, D. J.; Hess, G. P. *Proc. Nat. Acad. Sci., USA* **1980**, *77*, 842.
22. Gerencser, G. A.; Loo, S. Y. *Comp. Biochem. Physiol.* **1982**, *72A*, 727.
23. Gawade, S. P.; Gaitonde, B. B. *Toxicon* **1982**, *20*, 797.
24. Lind, P.; Eaker, D. *Toxicon* **1981**, *19*, 11.
25. Deakins, D. E.; Saunders, P. R. *Toxicon* **1967**, *4*, 257.
26. Schaeffer, R. C.; Jr.; Carlson, R. W.; Russell, F. E. *Toxicon* **1971**, *9*, 69.
27. Carlisle, D. B. *J. Mar. Biol. Assoc., U. K.* **1962**, *42*, 155.
28. Haavaldsen, R.; Fonnum, F. *Nature* **1963**, *199*, 286.
29. Russell, F. E.; Panos, T. C., Kang, L. W.; Warner, W. M.; Colket, T. C. *Am. J. Med. Sci.* **1958**, *235*, 566.
30. Calton, G. J.; Burnett, J. W. *Toxicon* **1975**, *13*, 399.

31. Thulesius, O.; Al-Hassan, J.; Criddle, R. S.; Thomson, M. *Gen. Pharmacol.* **1983**, *14*, 129.
32. Al-Hassan, J. M.; Thomson, M.; Ali, M.; Criddle, R. S. *Toxin Rev.* **1987**, *6*, 1.
33. Gawade, S. P.; Bhide, M. B. *Bull. Haff. Inst.* **1977**, *5*, 45.
34. Su, B.; Lao, Z.; Sho, Z.; Chang, M.; Zeng, J.; Pan, F.; Wu, S.; Xu, L.; Mo, Y. *Redai Haiyang* **1984**, *3*, 41.
35. Setoguchi, Y.; Morisawa, S.; Obo, F. *Acta Med. Univ. Kagoshima* **1968**, *10*, 53.
36. Uwatoko-Setoguchi, Y. *Acta Med. Univ. Kagoshima* **1970**, *12*, 74.
37. Carey, J. E.; Wright, E. A. *Trans. R. Soc. Med. Hyg.*, **1960**, *54*, 50.
38. Ibrahim, S. A.; Thompson, R. H. S. *Biochim. Biophys. Acta* **1965**, *99*, 331.
39. Fohlman, J.; Eaker, D. *Toxicon* **1977**, *15*, 385.
40. Tan, N. H. *Biochim. Biophys. Acta* **1982**, *717*, 503.
41. Durkin, J. P.; Pickwell, G. V.; Trotter, J. T.; Shier, W. T. *Toxicon* **1981**, *19*, 535.
42. Uwatoko-Setoguchi, Y.; Minamishima, Y.; Obo, F. *Acta Med. Univ. Kagoshima* **1968**, *10*, 219.
43. Uwatoko-Setoguchi, Y.; Obo, F. *Acta Med. Univ. Kagoshima* **1969**, *11*, 139.
44. Tu, A. T.; Passey, R. B.; Toom, P. M. *Arch. Biochem. Biophys.* **1970**, *140*, 96.
45. Yoshida, H.; Kudo, T.; Shinkai, W.; Tamiya, N. *J. Biochem.* **1979**, *85*, 379.
46. Nishida, S.; Kim, H. S.; Tamiya, N. *Biochem. J.* **1982**, *207*, 589.
47. Uwatoko, Y.; Nomura, Y.; Kojima, K.; Obo, F. *Acta Med. Univ. Kagoshima* **1966**, *8*, 141.
48. Mari, N.; Tu, A. T. *Arch. Biochem. Biophys.* **1988**, *260*, 10.
49. Mori, N.; Tu, A. T. *Biol. Chem. Hoppe-Seyler* **1988**, *369*, 521.
50. Maeda, N.; Tamiya, N. *Biochem. J.* **1977**, *167*, 289.
51. Frykland, L.; Eaker, D.; Karlsson, E. *Biochemistry* **1972**, *11*, 4633.
52. Liu, C. S.; Blackwell, R. Q. *Toxicon* **1974**, *12*, 543.
53. Fox, J. W.; Elzinga, M.; Tu, A. T. *FEBS Lett.* **1977**, *80*, 217.
54. Wang, C. L.; Liu, C. S.; Hung, Y. O.; Blackwell, R. Q. *Toxicon* **1976** *14*, 459.
55. Mori, N.; Ishizaki, H.; Tu, A. T. unpublished data, **1988**.
56. Maeda, N.; Tamiya, N. *Biochem. J.* **1976**, *153*, 79.
57. Maeda, N.; Tamiya, N. *Biochem. J.* **1974**, *141*, 389.
58. Tsernoglou, D.; Petsko, G. A.; Tu, A. T. *Biochim. Biophys. Acta* **1977**, *491*, 605.
59. Maeda, N.; Tamiya, N. *Biochem. J.* **1978**, *175*, 507.
60. Kim, H. S.; Tamiya, N. *Biochem. J.* **1982**, *207*, 215.

RECEIVED June 26, 1989

Chapter 27

Origin, Chemistry, and Mechanisms of Action of a Repellent, Presynaptic Excitatory, Ionophore Polypeptide

P. Lazarovici[1,9], N. Primor[2], J. Gennaro[3], J. Fox[4], Y. Shai[5], P. I. Lelkes[6], C. G. Caratsch[7], G. Raghunathan[8], H. R. Guy[8], Y. L. Shih[6], and C. Edwards[6]

[1]Section on Growth Factors, National Institute of Child Health and Human Development, National Institutes of Health, Bethesda, MD 20892
[2]Osborn Laboratories of Marine Sciences, New York Aquarium, New York, NY 11224
[3]Department of Biology, New York University, New York, NY 10003
[4]Department of Microbiology, School of Medicine, University of Virginia, Charlottesville, VA 22308
[5]Molecular, Cellular and Nutritional Endocrinology Branch, National Institutes of Health, Bethesda, MD 20892
[6]Laboratory of Cell Biology and Genetics, National Institute of Diabetes and Digestive and Kidney Diseases, National Institutes of Health, Bethesda, MD 20892
[7]Department of Pharmacology, University of Zurich, 8006 Zurich, Switzerland
[8]Laboratory of Mathematical Biology, National Cancer Institute, National Institutes of Health, Bethesda, MD 20892

Pardaxin, a marine neurotoxic polypeptide isolated from the secretion of the flatfish *Pardachirus marmoratus* or synthesized by the solid phase method, is a single chain, acidic, amphipathic polypeptide with the sequence: NH_2-G-F-F-A-L-I-P-K-I-I-S-S-P-L-F-K-T-L-L-S-A-V-G-S-A-L-S-S-S-G-G-Q-E. Pardaxin is secreted together with a family of steroid aminoglycosides by pairs of cylindrical, acinar glands through secretory ducts to the ocean water. Pardaxin repels sharks and is toxic to marine organisms at 10^{-6}–10^{-4} M. The target of pardaxin is the gills and the pharynx of aquatic animals. It is suggested that pardaxin interference with the ionic transport of the gill epithelium (10^{-6}–10^{-4} M) is the main reason for its toxic effects and that its presynaptic activity (10^{-7} M) is responsible for the initiation of the escape behavioral reflex underlying its repellent action. On a molecular level pardaxin forms voltage-dependent, cation- and anion-permeable pores at low concentrations (10^{-10}–10^{-7} M) and causes cytolysis at higher concentrations (10^{-7}-10^{-4} M). The ionic selectivity properties of pardaxin channels in artificial

[9]Current address: Department of Pharmacology and Experimental Therapeutics, School of Pharmacy, The Hebrew University, P.O. Box 1172, 91010 Jerusalem, Israel

0097–6156/90/0418–0347$06.00/0
© 1990 American Chemical Society

membranes suggest ion binding sites in the channel. Models of the pardaxin pore favor an antiparallel oligomer of the helical segments with a narrow, negatively charged entrance due to the carboxy terminal groups. Pardaxin seems to be a suitable tool to investigate the molecular structures underlying channel selectivity and voltage dependence, and the relationship between channel activity, cytotoxicity and repellency to marine organisms.

Moses Sole — Source of Shark Repellents

Certain ichthyocrinotoxic fish secrete toxic compounds that repel their predators. Among them, the Red Sea Moses Sole, *Pardachirus marmoratus* (Figure 1a,b) exudes a fluid from specialized glands (Figure 1) into the surrounding ocean water. This secretion repels sharks, and it may function as a naturally occurring weapon of defense against shark predation (*1,2*). The principle factors in the secretion, a polypeptide and a family of steroidaminoglycosides, responsible for both its toxicity and noxious effects in sharks, were isolated, identified, and named pardaxin (*3-6*) and mosesins (*5,7*), respectively. The secretion of *Pardachirus pavoninus*, Peacock Sole, in the Western Pacific, presumably contains similar factors: a family of steroid monoglycosides (pavonins) (*7*) and ichthyotoxic peptides (*8*).

The Morphology of the Moses Sole Toxins Secretory Apparatus

Pardachirus marmoratus toxins are secreted by a double row of small cylindrical simple acinar glands (Figure 1b), one located epaxially and the other hypaxially between each of the fin rays from head to tail (Figure 1b). The ventral member of the pair of glands releases its secretion from a pore located on the ventral side of the fish, more peripheral (lateral) to that of its dorsal partner (Figure 1c). Each cylinder contains a central channel (Figure 1c,f) which is part of the secretory duct and is lined with interdigitating epithelial cells of the type found in glands in which water is withdrawn from the secretion before it is released (Figure 1c-e). The outside periphery of the cylinder houses a prominent capillary network and the secretory acini lined by thin secretory cells which surround the acinus, a compartment filled with the secretion. Together these pairs of secretory units comprise an appreciable portion of the body mass of this small flatfish.

A light microscope image through a longitudinally sectioned *pair* of toxin secreting glands shows acini in each gland, although the epaxial gland contains more secreted material than the hypaxial one (Figure 1c). The secretory duct lies within the plane of the fin ray (the midaxial plane) whose articulation can be seen in the left of the image, which separates both glands.

An electron micrograph of one acinus shows a large mass of stored secretory material and, adjacent to this, is the thin cytoplasm of the almost flat secretory cell (Figure 1e). The secretion is released into the acinar pool from the secretory cell as globules (Figure 1d,e), some of which can be seen elevating the plasma membrane. The cytoplasm of an epithelial cell (Figure 1e) contains mitochondria, smooth endoplasmic reticulum, and an extensive and dense network of intermediate filaments, probably keratin, but lacks secretory granules. The acinus is surrounded by a system of satellite cells (cells with pleomorphic nuclei) applied closely to them with no intervening amorphous basement material. Instead, both amorphous and reticular components of the basement material are peripheral to the satellite cell. An electron micrograph of the epithelium lining the secretory duct shows irregular rounded apical elevations which protrude into the duct space and the complexity of the interdigitations along their lateral boundaries (Figure 1f). These structures

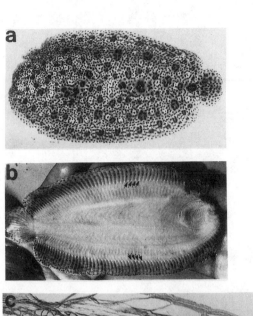

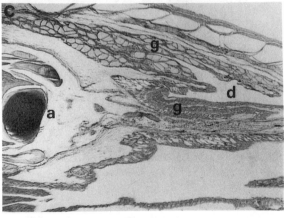

Figure 1. *Pardachirus marmoratus* fish and the morphology of the toxin secretory glands.

a,b - Lateral views of the fish. Arrows indicate white secretion around the gland openings.

c - Photomicrograph (20×) of the two toxin secreting glands in sagittal section. The glands (g), are right of the ray articulation (a) and are filled with secretion. The clear space within each gland lies in the secretory duct(s) (d) which end out of the image to the right. *Continued on next page.*

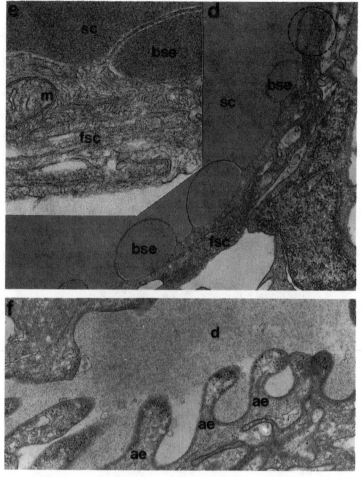

Figure 1. *Continued. Pardachirus marmoratus* fish and the morphology of the toxin secretory glands.

d - An electron micrograph (30,000×) of the glandular secretory epithelium and a portion of one acinus. Peripheral to the area of moderately electron dense material (sc) which is the toxic secretion(s), is the thin secretory epithelial cell (fsc). Several secretory vesicles (bse) can be seen on its surface, some separated from the contents of the acinar pool only by the thickness of the plasma membrane.

e - An electron micrograph (80,000×) showing the secretory cell cytoplasm in detail. The fibrillar nature of the cytoplasmic matrix is visible as well as the smooth (agranular) appearance of the endoplasmic membranes. Mitochondria (m) can be seen and a portion of one of the plasma membrane elevations within which is material of almost identical granularity and electron density to that in the acinar pool. No secretory granules are visible in the cytoplasm.

f - This electron micrograph (20,000×) is an image of the cells lining the secretory duct (d). Their rounded apical elevations (ae) project into the secretory space and come into contact with the secretion.

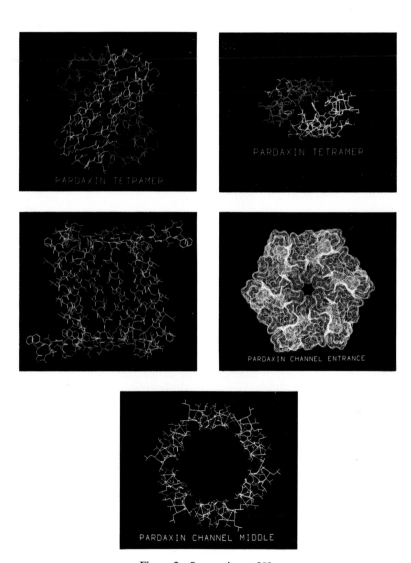

Figure 2. *See caption p. 352.*

- initial models indicated that adjacent α helices pack well next to each other when the C-helix extends to residue 29, and

- the segment 11-13 (Ser-Ser-Pro) is quite likely not to be in a helical conformation and the residues that precede a proline are not usually an α helical conformation.

The initial C-helix was created by a program that generated α helices with the backbone structure and side chain conformations most commonly observed for α helices in crystal structures (Cornette, Guy and Margalit, unpublished data). This program was not appropriate for an α helix that contains an interior proline. The initial backbone structure, the N-helix, was modeled from the α helical segment 120-128 in lactate dehydrogenase (13) with the sequence PHE-lys-phe-*Ile*-ILE-PRO-asn-ILE-*Val* in which most residues were identical or similar to those of the pardaxin segment 2-10 as indicated by capitals and underlining. The remaining C-terminus residues were assigned phi and psi backbone torsion angles commonly observed in random coil segments and that extended the Glu-33 carboxyl groups into a region postulated to be occupied by water and near Lys-8 and Lys-16 of adjacent helices. Initially, the conformations of Ser-11 and Ser-12 were adjusted manually using computer graphics so that N-helices could have apparently favorable interactions with the membrane lipid and with other protein segments.

Connolly (14) surfaces were added to the structures and monomers were manually docked to form channel structures using the Mogli program on an Evans and Sutherland computer graphics monitor. C-helices were packed so that Connolly surfaces formed a tight barrier between the inside and outside of the channel. Most hydrophilic groups not involved in internal hydrogen bonds were positioned to be exposed to water and most hydrophobic groups were positioned to be in contact with the hydrocarbon portion of the lipid membrane. Energies for individual dimers created this way were refined with the CHARMM program (15) using adopted basic Newton-Raphson method. A conversion criterion of .01 Å was used for the rms gradient during a cycle of minimization. These energy refined dimers were then used to reconstruct aggregates and a final energy refinement was performed on the complete aggregate. Each monomer was in an identical environment and had the same conformation.

Experimental data indicates that in aqueous solution, pardaxin appears to exist primarily as a tetramer (5) and shows a concentration-dependent oligomerization (5). On this basis, a tetramer (Figure 2) and "raft" models were developed using dimers that have the same packing for the C-helices as the channel models. The major differences among the monomer conformations of these models involve the C-terminus and residues Ser-11 and Ser-12 that control the relative positions of the N- and C-helices to each other. In the tetramer the dimers were packed next to each other so that the row of large hydrophobic side chains on the C-helices of each dimer packed next to those of the adjacent dimer (Figure 2A, B). This produced a structure similar to that of melittin tetramers (16). The tetramer structure had 2-fold symmetry about each orthogonal axis (Figure 2A, B). Helical interactions between C-helix dimers had "4-4 ridges-into-grooves" type packing. The N-helices were positioned so they fit into a hydrophobic groove formed by the C-helices. The N-termini of two N-helices approached each other on each side of the tetramer. All hydrophilic side chains were exposed on the surface and most hydrophobic side chains were buried. This model is in accordance with the unique property of pardaxin to reduce water surface tension (17) at concentrations more than 10^{-6} M, the critical micellar concentration of this peptide in aqueous solutions (Lelkes and Lazarovici, unpublished data).

Mechanism of Pardaxin-Produced Repellency

Location of the Pardaxin Target in Fish. In free-swimming sharks, mixing bait with the gland secretion stops the shark from completing its bite and causes it to leave with an abnormally open jaw (18). In sharks, the main site for chemical sensing is located within the lateral-line organ of the head. To determine whether the response to pardaxin was mediated through the lateral line system or via the pharyngeal cavity and the gills, an apparatus was constructed which prevented a mixing of the outflow from shark gills with water bathing its surface skin (19). Pardaxin administered to the medium bathing the skin surface of the shark's head did not elicit the behavioral responses set up by a repellent. Addition of pardaxin to the medium bathing the shark's pharynx and gills caused the shark to struggle violently immediately (19). It was rather unexpected to find that the shark's main sensory system, located at the head surface, did not respond to pardaxin (19). Furthermore, pardaxin elicited its repellent effect (10^{-6}–10^{-4} M) only when applied in sea water, and showed no effect when injected into the circulation (20). Previous studies have shown that the Na-K$^+$ ATPase of the gills of teleost fish is affected by pardaxin (10^{-6}–10^{-4} M) and histopathological effects have been described (21).

Pardaxin Impairment of Ion-Transport of the Opercular Epithelium. Pardaxin affects gills in both teleosts (21) and elasmobranchs (20), and its toxicity is higher in fish preadapted to a medium of high salinity (21). Therefore, its mode of action upon ionic transport by the opercular epithelium of the killifish *Fundulus heteroclitus*, whose ion-osmoregulatory properties closely resemble those of teleost gills (22), was examined. It is well established that fish adapted to sea water actively secrete NaCl by means of specific cells positioned in the gills and opercular epithelial skin (22). Administration of pardaxin to the mucosal (seawater) side of the isolated short-circuited opercular epithelium caused a transient stimulation of the active transport of ions (I_{sc}), followed by an inhibition (Figure 3). The stimulation was abolished by ouabain and/or removal of Na$^+$ from the Ringer (23). Pardaxin did not affect the I_{sc} when applied to the serosal (blood) side. On the mucosal side, pardaxin produced a net transient Na$^+$ current from the mucosal to the serosal side of 2.2 μequiv cm^{-2} h^{-1}. The sodium flux was 1000-fold higher than the passive permeability measured by the unidirectional inulin fluxes (23). It was concluded that the increased Na$^+$ influx underlay the stimulation, and this was suggested to be the mechanism responsible for the toxicity of pardaxin.

Pardaxin Evoked Increase of Intracellular Ca^{2+} in Chromaffin Cells. The previously described pore forming properties in artificial membranes (5) warranted examination of the effect of pardaxin on intracellular free Ca^{2+} concentrations, using chromaffin cells as a model and the fluorescent dye Fura-2 to measure intracellular [Ca$^+$]. The basal level of [Ca^{2+}]$_{in}$ for these cells is 50-90 nM (28). Stimulation of the cells with potassium triggers a rapid, transient increase of [Ca^{2+}]$_{in}$ (Figure 4). As expected, ionomycin, a calcium ionophore, causes a sustained rise in the free intracellular concentration to approximately 450 nM (Figure 4). In this system, pardaxin induced an increase in intracellular [Ca^{2+}] only in the presence of extracellular Ca^{2+} (Figure 4). These results indicate that pardaxin mediated a Ca^{2+} influx but did not release Ca^{2+} from intracellular stores. This influx is most probably mediated directly by pardaxin channels and possibly also indirectly by activation of the Ca^{2+} channels of the chromaffin cells by the depolarization produced by the pardaxin channels (data not shown). These observations further substantiate our hypothesis (10) that transmembrane fluxes of Na$^+$ and Ca^{2+} are involved in the pathological action of pardaxin.

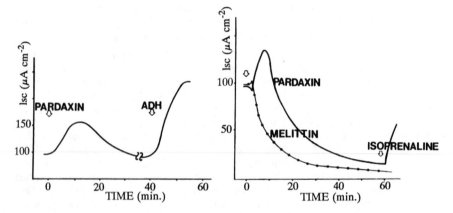

Figure 3. The effect of pardaxin on the short circuit current of the isolated epithelial preparations. (*Left*) Isolated skin of the frog *R. catesbeiana*. Pardaxin (6 x 10^{-5} M) was applied to the mucosal side. ADH (5 x 10^{-7} M) was applied to the serosal side. (*Right*) Isolated opercular skin of seawater-adapted killifish *F. heteroclitus*. Pardaxin (10^{-5} M) (full line) and melittin (3 x 10^{-5}) (circles) were added (first arrow) to the mucosal (seawater) side of the skin. Isoprenaline (2 x 10^{-6} M) was added to the serosal side (second arrow). The continuous tracing shows the short-circuit current (I_{sc}).

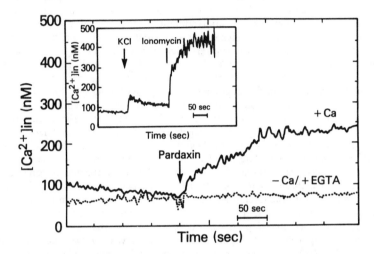

Figure 4. Effect of pardaxin on intracellular ionized calcium level in bovine adrenal chromaffin cells.

Isolated chromaffin cells were maintained in suspension culture and loaded with the fluorescent calcium indicator Fura 2 as previously described (28). 2×10^5 cells/ml were added into a cuvette containing standard buffer without (dotted line) or with (full line) 2 mM calcium. At the arrow, 10^{-7} M pardaxin was added. A rise in $[Ca^{2+}]_{in}$ was observed only in the presence of calcium. Inset, typical control experiment. Depolarization of the cells by 30 mM KCl resulted in rapid, partially reversible increase in $[Ca^{2+}]_{in}$. Further, addition of $0.1 \mu M$ ionomycin, a calcium ionophore, increased the intracellular calcium level.

Presynaptic Effects of Pardaxin. The epithelial skin isolated from the killifish oper-
culum contains an extensive supply of blood vessels and nerves on blood side (serosal) in close proximity to the epithelial cells; however the possible exposure of
nerve terminals on the mucosal (seawater) side and the detailed ultrastructural morphology are unknown. It is possible that the predator repellent action of pardaxin
might involve sensory and motor neuronal pathways in addition to ionic transport
systems. Therefore, we examined the possible effects of pardaxin on neuromuscular
transmission using the frog sartorius nerve muscle preparation. In this system, pardaxin (10^{-8}–10^{-6} M) produced presynaptic but not postsynaptic effects (29). It
increased the frequency of the spontaneous release of transmitter quanta in a dose-
dependent and temperature-influenced way up to more than 100 times the control
values. At the same time the quantal content of the evoked end-plate potentials
was greatly elevated (29). The glycosteroids isolated from the gland secretion were
relatively ineffective on neurotransmitter release; however, at high doses they had
postsynaptic effects, as shown by a diminution of the amplitude of the evoked
end-plate potentials. They did not reinforce the effect of the pardaxins.

At higher doses pardaxin depolarized the postsynaptic membranes, producing
muscle contractions which could not be blocked by (+)-tubocurarine or tetrodotoxin, and eventually also physical disruption of muscle cells. No effects on nerve
conduction were observed (29). It is suggested that the pardaxin effects on nerve
terminals are due to the fact that the absence of myelin makes the terminals more
susceptible to the toxin and the smaller surface/volume ratio of the terminal means
that a few channels can produce a large effect on the intracellular ionic contents
and so on the membrane potential. In addition, small changes in the intracellular
ionic contents and in the membrane potential of the nerve terminal could easily
produce large changes in the rate of release of quanta of transmitter. This profound presynaptic activity of pardaxin led us to suggest that pardaxin may act on
the seawater-facing cells of the gills and/or the pharynx by two mechanisms: (1)
repellency (10^{-8}–10^{-6} M) as a result of depolarization and activation of sensory-
motor neuronal pathways involved in the escape behavior, and (2) toxicity (10^{-6}–
10^{-4} M) as a result of the collapse of the osmoregulatory homeostasis of the fish.

Ionophore Activity of Pardaxin

Pore Forming Activity in Liposomes. When a pore forming factor inserts into the
hyperpolarized membrane of lipid vesicles, the existing ion gradients can produce
potential differences. A potentiometric cyanine dye (9) can be used to elucidate
some of the characteristics of pardaxin pore in phosphatidylcholine liposomes (9).
In this system, pardaxin acts at higher concentrations (10^{-10}–10^{-9} M), and in a
more complex way to permeabilize the bilayer, compared to gramicidin, a well-
known ionophore. This pardaxin channel does not discriminate significantly
between cations and anions and it is equally effective if sucrose or choline sulfate
replaces the Na_2SO_4 in the external buffer (9). From the fraction of depolarized
vesicles and the amount of radioactive pardaxin bound, it was estimated that 4-12
pardaxin monomers are required to form a pore, sufficient to alter the potential
differences across the membrane of one vesicle (9).

Aggregation of Phosphatidylserine Vesicles by Pardaxin. Besides forming pores in
lipid membranes, pardaxin induces aggregation of phosphatidylserine vesicles (26), a
phenomena which has been visualized with negative contrast electron microscopy
(5). The pardaxin-induced aggregation is very rapid and continues for an extended
period of time ($\geq$30 min). Furthermore, pardaxin-induced aggregation is strongly
modulated by the transmembrane and/or surface potential (26). In spite of the
extensive aggregation, no exchange of lipids between vesicles or fusion was observed

(26). We have proposed that vesicle aggregation is probably related to the disposition of pardaxin bound in the phosphatidylserine vesicle lipid bilayer (26). This conclusion is supported by the observation that phosphatidycholine vesicles are not induced to aggregate and that the pardaxin-induced phosphatidylserine vesicle aggregation is affected by charge polarization of the vesicle (26). This suggestion seems to be consistent also with the voltage dependence of fast "pore" activity of pardaxin, the channels which are open only at positive membrane potentials.

Single Channel Recordings and Selectivity of Pardaxin Channels on Planar Bilayers of Phosphatidylethanolamine at the Tip of Patch Pipets. To learn about the ion selectivities of the pardaxin channels, their behavior in lipid bilayers was examined. Bilayers of phosphatidylethanolamine were formed at the tips of micropipettes (Figure 5a) by the double-dip method (27). After the bilayer was found to be stable, pardaxin was added to the solution on one side, whose potential was set at +20 mV. The currents across the bilayer in the voltage clamp mode were monitored. At low concentrations of pardaxin (10^{-9}–10^{-8} M) single channel events (Figure 5b) usually appeared within 10-20 min. At higher toxin concentrations the latencies were shorter. A potential gradient was required for the toxin molecules to insert and/or express channel activity into the membrane; little or no activity was found in the absence of a potential gradient.

To determine the ionic selectivity of the pardaxin channels, various ion substitutions were performed and the bionic reversal potential, i.e., the potential at which the current across a bilayer with many open pardaxin channels changed sign, was determined. The relative permeabilities of the ions could then be determined from the general equation:

$$E_{rev} = RT \, (P_{C_1}[C_1] + P_{A_1}[A_1]) \, nF \, (P_{C_2}[C_2] + P_{A_2}[A_2])$$

C_1, C_2, A_1, and A_2 are the concentrations of cations and anions on sides one and two. In practice, it was found that Tris was an impermeant cation and HEPES an impermeant anion, and so salts of these compounds were used to simplify the experimental conditions. For cations the selectivity sequence was

$$Tl^+ > Cs^+ > Rb^+ > K^+, \, NH_4^+ > Li^+ > Na^+$$

which, except for Li^+, is the same as that of the relative hydrated sizes. For anions the sequence was

$$I^- > NO_3^- > Br^- > Cl^- > ClO_4^- > SCN^- > HCOO^-,$$

which is quite different from that of the relative hydrated sizes. Therefore, the mechanisms governing the permeabilities of cations and anions appear to be different.

Modeling Pardaxin Channel. The remarkable switching of conformation in the presence of detergents or phospholipid vesicles (5) suggests that pardaxin is a very flexible molecule. This property helps to explain the apparent ability of pardaxin to insert into phospholipid bilayers. In addition, it is consistent with the suggestion that the deoxycholate-like aminoglycosteroids (5,7) present in the natural secretion from which pardaxin is purified (5) serve to stabilize its dissociated conformation. The question of the mechanism by which pardaxin assembles within membranes is important for understanding pore formation and its cytolytic activity (5).

We have developed a series of models in which 8-12 monomers were packed in an antiparallel manner to form the channel. The kinetics of pore formation in liposomes (9) and the dependence of planar bilayer conductance on toxin

a

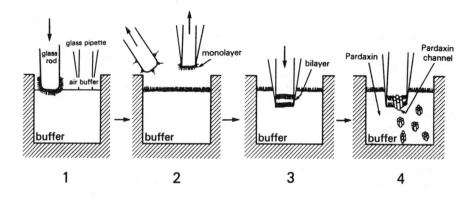

b

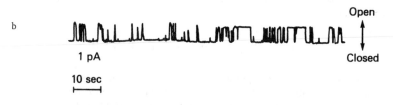

Figure 5. Scheme of the bilayer formation at the tip of the patch pipets (a) and single channel recordings of pardaxin pores (b).

Step 1 - Monolayer of phosphatidylethanolamine is prepared at the air water interface with a glass rod;

Step 2 - A patch pipet is removed from the solution, the polar head groups of the monolayer lipids are adsorbed to the interface while the fatty acid hydrophobic tails are exposed to the air;

Step 3 - The pipet is reinserted into the liquid, resulting in apposition of the hydrocarbon tails of the attached monolayer to those of the original monolayer, forming a bilayer (27);

Step 4 - Addition of the synthetic pardaxin (10^{-9} M) to the bath results after 10–20 min in tetramer insertion into the bilayer and pore formations as measured by the current fluctuations depicted in B at a positive potential of 100 mV. The single channel conductance can be estimated from the amplitude of the current steps divided by the applied voltage and was in the range of 10 pS.

concentration yield linear relationships with slopes around 8-12. The common interpretation of such results is that an aggregate of 8-12 molecules forms the ion conducting channel. Antiparallel models appear to be more probable because they allow:

1. the interaction of negative carboxyls of the C-terminus glutamate with positive charges of Lys-16 and Lys-8;
2. the favorable interaction of the backboned dipoles of adjacent C-helices;
3. the only negatively charged portion of the molecule, the C-terminus, to lie near to the radial center of the pore where it could make the channel selective to cations; and
4. large hydrophobic side chains on the C-helix to pack tightly next to each other.

Models in which the monomers are parallel, which we have previously considered (*10*), appear unlikely because: (1) all models with parallel α helices have highly positively charged regions that should strongly reduce cation permeability; and (2) we could find no parallel models in which side chains pack next to each other. Two ways were found to pack C-helices in an antiparallel manner to form dimers in which all the large alkyl side chains and the phenlyalanine side chain form a hydrophobic column on one side of the dimer and several serines form hydrogen bonds with each other on the opposite side of the dimers. In addition, the point of closest contact between α-helix backbones is Gly-23, the positively charged Lys-16 extends over the C-terminus of the adjacent α helix where it interacts favorably with the negative end of the helix dipole and carboxyls of the terminus Glu-33, and the helices cross each other at an angle of -15° predicted by '3-4 ridges into grooves' helix packing theory (*24*). These dimers for the C-helices were used to construct models of pardaxin in solution as a tetramer (Figure 2, top photographs), on the membrane surface as a "raft-like" structure (data not shown), and in the transmembrane orientation as a channel (Figure 2, bottom photographs).

These dimers were used to construct several channel models (Figure 2, bottom right) that have narrow regions formed by the C-termini near their entrances (Figure 2, bottom center), a large noncharged pore (Figure 2, bottom center) formed by the C-helices through the middle of the membrane, and a positively charged hydrophobic ring around the channel formed by the N-helices (Figure 2, bottom right). Figure 2 (bottom center) shows the entrance of the channel formed by residues 1-16 and 31-33 in one of these models. These segments contain all the charged groups. The N-helices form a hydrophobic and positively charged outer ring that surrounds the negatively charged C-termini. The narrowest portion of the channel is formed by a ring of six amide groups from the Gln-32 side chains. These form a network of hydrogen bonds with each other and with the carboxy termini. Lys-8 and Lys-16 form salt bridges to the two carboxyl groups of Glu-33. A sphere of diameter 4.8 Å could just pass through this ring. This is large enough to pass the known permeant cations but small enough to exclude Tris, which is impermeant.

The central region of this channel model formed by residues 17-30 is shown in Figure 2 (bottom right). Only serine side chains line the hexagonally shaped pore. The pore is about 20 Å wide. The exterior of the structure which would be exposed to lipid contains only alkyl side chains. The channel is large enough to contain an hexagonally shaped Ice I type water structure (*25*) in which each cross-sectional "layer" of water molecules contains 42 water molecules. The hydroxyl groups of the channel lining may interact with this or a similar kind of water structure that has 3-fold or 6-fold symmetry.

This pardaxin model is not unique. We have developed several similar models that are equally good energetically and equally consistent with present experimental results. It is difficult to select among these models because the helices can be packed a number of ways and the C-terminus appears very flexible. Our energy calculations are far from definitive because they do not include lipid, water, ions, membrane voltage, or entropy and because every conformational possibility has not been explored. The model presented here is intended to illustrate the general folding pattern of a family of pardaxin models in which the monomers are antiparallel and to demonstrate that these models are feasible.

The most likely way for pardaxin molecules to insert across the membrane in an antiparallel manner is for them to form antiparallel aggregates on the membrane surface that then insert across the membrane. We developed a "raft"model (data not shown) that is similar to the channel model except that adjacent dimers are related to each other by a linear translation instead of a 60° rotation about a channel axis. All of the large hydrophobic side chains of the C-helices are on one side of the "raft" and all hydrophilic side chains are on the other side. We postulate that these "rafts" displace the lipid molecules on one side of the bilayer. When two or more "rafts" meet they can insert across the membrane to form a channel in a way that never exposes the hydrophilic side chains to the lipid alkyl chains. The conformational change from the "raft" to the channel structure primarily involves a pivoting motion about the "ridge" of side chains formed by Thr-17, Ala-21, Ala-25, and Ser-29. These small side chains present few steric barriers for the postulated conformational change.

Conclusions

Pardaxin is the principal polypeptide toxic component of the secretion of the flatfish *Pardachirus marmoratus* (5,6). It is secreted into the water by a series of double, clindrical acinar glands. Although it is water soluble, at concentrations of 10^{-11}–10^{-7} M, this polypeptide spontaneously inserts into artificial or biological membranes to form voltage-gated pores which are permeable to cations and anions. At concentrations higher than 10^{-7} M pardaxin acts as a lytic agent. Both properties probably underly the toxicity and repellency to marine organisms. The primary sequence of pardaxin indicates a strong hydrophobic segment at the amino-terminal, followed by an α-helical amphipathic region and a hydrophylic carboxy terminal.

In analogy to a series of polypeptide channel forming quasi ionophores, a model of pardaxin tetramer in water and in the membrane is presented:

1. In aqueous buffer, pardaxin is comprised of four antiparallel monomers tightly packed with 2-fold symmetry of the "4-4 ridges into grooves" type; the hydrophobic amino-terminal segments of pardaxin monomers are shielded from the aqueous surface in the tetramer which most probably exposes the polar side chain to water.

2. Interaction with a lipid bilayer driven by a potential difference and by polar and/or hydrophobic forces between the amino acid side chains of the pardaxin tetramers and the polar membrane lipid head group triggers insertion from a "raft" like structure.

3. In the bilayer or upon interaction with detergent micelles, a structural reorganization of pardaxin aggregates takes place, in which the polar side chains interact with themselves and the hydrophobic residues are externally oriented in the pardaxin aggregate, therefore allowing interactions with the lipid backbone hydrocarbons.

4. In this pardaxin oligomer channel, the α helices cross at an angle of 25° and the model predicts a hydrophilic interior allowing for passage of ions.

5. The voltage induces the opening of pardaxin pores, or opens them indirectly, by affecting pardaxin oligomers, supposed to be stabilized by dipole-dipole interactions of monomer helices.
6. Most probably the presence of pardaxin pores alters the structure of the bilayer resulting in aggregation of phosphatidylserine vesicles mediated by contact but not by partial merging of their membranes.

Although for the moment this model is only partially supported by experimental data it offers the opportunity to design new experiments which will help to understand the mechanisms of pardaxin insertion and pore formation in lipid bilayers and biological membranes which at a molecular level are the events leading to shark repellency and toxicity of this marine toxin.

Acknowledgment

This work was supported in part by Research Grant N-00014-82-C-0435 from the Department of Defense, Office of Naval Research, United States Navy and Koret Foundation, San Francisco.

Literature Cited

1. Clark, E.; Chao, S. *Bull. Sea Fish Res. Stn. Israel* **1973**, *60*, 53-6.
2. Clark, E. In *Shark Repellents from the Sea*: *New Perspectives*; Zahuranec, B. J., Ed.; American Association for the Advancement of Science: Washington, DC, 1983; pp 135-50.
3. Primor, N.; Zlotkin, E. *Toxicon*1975/fl, *13*, 227-234.
4. Primor, N.; Parness, J.; Zlotkin, E. In *Toxins*: *Animal, Plant and Microbial*; Rosenberg, P., Ed.; Pergamon Press: New York, 1978; pp 539-47.
5. Shai, Y.; Fox, J.; Caratsch, C.; Shih, Y. L.; Edwards, C.; Lazarovici, P. *Febs. Letters.* **1988**, *242*, 161-166.
6. Lazarovici, P.; Primor, N.; Lelkes, P. I.; Fox, J.; Shai, Y.; Raghunathan, G.; Guy, H. R.; Shih, Y. L.; Edwards, C. *Biochemistry* (submitted for publication.
7. Tachibana, K.; Sakaitanai, M.; Nakanishi, K. *Science* **1986**, *226*, 703-5.
8. Thompson, S. A.; Tachibana, K.; Nakanishi, K.; Kubota, I. *Science* **1986**, *233*, 341-3.
9. Loew, L. M.; Benson, L.; Lazarovici, P.; Rosenberg, I. *Biochemistry* **1985**, *24*, 2101-4.
10. Lazarovici, P.; Primor, N.; Caratsch, C. G.; Munz, K.; Lelkes, P.; Loew, L. M.; Shai, Y.; McPhie, P.; Louini, A.; Contreras, M. L.; Fox, J.; Shih, Y. L.; Edwards, C. In *Neurotoxins as Tools in Neurochemistry*; Dolly, J. O.; Ellis Horwood Limited: Chichester, England, 1988, pp. 219-240.
11. Eisenberg, D. *Ann. Rev. Biochem.* **1984**, *53*, 595-623.
12. Garnier, J.; Osguthorpe, D. J.; Robson, B. *J. Mol. Biol.* **1978**, *120*, 97-120.
13. White, J. L.; Hackert, M. L.; Buehner, M.; Adams, M. J.; Ford, G. C.; Lentz, P. J., Jr.; Smiley, I. E.; Steindel, S. J.; Rossmann, M. G. *J. Mol. Biol.* **1976**, *102*, 759-68.
14. Connolly, M. L. *J. Appl. Cryst.* **1983**, *16*, 548-58.
15. Brooks, B. R.; Bruccoleri, R. E.; Olafson, B. D., States, D. J.; Swaminathan, S.; Karplus, M. *J. Compt. Chem.* **1983**, *4*, 187-217.
16. Terwilliger, T. C.; Eisenberg, D. *J. Biol. Chem.* **1982**, *257*, 6010-22.
17. Primor, N. *Gen. Pharmacol.* **1986**, *17*, 413-8.
18. Gruber, S. H.; Zlotkin, E.; Nelson, D. R. In *Toxins, Drugs and Pollutants in Marine Animals*; Bolis, L.; Zadunaisky, J.; Gilles, R., Eds.; Springer-Verlag:Germany, 1984; pp 26-41.

19. Primor, N. *Experientia* **1985**, *41*, 693-696.
20. Primor, N.; Zadunaisky, J. A.; Murdaugh, M. V., Boyer, J. L.; Forrest, J. N. *Comp. Biochem. Physiol.* **1984**, *78C*, 483-90.
21. Primor, N.; Sabnay, I.; Lavie, V.; Zlotkin, E. *J. Exp. Zool.* **1980**, *211*, 33-43.
22. Degnan, K. J.; Karnaky, J. J.; Zadunaisky, J. A. *J. Physiol. Lond.* **1977**, *271*,155-91.
23. Primor, N. *J. Exp. Biol.* **1983**, *105*, 83-94.
24. Chothia, C.; Levitt, M.; Richardson, D. *J. Mol. Biol.* **1981**, *145*, 215-50.
25. Eisenberg, D.; Kauzmann, W. *The Structure and Properties of Water*; Oxford University Press: New York, NY, 1969.
26. Lelkes, P.; Lazarovici, P. *FEBS Lett.* **1988**, *230*, 131-6.
27. Suarez, I.; Lindstrom, W.; Montal, M. *Biochemistry* **1983**, *22*, 2319-23.
28. Lelkes, P. I.; Pollard, H. B. *J. Biol. Chem.* **1987**, *262*, 15496-505.
29. Renner, P.; Caratsch, C. G.; Waser, P. G.; Lazarovici, P.; Primor, N. *Neuroscience* **1987**, *23*, 319-25.

RECEIVED August 18, 1989

Author Index

365

Affiliation Index

Subject Index

Production: Donna Lucas
Indexing: Deborah H. Steiner
Acquisition: Robin Giroux

Elements typeset by Hot Type Ltd., Washington, DC
Printed and bound by Maple Press, York, PA

Paper meets minimum requirements of American National Standard
for Information Sciences—Permanence of Paper for Printed Library
Materials, ANSI Z39.48–1984 ∞